KU-539-750

MULTIPHASE TRANSPORT AND PARTICULATE PHENOMENA

VOLUME 3

Edited by

T. Nejat Veziroğlu
Clean Energy Research Institute
University of Miami

◉ HEMISPHERE PUBLISHING CORPORATION
A member of the Taylor & Francis Group

New York   Washington   Philadelphia   London

EDITOR
T. Nejat Veziroğlu
Clean Energy Research Institute
University of Miami
Coral Gables, Florida, U.S.A.

EDITORIAL BOARD
Sadik Kakaç
University of Miami
Coral Gables, Florida, U.S.A.

Manoj Padki
Clean Energy Research Institute
University of Miami
Coral Gables, Florida, U.S.A.

John W. Sheffield
University of Missouri at Rolla
Rolla, Missouri, U.S.A.

MANUSCRIPT EDITOR
Jennie Myers
Clean Energy Research Institute
University of Miami
Coral Gables, Florida, U.S.A.

MANUSCRIPT ASSISTANTS
M. Akçin
Clean Energy Research Institute
University of Miami
Coral Gables, Florida, U.S.A.

M. T. Özgökmen
Clean Energy Research Institute
University of Miami
Coral Gables, Florida, U.S.A.

MULTIPHASE TRANSPORT AND PARTICULATE PHENOMENA: Volume 3

Copyright © 1990 by Hemisphere Publishing Corporation. All rights reserved. Printed in the United States of America. Except as permitted under the United States Copyright Act of 1976, no part of this publication may be reproduced or distributed in any form or by any means, or stored in a data base or retrieval system, without the prior written permission of the publisher.

Cover design by Sharon DePass.

1 2 3 4 5 6 7 8 9 0   E B E B   8 9 8 7 6 5 4 3 2 1 0 9

Library of Congress Cataloging-in-Publication Data

Multiphase transport and particulate phenomena / edited by T. Nejat
    Veziroğlu.
        p.   cm.
    Papers presented at the 5th Miami International Symposium on
Multiphase Transport and Particulate Phenomena.

    1. Multiphase flow—Congresses.   2. Heat—Transmission—
Congresses.   3. Mass transfer—Congresses.   I. Veziroğlu, T.
Nejat.   II. Miami International Symposium on Multiphase Transport
and Particulate Phenomena (5th)
TA357.M85   1990
620.1'064—dc20   80-19776
                                                              CIP

ISBN 1-56032-026-5 (set)
ISBN 1-56032-032-X (Vol. 3)

Contents

SUSPENSIONS

FLUIDIZED BEDS

COMBUSTION

Preface

Multiphase transport and particulate phenomena applications are found in a wide range of engineering systems, such as heat exchangers, boilers, evaporators, condensers, boiling water reactors, pressurized water reactors, particle separators, contamination detectors and controllers, filters, slurry transporters, and fluidized beds. Over the past three decades, problems in two-phase flow heat transfer and instabilities have challenged many investigators. Instabilities can induce boiling crises, disturb control systems, and/or cause mechanical damage. It is thus important to be able to predict the conditions under which a two-phase system will perform reliably. At the same time, the importance of the particulate phenomena science and other technology is growing in other areas, including particle sizing, liquid particle interactions, mechanics of suspensions and emulsions, and sedimentation.

Due to recent energy and environmental crises, many other multiphase transport and particulate phenomena problems have also become important. Some of these include the modeling of the loss-of-coolant accident in pressurized water nuclear reactors, scaling up of fluidized bed reactors for converting coal to clean gaseous and liquid fuels, design of heat exchangers for liquified natural gas and liquified petroleum gas, control of microcontamination, effect of aerosols, and air pollution control.

The Fifth Miami International Symposium on Multiphase Transport and Particulate Phenomena continued the tradition established by its four predecessors. It provided a high-level international platform in pleasant surroundings for the presentation of the latest research results and for the exchange of ideas on the important topics of multiphase transport and particulate phenomena.

The lectures and papers presented at the symposium and prepared in accordance with the paper guidelines, have been divided by their subject matter into 16 parts and 3 volumes. The reader should be advised that it was difficult to classify specifically some of the papers where there was an overlap in subject matter. In such cases, we tried to make the best possible choice. This three-volume set should serve as a valuable reference, covering the latest developments in the growing areas of multiphase transport and particulate science and technology.

T. Nejat Veziroğlu

Acknowledgments

The Organizing Committee gratefully acknowledges the assistance and cooperation of the International Association for Hydrogen Energy, the International Atomic Energy Agency, the International Solar Energy Society, the Florida International University, the Florida Solar Energy Center, and the Department of Mechanical Engineering, University of Miami.

We also wish to extend sincere appreciation to the keynote speaker, William L. Grosshandler, Thermal Sciences and Engineering Program, National Science Foundation, Washington, D.C., and to the banquet speaker, Robert T. Lahey, Nuclear Engineering Department, Rensselaer Polytechnic Institute, Troy, New York.

Special thanks are due to our authors and lecturers, who have provided the substance of the proceedings.

And last, but not least, our debt of gratitude is owed to session developers, chairpersons, and co-chairpersons for the organization and execution of the technical sessions. In acknowledgment we list these session officials on the following pages.

Organizing Committee

Acknowledgements

Organizing Committee

K. Akyüzlü
University of New Orleans
New Orleans, LA, USA

S. G. Bankoff
Northwestern University
Chicago, IL, USA

K. J. Bell
Oklahoma State University
Stillwater, OK, USA

A. E. Bergles
Rensselaer Polytechnic Institute
Troy, NY, USA

J. S. Chang
McMaster University
Ontario, Canada

X. Chen
Xian Jiaotong University
Xian, China

A. Duyar
Florida Atlantic University
Boca Raton, FL, USA

M. Gashgari
King Abdulaziz University
Saudi Arabia

D. Geldart
University of Bradford
Bradford, UK

D. Gidaspow
Illinois Institute of Technology
Chicago, IL, USA

N. Güven
Texas Tech University
Lubbock, TX, USA

S. A. Hoenig
University of Arizona
Tucson, AZ, USA

H. Hoffman
Oak Ridge National
 Laboratory, USA

W. S. Janna
University of New Orleans
New Orleans, LA, USA

K. Johannsen
Berlin Technical University
Berlin, FRG

V. Kakabadze
Academy of Sciences
Tbilisi, USSR

S. Kakaç
University of Miami
Coral Gables, FL, USA

R. T. Lahey
Rensselaer Polytechnic Institute
Troy, NY, USA

S. S. Lee
University of Miami
Coral Gables, FL, USA

R. Lyczkowski
Argonne National Laboratory
Argonne, IL, USA

A. Mertol
Science Applications
 International Corp.
Los Altos, CA, USA

A. S. Mujumdar
McGill University
Canada

V. J. Novick
EGG&G Idaho, Inc.
Seattle, WA, USA

M. N. Ozisik
North Carolina State University
Raleigh, NC, USA

M. R. Parker
University of Salford
Salford, UK

R. W. Peters
Purdue University
West Lafayette, IN, USA

J. T. Pytlinski
University of Puerto Rico
Puerto Rico

P. Ramakrishnan
Indian Institute of Technology
Madras, India

T. Raunemaa
University of Kuopio
Helsinki, Finland

T. M. Romberg
CISRO
Sutherland, Australia

C. Schweiger
University of Duisburg
Duisburg, FRG

N. Selcuk
Middle East Technical University
Turkey

J. S. Sheffield (Co-Chairperson)
University of Missouri
Rolla, MO, USA

S. Sideman
Israel Institute of Technology
Haifa, Israel

C. W. Snoek
Atomic Energy of Canada, Ltd.
Chalk River, Canada

I. K. Stephan
University of Stuttgart
Stuttgart, FRG

R. L. Sterling
University of Minnesota
Minneapolis, MN, USA

Y. K. Tan
South China Institute of
Technology
Guangzhou, China

Y. Ueno
University of Missouri
Rolla, MO, USA

T. N. Veziroğlu (Chairperson)
University of Miami
Coral Cables, FL, USA

J. H. Vincent
Institute of Occupational Medicine
Edinburgh, UK

J. Weisman
University of Cincinnati
Cincinnati, OH, USA

A. A. Zkauskas
Mokslu Akapemija
Vilnius, USSR

STAFF

Executive Secretary   Lucille Walter

Conference Coordinators   Javonne Gelineau
                             Aymara Schmidt

Graduate Assistants	L. Kazi
	N. Lutfi
	T. Özgökmen
	M. Padki
	T. Tekindur
Undergraduate Assistants	M. Akcin
	C. Blaisure
	K. Cerretti
	V. Yankowski

AEROSOLS

Correlations of Atmospheric Turbidity Parameters with Direct Normal Radiation under Clear Sky Conditions in Baghdad

A. AKRAWI, D. M. ABDULRAZZAK, and A. AZIZ
Scientific Research Council
Solar Energy Research Center
Jadiriyia, P.O. Box 13026
Baghdad, Iraq

Abstract

Hourly measurements of aerosol optical depth have been recorded at the research field observatory of the solar energy research center in Jadiriyia, Baghdad, Iraq (Latitude $33^\circ\ 21'$ N, Longitude $44^\circ\ 25'$ E, elevation 34 m above MSL). These measurements have been made in four wavelengths, (368nm, 500nm, 675nm and 778nm) using filter wheel fully automatic sunphotometer model Ms-110 manufactured by EKO in Japan.
A wide variety of additional meteorological and solar radiation data are also recorded using the automatic weather station located at the same site.
The purpose of this paper is to analyze the data set obtained under clear sky conditions for the period April 1987 to March 1988, and to assess the influence of atmospheric content on direct normal radiation.
The mean annual and seasonal atmospheric turbidity due to aerosols T_a, the linke turbidity factor T_L, the Angstrom Coefficient β, and the Schuepp coefficient B are derived. Linear regression correlations between the mean monthly direct normal spectral solar irradiance I_λ, and the aerosol optical depth T_a are established.
Similar correlations between the ratio of diffuse to global solar radiation and the aerosol optical depth at 500nm, and the ratio of diffuse to direct normal radiation with the aerosol optical depth at 500nm are developed.
Significont decrease in the attenuation of beam radiation by aerosols observed when the wavelength increases.

1. INTRODUCTION

The knowledge of the available solar irradiance is essential to many solar conversion systems in terms of their design selection, and performance efficiency as well as in building heating load calculations.
The direct irradiance is utilized in concentrating devices such as solar furnaces while total radiant energy is needed in solar heating applications [1].
The direct normal spectral solar irradiance is important for design of certain solar energy applications such as, photovoltaics, and in the analysis of the atmosphere with regard to its turbidity and constituents in addition to the study of the distribution of in coming, outgoing and net solar

radiation [2].

The strength of the solar beam as it enters the earths atmosphere is attenuated by absorption and scattering. Absorption by the molecules and the atoms is in discreet wavelengths. The main gaseous absorbers are O3 ,O2 , H2O and Co2. All atmospheric gaese and aerosols scatter solar radiation at all wavelengths. The aerosols also absorb radiation somewhat continuously in wavelength. However, absorption by the aerosols is much smaller than scattering by the aerosols [3].

The attenuation of the solar beam as it traverses from the top of the atmosphere and reaches the ground varies with the mass of gases and aerosols it encounters in its path. The concentration of N_2, O_2 and Co2 in the atmosphere remains more or less constant in time and space.

However, the total amount of O_3 in the vertical direction varies with latitude and season.

Ozone, which is distributed in a layer of 10 to 30 km above ground level, absorbs mainly in the ultraviolet and produces an abrupt termination of solar radiation below 290nm. The portion of solar radiation absorbed by ozone varies from 1.5 to 3% , depending on the atmospheric path length and the total ozone content in the atmosphere. [4].

The average monthly ozone values are tabulated in the literature [5] or can be estimated [6].

The total amount of water vapour in the atmosphere in the vertical direction is highly variable and depends on the instantaneous local conditions. However this amount, generally expressed as precipitable water thickness, can be readily computed through a number of atmospheric parameters, such as relative humidity, ambient temperature, dewpoint temperature or vapour pressure.

The precipitable water vapour thickness can vary from 0.0 to 5 cm. Some of the most commonly employed methods of computing the precipitable water thickness are summarized in [7]. Water vapour absorbs solar radiation selectively and has scattering properties as well. Values of 3 to 9% for absorption and 0.5 to 5% for scattering were reported [8].

Molecular, or Rayleigh scattering is caused by density fluctuations and molecular anisotropy by the air molecules, such as N_2 and O_2 and particles in size ranges smaller than the wavelength of solar radiation.

The observed reduction in total radiation due to molecular scattering is about 9 to 12% [9].

In the visible band (400-750 nm), where gaseous absorption is minimal under clear sky conditions, suspended particles within the atmosphere display considerable diversity in volume, size, form and material composition.

Aerosols are the major cause of depletion of direct beam solar radiation through scattering and absorption.

There are two dominant layers of aerosols in the atmosphere - one near the Earths surface (0.3km), which is affected by natural duststorms and man-made inputs to the atmosphere, and another stratospheric dust layer 15-25km above sea levels affected by volcanic eruptions and cosmic sources [10].

The extinction by atmospheric aerosols of intensity of direct

solar radiation for a clear sky is called atmospheric
turbidity. For a beam of direct solar radiation, the upper
layer absorbs 1 to 5% and scatters 0.9 to 6% whereas the lower
layer absorbs 0.5 to 5% and scatters 0.6 to 21% [11]. In desert
climates with sand suspension in the atmosphere, an absorption
of 4.8% was observed [12].
The amount of aerosols present in the atmosphere in the
vertical direction has been represented in terms of the number
of particles per cubic meter. However, it is more common to
represent the amount of aerosols by an index of turbidity.
The paper discusses the measurement of the aerosol aptical
depth T_a under cloudless skies, the deduction and analysis of the
monthly and seasonal T_a , the linke turbidity factor T_L , the
Angstrom coefficient β ,and the schuepp coefficient B.
Mathematical relationships are established between the direct
normal spectral irradiance and the aerosol optical thickness.
Most solar thermal power plants use reflectors to concentrate
solar radiation.
The concentration systems operate best under cloudless skies. A
knowledge of the atmospheric turbidity coefficients is very
important to predict the availability of solar radiation under
cloudless skies.
This ability to predict the direct solar irradiance is
essential for the design of solar thermal power plants and
other solar energy conversion devices with concentration
systems.
The turbidity parameter is also required in order to determine
the amount of spectral global irradiance for the design of
photovoltaic systems, and calculation of the photosynthetic
energy for plant growth.
Several atmospheric turbidity indices have been proposed.
Abrief description of the most currently used is described as
follows:
(a) The Link Turbidity Factor T_L
This is an index of the number of standrd atmosheres of pure
and dry air which produces the same total depletion or give
rise to the same intensity of direct solar radiation as in the
given turbid atmosphere. This was based on the consideration
that scattering and absorption by pure, dry air is the basic
atmosphric effect. Its value can vary from 1 to 10. It is a
useful parameter for comparison of cloudless atmospheric
conditions, however, it has one serious disadvantage. Even
under constant turbidity conditions, T_L, was found to exhibit a
diurnal variation, or the so called virtual variation with
airmass. This was ascribed to the dependence of the extinction
coefficient on the wavelength of radiation [13]. It may also be
noted that T_L is a measure of total turbidity, including the
contribution by water vapour, since it is based on solar
irradiance measurements without filter.
(b) Angstrom Turbidity Coefficient β
The amount of aerosols present in the atmosphere in the
vertical direction can also be represented by an indx called
Angstroms turbidity coefficient , proposed by Angstrom [14],
where β is a constant characteristic of the aerosol content,and α
is anumber between (0-4), characteristic of the size

5

distribution.
On the basis of spectral investigations, and for most natural atmospheres,the exponent was found to be(1.3 - 0.5)has been in wide use by meteorological organizations all the world as a rough guide to the measure of dust, smoke and haze of the local atmosphere.
The principal drawback of β is the assumption of a constant value of α which may widely differ from 1.3, thereby giving rise to the virtual variation of β with air mass under steady turbidity conditions of the atmosphere. This is particularily true in tropical and subtropical countries, where particle size distribution has wide seasonal variations .
(C) Shueppes generalized turbidity coefficient B
shueppe [15] has defined a turbidity coefficient which refers to the measurement of direct spectral solar irradiance at 500nm in the central part of the visible spectrum and is related to Angstrom Coefficient by:

$$at \ \alpha = 1.3 \ , \ B = 1.069 \beta .$$

2.MEASUREMENTS

Hourly measurements of the aerosol optical thickness on clear and cloudless days are made at the solar research observatory located in the solar energy research center in Jadiriyia Baghdad (Latitude $33^{\circ} \ 21' \ N$, Longitude $44^{\circ} \ 25E$). Measurements for 4 wavelengths (368nm, 500nm, 675nm and 778nm) were taken during the period April 1987 to March 1988 using a fully automatic sunphotometer mode ms-110 manufactured by EKO of Japan. The system is composed of sunphotometer sensor tube containing the interference filters, controller, equatorial mount, convertor , a data logger type solar Mp-080 and a micro computer consisting of personal computer Pc-9801, two floppy discs one containing the complete computer programme to evaluate the turbidity data while the other is used to store the data. The system also contains a display unit and a printer. The direct normal solar radiation passing through the incidence part enters the spectral system. This system has interference filters which take out a given wavelength.
The signals detected by the filter sensor indentify each wavelength.
The output signals from the sensor corresponding to each wavelength are stored in the sample holder electrically at inlervals of 10 ms.
Those discontinuous signals are converted to analog signals by low pass filters, and the outputs corresponding to each wavelength are taken out independently . The signals are displayed on the controller and recorded on a data logger. The voltage output from the sunphotometer is proportional to the intensity of direct normal radiation incident upon it. The system configuration is shown in Figure (1), and is described fully in [16].
Hourly measurements of the direct beam spectral irradiance are evaluated from the sunphotometer measurements, while the global solar irradiance on a horizontal surface in the wavelength

6

range 0.3 to 3 μm are made by a star shaped pyranometer according to Dimhirn.
The sensors convert the physical variables into electrical signals which are transmitted to an automatic data acquisition system and lead to the corresponding tranducers. The microprocessor interrogates every second the instantaneous values and calculates the mean values every hour. When the system is activated the output is made automatically on the printer.

3. MATHEMATICAL ANALYSIS

For a parallel beam of monochromatic radiation, I , falling perpendicularily on a medium of layer thickness x, the transmitted radiation, I_λ , is given by Beer-Lambert relation (Bouguer law) [17]:

$$I_\lambda = Io\lambda \exp - (\gamma_\lambda . x) \quad \ldots\ldots(1)$$

where γ_λ is the volume extinction coefficient with dimensions L^{-1}
The product, γ_λ. x, is called the total optical depth T_λ, and is dimensionless, since the solar radiation arrives at the earth surface at an inclined path, equation (1) is corrected for the increased path length by multiplying by the optical depth the optical air mass, m, thus:

$$I_\lambda = \frac{Io\lambda}{s} \exp - T_\lambda . m \quad \ldots\ldots\ldots(2)$$

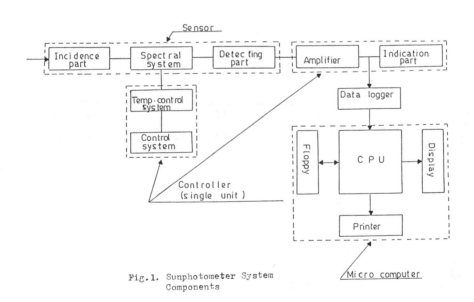

Fig.1. Sunphotometer System Components

Where
I_λ the measured spectral solar irradiance at the ground.
Io_λ the extraterrestrial spectral solar irradiance at zero
   air mass
s   the correction factor for mean sun-earth distance
T_λ the total optical depth
m   sec z
z   Solar zenith angle
λ wavelength nm

The total optical depth results from a combination of effects
which are illustrated mathematically by:

$$T_\lambda = T_{R\lambda}\frac{P}{Po} + T_{a\lambda} + T_{o\lambda} \quad \ldots\ldots (3)$$

where
$T_{R\lambda}$ Rayleigh or molecular scattering spectral optical depth for
   a standard atmosphere
P   atmospheric pressure at observers site
Po   atmospheric pressure at sea level for a standard atmosphere
$T_{a\lambda}$ aerosol spectral optical depth resulting from scattering
   and absorption by airborne atmospheric particulates
   contained in the air
To_λ Optical depth arising from absorption by ozone

The sunphotometer does not measure the spectral irradiance but
rather provides a meter reading E_λ which is proportional to I_λ ,
similarily, the extraterrestrial, or calibration voltage Eo is
the meter reading when the irradiance is Io .
Thus equation (3) takes the following form:

$$Ta(\lambda) = \frac{In\ \dfrac{Eo_\lambda}{E_\lambda \cdot S}}{m} - (\frac{P}{Po}T_{R\lambda} + To_\lambda) \quad \ldots\ldots\ldots\ldots (8)$$

It should be noted that measurements were made at wavelengths
where no selective absorption of CO_2 or water vapour takes
place.
Rayleigh scattering is defined as the scattering light from
small particles with dimensions smaller than the radiation
wavelength.
The Rayleigh Spectral optical depth is given by Frohlich and
shaw [18]:

$$T_{R\lambda} = 0.00838 . \lambda^{-(3.916+0.074.\lambda+0.050/\lambda)} \quad \ldots\ldots\ldots (9)$$

The ozone transmittance $T_{o\lambda}$ was evaluated from leckners equation
[19].

$$To_\lambda = exp\ (-a_{o\lambda}\ 03\ Mo) \quad \ldots (10)$$

where:
$a_{o\lambda}$ is the ozone absorption coefficient,
03, the ozone amount (atm-cm),

and Mo the ozone mass.
The ozone absorption coefficients given in [20] are used, the
ozone mass expression of Robinson as given by Iqbal
(16), has been adopted. The ozone mass is given by

$$Mo = (1+ ho/6370)/[cos(z)+2ho/6370]^{0.5} \quad \ldots\ldots (11)$$

The parameter ho is the height of maximum ozone concentration,
which is approximately 23 km. The ozone height varies with
latitude and time of year.
The ozone amount for Baghdad has been estimated [12].
Leckners expression for uniformly mixed gas transmittance is
used and is expressed as :-

$$T_{o\lambda} = exp [-1.41\ au\ M/(1+118.3\ au_\lambda\ M)^{0.45}] \quad \ldots\ldots (12)$$

where au_λ is the combination of an absorption coefficient and
gaseous amount, leckners values are used for au_λ[19], M is the
ozone mass.

The values of $T_{R\lambda}, T_{o\lambda}, I_o$ and $E_{o\lambda}$ are given in the following
table:-

wavelength nm	368	500	675	778
$T_{R\lambda}$	0.4945	0.1391	0.0410	0.023
$T_{o\lambda}$	0	0.0114	0.0144	0
$I_{o\lambda}$ (Wm^{-2} nm^{-1})	1.088	1.859	1.512	1.188
$E_{o\lambda}$ (mv)	9.03	10.96	8.08	7.64

The linke turbidity factor can be expressed as:

$$T_L = \frac{lnIo - lnI_\lambda - lns}{T_{R\lambda}\ m\ lne} \quad \ldots\ldots (13)$$

where
e the water content of the atmosphere, the Angstrom turbidity
coefficient is calculated from the Angstrom turbidity formula.

$$T_{a\lambda} = \beta\lambda^{-\alpha} \quad \ldots\ldots\ldots (14)$$

The values for a pair of wavelengths is determined from:

$$\alpha = ln\frac{Ta\lambda_1}{Ta\lambda_2} / ln\frac{\lambda_2}{\lambda_1} \quad \ldots\ldots (15)$$

The schuepp turdidity coefficient B is calculated from:

$$Ta\lambda = B\ ln\ 10 \quad \ldots\ldots (16).$$

The spectral solar irradiance I_λ is evaluated from:

$$I_\lambda = \frac{E}{Eo} \cdot I_{o\lambda} \quad \ldots\ldots (17)$$

4 RESULTS AND DISCUSSION

4.1 Analysis of Turbidity Factors

Spectral aerosol optical values were determined for 1200 set of observations, cloudless hours were selected according to the following conditions:

(a) The relative sunshine duration $\frac{S}{3_D}$ =1
where S is the actual sunshine duration and is the daylength.
(b) The total cloud amount N $\leqslant$ 1 okta at both the begining and the end of the respective hour.

Some typical values are presented in tables I and II, the tables show the measured values of the spectral aerosol optical depth and the calculated values of the spectral solar irradiance (Wm nm) representing Seasonal variation on four different days. The highest values of the aerosol optical depth are recorded around noon for the four wavelengths. The maximum aerosol optical depth values are obtained in the summer season at the wavelength 368 nm a value of 0.49 is recorded on 6.7.1988 Extreme.weather conditions in terms of aerosol optical depth are not uncommon in Baghdad during the summer. An average summer day starts fairly clear, but as solar intensity increases and heats the ground, thermal convection develops within the boundary layer. This, in turn causes dust to rise and increass atmospheric turbidity.

The hourly variation of the aerosol optical depth on a clear winter day 29.12.1987 in the wavelength 368 nm is shown to vary between 0.19 to 0.37.

Similarily , the monthly mean values of the spectral direct normal solar irradiance (Wm nm) are evaluated from equation (17), and presented in table III for the months of March, August, October and December representing the Spring, Summer, autumn and Winter seasons.

The mean annual and seasonal values of Ta368, Ta500, Ta675, Ta778, $T_L 368$, $T_L 500$, B368, B500 and $\alpha 1$, $\alpha 2$, $\alpha 3$, $\alpha 4$, are evaluated from equations (8), (13), (16) and (15) respectively, and presented in table (IV).

The values are obtained in the four bands:

$$1 \longrightarrow 368 - 500 \text{ nm}$$
$$2 \longrightarrow 500 - 675 \text{ nm}$$
$$3 \longrightarrow 368 - 675 \text{ nm}$$
$$4 \longrightarrow 368 - 778 \text{ nm}$$

Knowing the values, the corresponding $\beta 1$, $\beta 2$, $\beta 3$ and $\beta 4$, values are calculated from equation (14).

It is clear from table (IV) that the highest mean annual and seasonal turbidity parameters occur during the summer season due to the moderate wind velocity which under favourable conditions of insolation and instability help in raising large amounts of dust which can be very extensive and persistent for many hours, in addition the persistance of haze under stable weather conditions in this season contributes to the high turbidity values and redues visibility.

The least turbidity values occur in winter which is normally characterized by clean atmosphere resulting from the continuous scavenging of aerosols and pollutants by rainfall.

TABLE I: MEASURED VALUES OF SPECTRAL AEROSOL OPTICAL DEPTH AND CALCULATED SPECTRAL SOLAR IRRADIANCE $Wm^{-2}\mu m^{-1}$

DATE 6.4. 1988

hour	solar elevation h	Airmass m	$T_{a}368$	$T_{a}500$	$T_{a}675$	$T_{a}778$	I 368	I 500	I 675	I 778
8	27.4	2.16	0.36	0.17	0.07	0.08	179.4	924.4	1118.2	958.8
9	39.4	1.57	0.38	0.16	0.04	0.04	288.8	1145.3	1285.8	1069.5
10	50.4	1.30	0.43	0.17	0.04	0.05	346.0	1216.7	1319.0	1082.3
11	65.3	1.1	0.44	0.18	0.03	0.04	365.0	1255.2	1335.5	1045.6
12	63.1	1.12	0.46	0.19	0.02	0.05	391.9	1266.0	1370.5	1099.5
13	60.2	1.15	0.43	0.17	0.01	0.03	393.8	1279.6	1381.6	1123.0

DATE 6.7. 1988

hour	solar elevation h	Air mass m	$T_{a}368$	$T_{a}600$	$T_{a}675$	$T_{a}778$	I 368	I 500	I 675	I 778
8	35.3	1.37	0.37	0.17	0.08	0.09	263.3	1031.5	1144.0	949.4
9	47.8	1.35	0.40	0.20	0.07	0.09	343.5	1187.0	1180.8	1054.2
10	60.3	1.15	0.46	0.24	0.10	0.12	367.0	1145.3	1210.8	979.1
11	72.0	1.05	0.48	0.25	0.11	0.11	396.9	1182.7	1212.1	1002.1
12	79.4	1.02	0.46	0.22	0.10	0.07	418.6	1238.7	1223.2	1047.9
13	74.3	1.04	0.49	0.24	0.12	0.10	399.1	1198.0	1201.1	1010.4

TABLE II: MEASURED VALUES OF SPECTRAL AEROSOL OPTICAL DEPTH
AND CALCULATED SPECTRAL SOLAR IRRADIANCE Wm^{-2}μm^{-1}

DATE 14.10. 1987

hour	solar elevation h	air mass m	$T_{a}368$	$T_{a}500$	$T_{a}676$	$T_{a}778$	I 368	I 500	I 675	I 778
8	27.9	2.13	0.30	0.13	0.02	0.03	187.2	1025.7	1207.5	1050.2
9	29.2	2.04	0.32	0.14	0.04	0.04	218.8	1046.8	1245.3	1054.0
10	37.4	1.64	0.33	0.12	0.01	0.02	300.2	1201.4	1361.3	1129.4
11	42.6	1.47	0.38	0.13	0.01	0.03	321.4	1201.4	1363.2	1118.4
12	43.8	1.44	0.35	0.13	0.01	0.03	342.2	1247.3	1327.9	1124.7
13	40.8	1.53	0.34	0.13	0.01	0.02	321.9	1218.4	1387.1	1110.9

DATE 29.12. 1987

hour	solar elevation h	airmass m	$T_{a}368$	$T_{a}500$	$T_{a}675$	$T_{a}778$	I 368	I 500	I 675	I 778
8	8.9	6.24	0.19	0.09	0.04	0.04	16.54	428.2	823.4	840.4
9	19.4	3.14	0.25	0.10	0.03	0.03	114.51	880.2	1166.1	1055.6
10	26.2	2.26	0.26	0.09	0.01	0.01	217.7	1119.8	1315.3	1143.6
11	31.4	1.91	0.29	0.09	0.01	0.01	267.2	1215.0	1377.9	1179.7
12	33.5	1.81	0.34	0.12	0.02	0.02	259.5	1167.4	1339.2	1145.2
13	32.0	1.88	0.37	0.14	0.03	0.02	234.1	1109.6	1306.1	1131.0

4.2 Deveopment of Correlations

The hourly sums of global radiation G(h) and diffuse solar radiation D(h) during cloudless hours were used to compute the hourly sums of direct solar radiation $I(\gamma)$ by means of the equation:

$$I(\gamma) = \frac{G(h) - D(h)}{\sin(\gamma)} \qquad \ldots..(18)$$

where γ is the solar altitude from the hourly values of $I(\gamma)$, the following linear correlations were established between hourly spectral aerosol optical thickness and the hourly direct solar radiation.

The Root Mean Square Error, (RMSE), the Mean Bias Error (MBE) and the mean percentage deviation between measured and calculated values were evaluated thus:-

$$Ta368 = 0.6 - 0.12 \sin(\gamma) \, \ln \frac{Io}{I(\gamma)} \qquad \ldots.(19)$$

RMSE = 0.054, MBE = 0.0082
Mean deviation = 7.3 per cent

$$Ta500 = 0.38 - 0.10 \sin(\gamma) \, \ln \frac{Io}{I(\gamma)} \qquad \ldots..(20)$$

RMSE = 0.09 MBE = 0.051
Mean deviation = 7 per cent

TABLE III: MONTHLY MEAN SPECTRAL DIRECT NORMAL SOLAR IRRADIANCE $Wm^{-2}\mu M^{-1}$

Wavelength (nm) Month	368	500	675	778
March	255.16	1005.57	1166.21	942.16
August	230.80	949.15	1061.84	832.81
October	262.59	999.51	1140.46	929.78
December	171.05	918.26	1101.77	964.06

TABLE IV: MEAN ANNUAL AND SEASONAL TURBIDITY PARAMETERS

	$T_{a}368$	$T_{a}500$	$T_{a}675$	$T_{a}778$	$T_{1}368$	$T_{1}500$	$B368$	$B500$	1	2	3	4	1	2	3	4
Mean Annual	0.52	0.30	0.16	0.19	2.08	3.13	0.22	0.13	0.08	0.08	0.09	0.10	2.06	2.39	2.37	1.80
Autumn	0.49	0.28	0.17	0.17	2.04	3.19	0.22	0.13	0.08	0.08	0.10	0.16	2.07	1.97	2.28	1.88
Winter	0.45	0.22	0.09	0.10	1.94	2.66	0.20	0.10	0.05	0.02	0.07	0.09	2.35	3.00	2.49	2.33
Spring	0.47	0.25	0.12	0.21	1.99	2.84	0.21	0.10	0.03	0.04	0.04	0.09	2.28	2.30	2.85	2.38
Summer	0.58	0.36	0.20	0.24	2.27	3.65	0.25	0.15	0.12	0.09	0.09	0.17	1.65	1.70	1.87	1.24

$$Ta675 = 0.18 - 0.03 \sin (\gamma) \text{ In } \frac{Io}{I(\gamma)} \quad \ldots\ldots (21)$$

RMSE= 0.019 MBE = 0.004
Mean deviation = 9.8 per cent

$$Ta778 = 0.30 - 0.10 \sin (\gamma) \text{ In } \frac{Io}{I(\gamma)} \quad \ldots\ldots\ldots (22)$$

RMSE = 0.03, MBE = 0.008
Mean deviation = 8.5 per cent.
A value of 1367 Wm^{-2} was assumed for the extraterrestrial solar irradiance Io
The following correlations between the ratio of the mean daily direct normal spectral solar irradiance $I\lambda$ and the extraterrestrial solar spectral irradiance $Io\lambda$ with the aerosol optical thickness Ta and the airmass, m, are deduced the correlation coefficients, (r), and standard deviations (S.D), are calculated thus:

$$I368 = 0.789 \text{ Io}(368) \exp - 1.603 \text{ } T_{a368} \text{ m} \ldots\ldots (23)$$
r= 0.85, S.D = 9.6%
$$I500 = 0.771 \text{ Io}(500) \exp - 0.930 T_{a500} .\text{m} \ldots\ldots (24)$$
r= 0.916, S.D = 2.8%
$$I675 = 0.934 \text{ Io}(675) \exp -1.045 \text{ Ta675. m} \ldots\ldots (25)$$
r = 0.997 S.D = 1.1%
$$I778 = 1.004 \text{ Io}(778) \exp - 1.085 \text{ Ta778 . m} \ldots\ldots (26)$$
r = 0.997 S.D= 1.1%

By applying linear regression techniques the following four correlations were deduced computer to estimate the mean daily diffused solar radiation using the mean daily global solar radiation G, the direct normal spectral solar irradiance I500 and the aerosel optical thickness Ta500.

$$\frac{D}{G} = 0.1332 + 0.4044 \text{ Ta500} \quad \ldots\ldots (27)$$

r=0.886

$$\frac{D}{I500} = 0.1241 + 0.7462 \text{ Ta500} \ldots\ldots (28)$$

r= 0.891

In order to predict the aerosols effect on the attenuation of the direct normal solar irradiance, an investigation was made of the correlation between the mean daily values of $I\lambda$ and the aerosol optical depth at 368nm and 500nm, the following least square regressions are deduced:

$$\text{In } I368 = 5.87 - 0.72 \text{ Ta368} \ldots\ldots (26)$$
r= 0.95
$$\text{In } T500 = 7.16 - 1.04 \text{ Ta500} \ldots\ldots (27)$$
r= 0.96

15

Furthermore, the ratio between the mean daily amount of the global solar irradiance on a horizontal surface, $(W\ m^{-2})$ and the spectral direct normal solar irradionce G

$\dfrac{}{I\lambda}$ with various atmospheric turbidity coefficients is investigated, the following correlations are faond:-

$$\frac{G}{I368} = 1.03 + 2.53\ Ta368 \qquad \ldots\ldots (28)$$

$$(r=0.90)$$

$$\frac{G}{I500} = 0.27 + 0.99\ Ta500 \qquad \ldots\ldots (29)$$

$$r=0.86$$

$$\frac{G}{I675} = 0.33 + 0.95\ Ta675 \qquad \ldots\ldots (30)$$

$$r=0.90$$

$$\frac{G}{I778} = 0.40 + 1.08\ Ta778 \qquad \ldots\ldots (31)$$

$$(r=0.92)$$

$$\frac{G}{I368} = 1.04\ T_{L}368 \qquad \ldots\ldots (32)$$

$$(r=0.91)$$

$$\frac{G}{I500} = -0.44 + 0.17\ T1500 \qquad \ldots\ldots (33)$$

$$(r=0.96)$$

5. CONCLUSIONS
1. The results indicate that there is a significant decrease in the attentuation of direct normal solar irradiance due to aerosols and particulate matter suspended in the atmosphere, the extinction effect of the aerosols have been found to be dependent on the particle size.

2. The summer values of the T_a, T_ω, B and $\beta's$ were found to be larger than their winter values, due to the combined effect of solar radiation and wind speed. A sharp drop of turbidity takes place after August though it remains relatively large in comparison with clear winter days. Turbidity is particularily large during the summer because of frequent dust phenomenon.

3. The atmospheric turbidity for a location is highly dependent on weather and climate in that location.

4. Correlations between the mean daily direct solar irradiance and atmospheric turbidity parameters have been developed with correlation coefficients ranging between (-0.85 - 0.997).

5. Linear correlations of hourly atmospheric turbidity parameters with hourly direct solar irradiance have been derived.

6. Correlations between the ratio of the mean daily global solar irradiance and direct normal spectral irradiance with various turbidity parameters have been deduced, the correlation coefficients are practically always greater than 0.86.

7. The direct normal spectral solar irradiance at (368 nm and 500nm) recorded its lowest values during winter, however its values at (675 and 778 nm) reach their minimum in the summer.

REFERENCES

1. R.King and R.O. Buckius, Direct Solar transmittance for a clear sky,
   Solar energy, Vol(22),PP 297-301 (1979).
2. IGY Instruction manual, part V, Radiation instruments and manuals.
3. A. Louche M. Maurel, G.Simonnot, G.Peri and M.Iqbal Determination of Angstroms turbidity coefficient from direct total solar irradiance measurements solar energy Vol (38), No.2.PP.89-96,(1987)
4. L.Ya.Kondratyev, Radiation in the atmosphere Academic press, New York (1969)
5. N.Robinson (Ed.), solar Radiation.
   American El-sevier, New York (1966).
6. T.K.Van Heuken, Estimating atmospheric ozone for solar radiation models. Solar energy ,Vol (22) PP 63-68 (1979).
7. M. Iqbal, An Introduction to Solar Radiation, Academic press,New York (1983).
8. D.V. Hoyt, Theoretical calculations of true solar noon atmospheric transmission U.S. Dept of Commerce, NOAA (1977).
9. A.Mani and O. Chacko, Attenuation of solar radiation in the atmosphere. Proceedings of ISEE conference, Atlanta, GA (1979).
10. On the nature and distribution of Solar radiation. U.S. Dept of commerce, National Techical Information service, HCP/T 2550-01, March (1978).
11. E.C. Flowers R.A. M cormick and K.R. kurfis, Atmospheric turbidity over the united states. Journal of Applied meteorology, vol (8), (1969).
12. M. Ya. Kondratyev, spectral radiative flux divergence and its variability in the troposphere, 0.4m to 2.4 m region. Applied optics vol (13), (1974).
13. World Meteorological Organization Technical Note No. 172 meteorological Aspects of the utilization of Solar Radiation As An Energy Source Wm0.557 PP 13-14, (1981)
14. A.Angstrom, Techniques of determining the turbidity of the atmosphere Tellus Vol (13), PP 214-223, (1961).
15. W. Schuepp,Die Bestimmung der Komonenten der atmospherschen Trubung aus Aktinometer Messungen, Arch. Meteorol. Geophy. Bioklimatol., ser.B, 1:25 (1949).

16. A.Akrawi, "Automation of atmospheric turbidity measurements at Baghdad", Proceedings of the fourth international Symposium on multi-phase transport and particalate phenomenon 15-17 December, Miami Beach, Florida, U.S.A
17. G.F. Lothian, Beers low and its use in analysis, Analyst Vol (88), PP 678-685, (1963).
18. C.Frohlich and G.E. Shaw, "New determination of Rayleigh scattering in the terrestrial atmosphere, Applied optics Vol (19), PP 1773-1775, (1980).
19. B.Leckner, The spectral distribution of solar radiation at the earths surface, elements of a model, Solar Energy Vol (20), PP 143-150, (1978)

The Application of the General Energy Theory to the Studies of Chemical Engineering: Some Fundamental Aspects of the Properties and Dynamics of Aerosols

K. PARTHASARATHY
Department of Electrical Engineering
Annamalai University
Annamalai Nagar 608 002, India

ABSTRACT :

In this paper, the optical properties , Brownian motion and other related physical characteristics of the fine particles of sols are studied in detail through the principles of the General Energy Theory. This is extended to the studies on the mechanism of adsorption at the solid-solution and the solid-gas interfaces. These studies will reveal that the buffer clouds of the colloidal particles play an important role. Colloids maintain stability in solution or coagulate depending upon the polarities of the energy particles involved in the buffer clouds. On this score, many of the details of the physical and chemical adsorption phenomena are discussed in detail. These studies enable us to investigate into many of the properties of the aerosols.

1. INTRODUCTION :

In this paper, some of the fundamental aspects of the properties and dynamics of the aerosols are considered. In as much as the aerosols are forming a particular type of colloids, many of the following discussions can be generalised so that they are applied to the colloids also(cf: companion paper on colloids). In addition, the properties of the sedimenting particles would also be inter-related with these discussions and hence would enable the better visualisation of the various aspects of the theory of sedimentation. (cf: companion paper on sedimentation theory). Thus these papers will become mutually complimentary.

2. OPTICAL PROPERTIES OF FINE PARTICLES :

The light scattering properties of the colloids and related microhetrogeneous particle aggregates had been investigated by Mie et al., (49). Such studies have helped to interpret the diffu - sion, Brownian motion, sedimentation and coagulation of colloids.

Light when passed through colloidal solutions suffers

refection, refraction and often opalescence(scattering). While true
solutions are highly transparent to light, colloidal solutions
are always characterised by opalescence caused by the light scattering exhibiting the typical Tyndall cones. Accordingly, it has
been found that light scattering occurs when the size of the particle is much smaller than the wavelength of the incident light(λ)
while reflection would always occur when the particle size becomes
larger than $\lambda/4$. When the size of the colloidal particle is very
small then the light is scattered predominently at the 0° and 180°
to the direction of the incident light. (figure-1a;i). If the particle size is comparitively bigger(but still smaller than $\lambda/4$), then
the light scatters essentially along the 0° to the incident light
(figure-1a;ii). While the scattering light is not polarised at 0°
and 180° for small particles, it is highly polarised at 90° . For
larger particles the direction of the maximum polarisation is at
an angle to the 90°- direction. Rayleigh deduced an expression for
the intensity of the scattering light which is proportional to the
sixth power of the radius of the particles. Also, since the intensity of the transmitted light is proportional to $1/\lambda^4$, it is found
that while the transmitted light through colloidal solutions exhibited red colour, the scattered light is always blue. The above analyses apply to particles of non-conductor of electricity. When conducting particles are involved, electromagnetic radiations are usually
completely absorbed causing opacity of the particles. In such particles as the experiments would show, the intensity of the opalescence
does not grow as the λ decreases, but passes through the maximum
charecteristic of every metal. As the degree of dispersion is varied,
a lower value for such sols would tend to shift the maximum towards
blue spectral end .

Particles adsorb light to a great extent . The Bougeur-Lambert
law which connects up the intensity of the transmitted light (I_t)
with the intensity of the incident light (I_0) is

$$I_t = I_0 e^{-kt} \tag{1}$$

where k- absorption coefficient and t- thickness of the absorbing
layer. Accordingly, it will be seen that as the thickness of the
layer increased arithmetically , the transmitted light decreased
geometrically. Beer showed that the absorption coefficient of the
solutions with colourless and transparent solvents is proportional
to the molar concentration of the solute.

It has been experimentally found that the Bougeur-Lambert law
is more applicable to sols having high dispersion but not with high
thickness of layer or concentration . However, its applicability
to very low dispersed sols with high opalescence is very difficult.
Often much of the incident light is scattered and sometimes this is
considered as imaginary absorption. Such an absorption will give
the sol orange colour in transmission and bluish in scattering.
Metallic sols exhibit anomalous behaviour in absorption. Like light
scattering , absorption by metallic sols reaches the maximum at
definite values of the wave lengths and the particle radius .As the
dispersion degree increases , the maximum shifts towards the short

wave lengths and the value of this maximum first grows and then decreases . This maximum is attributed to the imaginary absorption. Also the metallic sols absorb heavily and hence the degree of screening increases with dispersion.

Colloidal systems are often coloured. Such colours are often very intense and depend upon the size of the particles , the degree of dispersion etc., . Non metallic sols which do not have any selective absorption usually exhibit orange transmitted light and bluish opalescence in refelected light. The problem of the colour of the metallic sols is more difficult to understand and corres – ponds to the case of true absorption. Sols of the same metals may often be differently coloured as the particle size and λ are varied. For example, coarsely dispersed gold sol will exhibit blue transmission and red scattering light. When the dispersion is made high the transmission becomes highly red while the scattering is highly blue. In the former the true absorption maximum is in the red and in the latter it is in the yellow-green.

The application of the principles of the GET will now show that all the above mentioned phenomena can be systematically deduced. To this extent, we shall first consider the reflection and transmission of light through typical particles. Referring to figure-1b, we find that when light falls on a body of radius R ,the area of cross section which would receive the light is πR^2.Light photons have typically an area of cross section of πr^2, where r is the radius of the light photon. Hence the number of beams of light that will be involved in the excitation is R^2/r^2 .

Consider for simplicity, the wavelength of the incident light as 4000A° and the size of the particle as R=1000A° and hence the number of beams intercepted will be $[(1000 \times 10^{-8}/3.3 \times 10^{-11})]^2 =$ 9.2×10^{10} . With the wavelength at 4000A°, the frequency will be 0.75×10^{15} Hz. In as much as each wavelength contains one photon of h erg sec(the energy particle equivalent of h being 0.88×10^5), the number of particles received per second will be equal to $(0.75 \times 10^{15}) \times (9.2 \times 10^{10}) \times (0.88 \times 10^5)=6.21 \times 10^{30}$. The volume of the body is $(4\pi/3) \times (1000 \times 10^{-8})^2 = 4.2 \times 10^{-15}$ cc so that the density will become $6.21 \times 10^{30}/4.2 \times 10^{-15} = 1.24 \times 10^{45}$ p/cc. On the same lines, when the radiation corresponds to 7000A° and the size of the body becomes 1800A°, the final density generated will be 0.45×10^{45} p/cc. In general, with the number of beams received being R^2/r^2 , the particles received per second will be proportional to $(R^2/r^2)\nu$ and the density proportional to $(R^2/r^2)\nu \times (1/R^3) \propto (\nu/r^2) \times 1/R$ For a given ν (with r being constant), the density is inversely proportional to R.Hence when light falls on a body of given radius it will show reflection(figure-1b;iii) so long its size is large when compared with $\lambda/4$ and at R$<\lambda/4$, the density of energy particles received by the body exceeds that of heat photons. It should be seen that the critical case $\lambda/4 = R$ is worked as above for numerical values of 4000A° and 7000A° which are the limits of visible light. In as much as the heat photons scatter in all three dimensions, their existence in a body will hinder the coherency of the incident light so that the body no longer reflects light .

Therefore when a microscope employs a light of wavelength λ, in the region of 4000A°-7000A its resolution highly fails to visualise bodies of sizes less than about 1000A° which is found experimentally. On the same basis, we can show that when an electron beam is used, the density level of electron accretion being 4×10^{60} p/cc it would permit much smaller particles to be resolved . The computations will show that the size of the particles involved can be down to about 1A°. In this way we understand as to why electron microscopy can be much superior to conventional optical microscopes.

The above will lead to the analyses of the Mie's experiments on the light scattering. We will notice that when a light beam pumps the energy particles into the buffer of the given particle, there is a rapid accumulation which will tend to enlarge the buffer However, for small particles, the buffer size itself is much smaller and hence the addition of the photons would tend to grow the buffer preferentially along the axis of the light radiation. The cylindrical buffer formation is obvious since in this way , the buffer can stil tend to maintain its value at around 10^{45} p/cc. Thus the chain(cylindrical) formation is the easiest to build up and it can be shown that the central core of the given particle will not be greatly stressed due to the accumulation of the energy particles in the given linear(axial)direction only. On this score, the p+ contained in the incident light will tend to scatter preferentially along this axis. This is greatly brought in by the A-circle (p+ in character). The A-circle particles will tend to scatter the incoming p+ of the light photons in the forward direction. At the same time on the frontal regions, some p+ can manage to enter into the A-circle (from the incident light). This will only enhance the density of the A-circle which will now redistribute the p+ particles in such a way that the excess p+ (that entered in) will be pushed out along the diametrically opposite end to the entry point. The prominent lobes of the scattering diagram would then be understood. The quick redistribution of the p+ by the A-circle would prevent the scatter of the p+ along the perpendicular direction and hence the Mie diagram would show minimum scattering in this way. Any such light photon that gets a 90°-rotation (polarisation effect) will be easily moved out with out any hindrance from those moving along the axial direction. Hence we have more polarised light scattered in the perpendicular direction. Such a 90°-rotation for the light photon is not possible in the axial direction. When the size of the particle is relatively large , its buffer also will be big . Hence the light photons entering will not be able to distort the buffer of the particle along the axial direction. The growth will tend to be spherical as the accumulation is more in the frontal zone . The scattering will be more in the forward direction due to the A-circle scattering of the p+ . In both these cases, the essential difference comes in from the fact that when an incoming spherical unit (light photon)has comparable size to that of the receiver(particle) the shape of the combined object will tend to be a dumb-bell and when the receiver is sufficiently big and spherical, the addition will not materially alter the spherical shape. The scattering of the p+ (blue) being effected by the A-circle, the p- pass along with out any hindrence and hence the transmitted light will always

be red. The scattering of p+ by the A-circles will be apparent
in the case of bigger particles when the moderately bulged lobe
is made at 180°. In view of the unsymmetrical scattering of the
p+ (along the 0° and 180°)it is logical that the polarisation
effect must be more tilted from the 90°-plane towards the 0°-end
(incident direction).

For non metallic particles light scattering effect always
shows a maximum at a particular λ . This is easily visualised .
For a given particle dimensions, we had seen that as the λ is
varied, there will be light reflection so long the dimensions of
the particle is quite different from $\lambda/4$ (i.e., $R \gtrless \lambda/4$). But at
$R = \lambda/4$, the heat photon generation occurs , which is the cause
for the light scatter as explained earlier. When the density of
energy particles are different from 10^{45} p/cc, no scattering occurs
since the heat photons only are highly p+ so that the A-circle
repulsion can be operative. On this score, it can be seen that as
the sizes of the various atoms and their spectral absorption (p+)
would change , even for the same size of particle (of the sol),the
effective λ for maximum scattering must change. Also as the dispersion increases, the A-circles of the adjacent particles would be
made to come closer that would stress them to greater extent. This
will be equivalent to enhancing their density values and reducing
the size of the A-circles(equivalent to reducing the particle size)
and so the corresponding λ must be smaller (for the $R = \lambda/4$ condition). In other words, the maximum shift towards the short-wave
lengths as the dispersion increases.

The Bougeur-Lambert law for the transmitted light can be understood from the fact that the relationship $(dI_t / dt) = - kI_0$ so
that we obtain $I_t = I_0 e^{-kt}$. In as much as the exponential variation would show that for successive layers of thickness t which
increases arithmetically, the intensity would decrease geometrically. This would imply for example, that if the intensity of the
transmitted light at the top layer of a sol is 100 % , for the next
layer it will be 50 % and so on .From the physical aspect,this variation must be readily understandable.

We notice that in as much as the intensity of light is mainly
contributed by the number of light beams striking an unit area ,
it amounts to the fact that for the geometric variation mentioned
above, the density of the energy particles in the successive layers
will be $\sigma_1 = 2\sigma_2$; $\sigma_2 = 2\sigma_3$ and so on . Referring to figure-1c, we find
that as the light falls on the given particle it tends to immedia-
tely build up its density σ_1 which will be the same in all the plane
(perpendicular to the incident light). Such a particle will tend
to act as the core governing the buffer zone of radius R all around
This implies that with R=t, the number of particles in the circle
will be $\pi R^2 \sigma_1$ and if this is effectively radiated in all the direct-
ions(with respect to the central particle as the controlling factor)
the radiation envelop will be $2\pi R^2$ and the density σ_2 builds up
such that $\pi R^2 \sigma_1 = 2\pi R^2 \sigma_2$ from which $\sigma_1 = 2\sigma_2$. The next layer is formed
by an appropriate particle on the spherical shell of the above
(first)particle along the direction of the incident light generating

its own compliment $\delta_3 = \delta_2/2$ etc., . In other words, equal successive
layers will show up the required geometric variation. The control
spheres generated in a given layer will obviously depend upon the
molar concentration (that will decide the number of such particles
of the dispersed phase)and the absorption coefficient (which stands
for the measure of the number of energy particles confined within
the volume of the hemispheres) must therefore be proportional to
the molar concentration as is postulated by the Beer's law. The
applicability of the Bougeur-Lambert-Beer law to sols of high dispersity but low thickness and concentration is evident from the above
since, the smaller the size of the particles, the more freely the
buffer zones would interact effectively leading to the density law
$\delta_1 = \delta_2/2$ etc., . (the smallness of the size increases the dispersity,
reduces the thickness (R=t) and a lower concentration gives greater
gaps for full establishment of individual buffers of particles).For
low dispersity the particle size is larger and should call for a
thicker layer equivalent . But in this condition, the bigger size of
the particle would generate the front-end (in the direction of the
incident light)opalescence zone which implies that the incident light
is not fully transmitted to the next hemisphere(of the second layer
particle)along the forward direction , but scattered back considerably there by reducing the light available for the layer-to-layer
onward transmission. This means that the Bougeur-Lambert-Beer law
must fail to hold good. Incidentally, as long as the scattering is
present, the transmitted light must obviously be reddish and the
scattered light bluish as explained earlier. The metallic particles
of a sol exhibiting absorption maxima and the corresponding shift
in the value of the maximum towards the short wavelengths will also be
obvious from the previous considerations.

The gold sol showing basically a reversal of scattering property
will be quite interesting in as much as it gives reversed colours
of transmitted and scattered lights as the dispersion is varied. Thus
referring to figure-2, when the dispersion is low, the size of the
sol particles will be large and the associated buffer will be more
firmly confined to the individual particles. As the incident ray pumps
energy particles into the system, the large interparticle gaps and
the consequent feable interaction between adjacent A-circles will
permit the p+ of the rays to pass freely in the transmission. The
p- will now be greatly reflected by the strong buffer around each
particle so that a great proportion of it appears as scattered light.
The colours of the two lights will be thus accounted correctly. When
the dispersion is high, the particles come closer to each other and
the A-circles and buffers coalesce greatly tending to block the energy
particles . The coalesced A-circles will now effectively prevent the
the p+ and reflect them as scattered blue light. The buffer, because
of coalescence becomes much weaker and hence the p- can easily pier-
ce through and show up the red transmission. The exhibition of the
maxima in the intensity -wavelength curve and its shifting towards
shorter wavelengths has already been explained as above.

3. BROWNIAN MOTION IN COLLOIDAL SYSTEMS :

While crystals have long range ordering , the molecules in a

liquid show a definite short range ordering. However, in a liquid
the ordering is highly unstable and are highly temperature dependent. As per the modern concepts , the short range order in liquids
is the manifestation of voids in the molecular packings. These
holes increase in number and size as the temperature is increased
causing melting etc., . The ability of a molecule to jump over the
potential hole (void) is a measure of its self diffusion. The energy
it requires for such an act is called the activation energy. The
thermal effect is to increase the energy of a given molecule in
that it tends to vibrate around a mean stable position and also
drift from one place to another . Such translatory motion is usually in jumps . These motions are largely known as the Brownian
motions.

   Brownian motion does not depend upon the external energy sources such as light , shock etc., . Temperature increase intensifies
it while for particle sizes exceeding 50,000A°, the motion completely stops. Many attempts were made to account for the motion
through recourse to the concept of convection currents, the effect
of the electric field , the property of wetting etc., , but,they
are all not satisfactory. Gouy and Exner attributed the effect
as due to the molecular kinetic forces and this was later confirmed by Einstein etal., theoritically and by Perrin etal., experimentally (49).

   The diffusion is the spontaneous equilisation of the concentration of the molecules and is irreversible. The osmotic pressure
is often assumed to be the cause of diffusion. Fluctuations associated with the mean stationary location of a given molecule will
be opposite of diffusion, although both are caused by thermal
motion.

   Osmotic pressure is always associated with semipermeable membranes and the pressure existing between two systems composed of
particles of radii r_1 and r_2 is found to be inversely proportional to the cube of their radii ($\pi_1/\pi_2 = r_2^3/ r_1^3$) . Hence the osmotic pressure of lyosols can be considerably altered by extraneous
forces.

   By applying the principles of the GET, we find that all the
associated features of the Brownian motion can be completely visualised. The molecular ordering that is envisaged for liquids is
primarily the result of the buffer interaction. As the molecules
tend to come closer together, this interaction will lead to coalescence of the buffers and the consequent aggregation of the parent
masses(cf; companion paper on colloids). This situation is clearly one of improving the ordering in molecules and is of short
range in character. Simultaneously when the very buffer strength
can be increased due to temperature (heat photons of p+ character) , the mutual repulsion will tend to separate the two bodies
(as the cores of the parent bodies will always absorb heavily p+
because of their nuclear character). Thus disordering tends to
prevail. In a given situation, the two oppositely directed effects
will be largely governed by many factors like the temperature ,

the modes of energising the parent bodies otherwise etc., . In as
much as the general character of the buffers tending to coalesce
into a common envelop always predominates , the separation of them
into individual units must be at the cost of injected energy particles directly into the parent body(within the nuclear -electronic
interspace as the spectral, magnetic etc., forms of excitation)
that would propmote such an act and hence the injected energy
becomes the activation energy .

When the activation energy is injected into the given molecule
it will tend to enhance the internal density of the nuclear zone
which will reflect, by virtue of the density gradation effect, in
the form of an increased buffer size. This implies that the adjacent
molecules(which may also exhibit similar expansion)would tend to
move outward radially to the first. However, the enhancement of
the density within the original molecule is not three deimensiona-
ally stable for the given molecule and unless further molecular
aggregation could be effected(as for example, by the simultaneous
compression besides the excitation), the activation energy (particles) must be dissipated away by the original molecule. This implies
that the molecule contracts permitting motion of its neighbours
now in the opposite direction. This process can now be again repeated leading to the molecular vibration around a mean position for
any given molecule. If however, the excitation (through activation
energy) exceed a specified threshold value , continuous motion will
be the result to such an extent as required by the expanding buffer.
In this regard, we notice that in a given system, the excitation
energy is picked up at random always by any of the molecules and
once this happens, such a molecule becomes the core molecule which
tends to establish its density gradation effect over the others.In
this respect, the surrounding molecules act as slaves and be merely pushed out leading to the characteristic Brownian motion(along
zig-zag paths). Such a slave molecule can become subsequently by
itself as another master thereby tending to generate its own slaves.

The Brownian motion will thus be enhanced due to the increase
in the temperature of the system as every molecule will now receive
greater quantities of energy particles($p+$ of the heat photons)
to expand their buffers. In addition, the Brownian motion will also
be characterised by typical scattering generated by particles when
illuminated by light(this shall not however be construed as speci-
fying the influence of excitation by light on Brownian motion which
is contrary to experimental results but the proper perspective
will become clear through the exposition to follow).

For sizes of particles sufficiently small in comparison with
the wave length of incident light ($R \doteq \lambda/4$), we had already seen
that the density level within the particle will reach $10^{4.5}$ p/cc
and promote scattering(cf:companion paper on colloids). In this
case the given molecule(or particle) first receives the excitation
energy particles (there is automatically a storage developed inside
and the consequent raise in the density) and subsequently radiates
them back(scattering). The discharge of these energy particles
will be governed by the molecular arrangement in a given zone. We

26

had seen above that in a given group of molecules , always a core
(target) molecule originates to receive the excitation and has its
zone of influence (cf: Bougeur-Lambert-Beer law). The molecules
surrounding need not be all at the same energy level and infact
there is always a dissymmetry that would result due to the nature
and size(of the atoms) of the molecules. This will imply that the
radiation from the given target molecule will often be unsymmetrical
(refer to figure-3a) . This will always lead to the unsymmetrical
movement which will become clear by the application of the force-
energy equation of the GET and as discussed further on the Brownian
movement (21). Presently, in as much as the scattering generates
a symmetrical discharge of energy particles (at 0° and 180° with
respect to the incident light)the actual activation energy discharge
can occur in planes at 90° to the above scattering -discharge plane
and preferentially in one direction or the other due to the neighbouring molecular dispositions leading to the final motion of the
particle(figure-3b).

When the particle size increases , the scattering becomes
unsymmetrical (Mie's experiment)and thus the particle motion again
can then result, this time inclusive of the 180°-direction as well.
Symmetry would demand that the particle would move more along the
180°-plane than along the 90° plane as for smaller particles.

In deducing the above conclusions, we observe that the motion
imparted to the particle is primarily because of the incident energy.
This has been assumed to be derived from the light source. This
is logical, since under standard temperature and pressure conditions, the atmosphere is already well-illuminated by the light derived
from the sun and in this respect, becomes the perennial and omni-
present source of excitation. In as much as the light beams had
already been considered to the fullest crowding possible (worked
out at 6.6×10^{-11} cm as diameter for the incident light ray-cf: section
on optical properties above), any variation in the light intensity
as for example, by further augmenting the natural light falling
upon the particles of the sols will have no effect appreciably
noticeable. Also the nature of the addition of energy to the given
molecule will be the same for both the light photons and thermal
sources(except that for the latter , the number of heat photons
can be considerably increased), we must expect the Brownian motion
to be present under the STP conditions. This is well borne out in
practice by the colloidal particles exhibiting the Brownian motion
copiously .

Brownian motion is found to completely stop for particles
exceeding the size of 50,000A°. This can be easily seen .Following
the computations of density made in the earlier section , here, the
number of beams that will fall upon the particle will be $(50,000 \times 10^{-8}/$
$3.3 \times 10^{-11})^2 = 2.3 \times 10^{14}$. Assuming the mean average wavelength of the
incident light (normal terrestrial)light at 5500A° ($\nu =5.5 \times 10^{14}$ Hz),
the number of particles received by the body will be $(2.3 \times 10^{14}) \times$
$(5.5 \times 10^{14}) \times (0.88 \times 10^5) = 1.11 \times 10^{34}$. The volume of the body being
$4 \pi /3$ $(50,000 \times 10^{-8})^3 = 6.25 \times 10^{-10}$cc , the density will become $1.11 \times 10^{34}/$
$6.25 \times 10^{-10} = 1.66 \times 10^{43}$ p/cc. This density is sufficiently low enough

to cause scattering and the consequent Brownian motion . Even if
the incident light improved in frequency toward short wavelengths
(by a factor of about 30, to X ray regions -180 to 200 A°), the
resultant density will be lower than that of the threshold value
for scattering and hence the Brownian motion is not observed. However, if the particles-size could be reduced even at the same
5500A° (which is much easily accomplished)in practice, say to about
1000A° (R$= \lambda/4$), the density value will become in the above case as
$50x1.66x10^{43} = 0.83x10^{44}$ p/cc(almost $= 10^{45}$ p/cc), setting up the
Brownian motion . This will be in accord with the size of the
colloidal particles as considered earlier. The properties of diffusion , the development of the osmotic pressure etc., have already
been brought out by the GET elsewhere to which reference must be
made(21). The characteristics of fluctuation has already been incorporated in the discussion of the Brownian motion given above.

4. ADSORPTION AT THE SOLID-GAS INTERFACE :

While Scheele studied the adsorption of gases, Lowitz studied
the adsorption of the substances from solution. Saussure found that
when porous bodies adsorb, heat is usually liberated . Investigations
would show that there is the physical as well as the chemical adsorption . While the former is reversible, the latter is always irreversible. Physical adsorption is of more interest to colloidal chemistry. Physical adsorption occurs spontaneously. The amount of the
adsorptive involved is given by the Langmuir relation , which describes the experimental results fairly accurately. The adsorption
forces are rather complex. However for nonpolar adsorbents and adsorptives , the Lennard-Jones equation will be given by :

$$\theta = A\, r^{-6} + B\, r^{-12}$$

where θ - potential that determines the interaction between the molecules , A and B are constants characterising the energy of attraction and repulsion respectively and r is the radius of the molecule. The quantity Br^{-12} had been added to the above equation as an
empirical approximation. In a given situation , the interaction of
all neighbouring atoms over a given core atom must be considered
and this would usually be presumed for about 100-200 neighbours.

When the adsorbent consists of ions, then besides the action
of dispersion forces, there will also be induction forces. The polarizability of the adsorptive molecule and the electrostatic field
of the adsorbent will become important. If polar molecules are
adsorbed on ions or dipoles of the surface, molecular orientation
will result. The simple exposition of the Langmuir adsorption theory
does not account for more complicated cases. In such cases the concept of equipotential surfaces was introduced by Polanyi etal.,
(49).This had been further modified by the Brunauer-Emmett-Teller
(BET) theory. This latter will presume in particular the existence
of a definite number of active centers on which adsorption can be
initiated. Also the interaction between adjacent molecules will be
neglected in this theory(these are rather highly conditional aspects).

28

However, not withstanding these, the BET theroy gives the best
results for physical adsorption phenomena.

Usually capillary condensation would occur when fine pores
exist in the adsorbent. The adsorptive vapours usually condense
into these capillaries. This effect had been largely studied by
Deriyagin etal., . Experiments show that there is a sharp transition from the adsorption layer to the liquid phase for polar substances like water, alcohol etc., . In adsorption accompanied by the
capillary condensation, hysteresis is often observed(adsorption
and desorption isothermals not coinciding). It has been found that
the removal of air from a porous adsorbent usually greatly reduced
the hysteresis.

In many cases physical adsorption preceeds the chemical adsorption or chemisorption. Physical adsorption are exothermal and have
usually low heats of condensation(2-8 kcal. per mole)while chemisorption involves larger heats(200 kcal.per mole). Elevated temperature reduces physical adsorption but enhances the chemical adsorption. The desorption is more difficult in chemisorption , in which,
usually another substance other than the adsorbate is usually desor-
bed. Chemisorption is also specific and its activation energy grows
with the number of molecules.

Physical adsorption occurs very quickly . About 90-95 % of
adsorption occurs in a comparitively short time of 1-2 seconds.
There is however retarded physical adsorption brought in by the
porous adsorbing surface, simultaneous chemisorption, presence of
air etc., . Amorphous adsorbents adsorb gases much better than crystalline ones. Saussure established that the gases adsorb better the
more readily they are liquified-i.e., higher the critical tempera-
ture(also when their boiling point is greater). It has been found
that the boiling point , critical temperature and adsorption are
interconnected.(the boiling point is usually two-thirds of the critical temperature). Such relations are not exhibited in the
chemisorption.

5. ADSORPTION AT THE SOLUTION-GAS INTERFACE :

In this instance, owing to the homogeniety and smoothness of
the surface of any liquid, the concept of active centers will not
be applicable. The property of the surface tension will be useful
in this regard. Under the action of the surface tension, and when
the external forces are absent, the liquid always assumes a spherical shape. Increasing temperature reduces the surface tension propor-
tionally. When an interface is formed between two liquids, the interfacial surface tension is more when the difference in polarities of
the liquids is small. Hence liquids similar in their polarities
mix well in all proportions and hence the interfacial surface tension is zero. Adsorption strongly affects the surface tension of the
solution. All substances can be classified either as surface active
or surface inactive. The surface active substances (surfactants)
are capable of being accumulating in the surface layer. Surfactants

have a positive adsorption and their surface tension is less than
for the liquid. Typical surfactants for water are organic compounds
and fatty acids with large hydrocarbon radical etc., . These surfactant molecules have the polar groups which have a considerable dipole
moment and well hydrated(and engenders the affinity of the surfactant for water)and a non-polar hydrocarbon radical which is highly
hydrophobic. The surface active substances have negative adsorption
and their surface tension is larger than for solvents. This enables
them to go into the solution. Such substances include all inorganic
electrolytes, acids, alkalis etc., . These do not have the hydrophobic moiety and disintegrate in water into well hydrated ions.
Substances which do not affect the surface tension of the liquid
will be distributed uniformly between the surface and volume of the
solution.

Duclaux and Traube found that the activities of the surfactants
increase at the solution-air interface, on the average by 3.2 times
per CH_2 group.(In otherwords, when the activity increases geometrically due to the length of the fatty acid chains growing arithmetically). This was attributed to the solubility of the fatty acid
decreasing as the length of the hydrocarbon chain increasing. As
the temperature is increased from the ambient (for which the factor
of 3.2 is specified as above) , the activity factor is reduced to
unity. This is attributed to the desorption effect . However in all
the above, the rule is observed only for aqueous solutions of surfactants. In nonpolar solvents,the solubility of the surfactant
increases as the size of the hydrocarbon grows and the surfactant
gets into the solution thus reversing the Duclaux-Traube rule.

The structure of the adsorption layer at the solution-gas
interface can be studied by forming a thin film of poorly water
soluble liquids on water. When the molecules of the liquid attract
each other strongly than what water molecules could do with them,
then a thin lens shaped droplet of the liquid will form on the water
surface . If the water strongly attracted , then a thin film(mono
molecular layer)could be formed . When non-polar liquid is involved,
drops are formed. Diphilic molecules produce monomolecular -thick
films. These monomolecular films on water can be gaseous , liquid
and solid. The latter two are known as condensed films. When the
gaseous surface film is made of the water surface, generally the movement of the gas molecules will be confined to the surface of the
water in that the film is highly two dimensional. The gas films on
water surface with typical surfactants like organic compounds with
diphilic character. The hydrocarbon chain usually has a range of
carbon atoms from 12-22 only. The condensed films usually have
molecules grouped as islands that tend to move together. In such
islands, the molecules are oriented in parallel to one another and
perpendicular to the water surface making a palisade. Condensed
films are usually liquids and the molecules in them move freely.
As the temperature is increased , in most cases the condensed films
are converted into gaseous films.But when the size of the hydrocarbon chain of these liquid molecules become large (24 carbon
atoms) the films become corresponding to that of a solid (no move-
ments between the molecules).

6. APPLICATION OF THE GET TO THE STUDIES OF ADSORPTION :

Many of the details enumerated in the last two sections can be visualised in greater detail through the principles of the GET. Thus when we consider the basic phenomenon of adsorption we find that it would be defined either as physical or chemical in character. Some of the basic aspects of these have already been discussed elsewhere (28,33,34) .

In essence, when an adsorbing surface is in contact with a typical adsorptive, the firm adherence of the latter molecules on to the former can be purely physical. In this case the buffers of the two atoms or molecules will coalesce in as much as they are of the same polarity . Then desorption by the breaking up of this coalesed buffer will be also possible. But the chemical adsorption occurs because of the two coalescing buffers are made of opposite charges and hence once they are mixed up, (due to the otherwise electrostatic attraction as per the conventional parlance), they cannot be easily separated . On this score, while the physical adsorption can be created , for example, by increasing the pressure of the system; it can also be destroyed by reducing the pressure as found experimentally. However, the chemical adsorption would require the engagement of the atoms by their electronic circles and hence when once this is done , breaking the bondage is not possible unless some other chemical reaction involved will transform the parent substances into newer combinations.

In physical adsorption, it is often presumed that the two molecules approaching each other would exhibit the long range attractive forces and short range repulsive forces. This is obviously one of the inherent misconceptions of the conventional theories. In as much as two such molecules when they come closer are being bonded by the coalescence of their buffers must be clear from the concept of the GET. In fact, based on this, we had already deduced the energy of interaction for the system of two molecules(as given by the London equation otherwise) in connection with the studies on colloids. In the present case, the determination of the potential for a system through the Lennard—Jones equation will be complimentary. Roughly, as the energy of such a system can be imagined to be the product of the potential and the relevant charges , we could consider charges which are proportional to r^{-6} to be associated with each moleculeand hence by Coulomb's law a repulsive term could be formed as proportional to r^{-12} for use with the Lennard—Jones' equation. However the untrue state of these aspects will be evident in the light of the GET. It can also be seen that an equation as this had given good results in as much as it is confined to the general buffer zone of an atom which is otherwise accounted by considering the field of influence that is extended upto 100-200 atoms only (about 36.8 R , by the GET).

The forces of physical adsorption are easily recognised when an adsorber is introduced into a gas medium , for example, due to the Brownian movementa large number of gas atoms will come into contact with the atoms of the adsorber. This will promote immediate

buffer coalescence and once this has occoured , the two atoms would
be closely engaged. In this endeavour, the A-circles of the two
atoms will also tend to coalesce. If this occurs, physical adsorption becomes highly stable. It should be seen that this can occur
simultaneously over the entire surface of the adsorber and will be
monomolecular layer in thickness. Subsequent layer formations may
tend to improve the density of the outermost layer of atoms of the
adsorber considerably from those of the atoms within the bulk of
the adsorber solid. This is clearly not possible . This imples that
on the outside, the building up of the thick layers of adsorbate
atoms is not possible. -or, in otherwords, desorption phenomenon
will also set in to equalise the density gradation effect. This is
clearly the equilibrium condition.

    In the above process, the effect of increasing the temperature
will become obvious. In as much as the heat photons added to the
surface atoms at the interface of solid-gas or liquid-gas will enhance the A-circle activity . When a polar substance is considered,
it can be seen from the principles of the GET that the atoms involved will have overall spectral absorption of p+ and p- so that
in relation to the molecular disposition, spatial concentration of
these p+ and p- will simulate the equivalent dipole moments. On
this score, their buffer will also be complimentary and charged
accordingly . On the otherhand, the ionised atoms or molecules will
also be either in excess of p+ or p- and should therefore shaw
buffers of similar nature. Thus when the polar molecules are brought
into contact with the ionised or dipole(surface) atoms and molecules
orientation of the fields will become evident. In this respect, the
electrostatic field,if any, exhibited by the surface atoms will also
be contributory. In all these cases, we can therefore expect molecular orientation.

    In the investigation of the adsorption phenomena, the BET theory
will presume the existence of the activation centers . While this
is rather curious from the conventional angles, the success of the
BET theory will be seen to orient from this particular aspect which
otherwise is the same as the locating of the core atoms of the GET.
We had seen that in a given group, only a particular atom can be
the core-one(and although it is randomly oriented initially)and
once it is fixed, the subsequent matter building (or equivalently
the adsorption layer formation) will be fixed. It will be seen that
in this respect, there can be no interaction between this core atom
and the surrounding slave atoms which is also another valid(but
from conventional angles, a curious) assumption of the BET theory.
Hence it is logical to expect good correlation between the BET
theory and observation.

    Adsorption of the gas and vapours into the capillaries of the
adsorber resulting in the condensation is logical from the GET. It
will be seen that the surface atoms of the capillary walls will now
have their A-circles highly compressed due to closeness to each
other(for atoms located diametrically to each other oppositely on
the capillary surface). This will result in the ready squeezing of
the incoming gas or vapour molecule so that such a molecule will

easily condense into a liquid. On this account the sharp transition between the gaseous and liquid phases will become evident. The stressed A-circles of the capillary atoms will tend to push in turn the atoms of the condensed liquid especially when polar(charged)liquid atoms are concerned. The adsorbed atoms will either tend to cling to the surface atom of the capillary by buffer coalescence or will be compacted to such an extent that it condenses into a liquid and drawn into the(capillary)bulk liquid by surface tension forces. In this way the generation of the sharp transition will become obvious.

The property of the hysteresis associated with the capillary condensation can now be investigated. The gas atoms, when introduced into the capillary space would initially have free and independent buffers. Hence it would require compression for coalescence to occur between the buffers of the adjacent atoms of the gas, which is brought about by certain pressure p_1 . The presence of air atoms inside the capillary would be only complimentary to the process of buffer compaction. When the pressure is released,the coalescence of the buffer is not quite destroyed since the density gradation law would have modified already the coalesced buffers and tended to produce a spherically symmetrical cloud.Once this has neen carried out, breaking the spherical structure becomes difficult, at the same initial pressure p_1 . In this regard, the air atoms will tend to generate the most compact nucleation masses($N_2 : O_2$ will be 4:1 in the air, which tends to form a typical tetrahedral structure that is highly stable) on to which the gas atoms can stably adsorb (even at a much reduced pressure p_2) and generate the overall buffer that is density wise graded. If the pressure is reduced still below , it amounts to greater gaps existing between the given number of atoms present in the pressurised vessel. This gap has to be filled up by the expansion of the buffers of atoms. In this way the separation of the buffers would occur at a much reduced pressure p_2 ($p_2 < p_1$). On this score, it will become apparent that if the air atoms are removed , such nucleation centers will be available for the gas atoms and by virtue of their like character(of the buffers), the gas atoms can easily dissociate consequent on the reduction in pressure thereby eliminating hysteresis.

By virtue of the fact that the buffer coalescence is not very intimate for the physical adsorption where as it is complete in the chemical adsorption (the coalescence occurs to the extent that the electronic circles of atoms must touch one another), it is apparent that the physical adsorption will always be observed to occur first . On this score, the desorption of the chemically adsorbed atoms will also be very difficult. Since the partial coalescence of the buffers of the two adsorbing atoms would involve the rejection of only a smaller quantity of the buffer clouds , the heat of condensation will be very low for physical adsorption. But large quantities of buffer rejection or addition will occur in chemical reactions and the chemisorption so that the heat of condensation will be very high. In this connection, the method of computing the free energy, enthalpy etc., associated with the chemical compounds through the GET have already been presented (21 22), to which

reference should be made. In chemisorption, often, the adsorbing
molecule will anchor itself so firmly with the surface molecule
in that the spherical or cylindrical envelop forming buffer will
hold the atoms (or molecule)of the adsorptive relatively with
those of the adsorber (chargewise p+ and p- buffers will tend
to hold each other firmly). In the process, some one of the initial
adsorber atoms can be expelled . For example , consider the adsorber
molecule being made of a p+ atom and a p- atom(classified spectrally). Normally these will be in firm contact and constitute the
adsorber molecule. If now the adsorber atom(either p+ or p-)
brought into coalescence, depending upon its polarity. one of the
adsorber atom will be released(the released atom will have the
same charge as that of the approaching one but strength wise of a
lower p+ or p- so as to be preferentially rejected by the oppositely charged original adsorber atom).

    The physical adsorption, by virtue of the low energy condition
(6-8 kcal. per mole) would naturally require only a small time for
for adsorption. Thus referring to figure-4a, this can be computed
for instance, in the case of an O-atom, adsorbing to a C-atom,(both
of almost of equal size) . In this case(and in general) the physical adsorption is manifesting because of the contact of the two
atoms over a small volume at the point of contact. For a given
atom this can be computed as $4\pi R^2 t$ (θ /360) where R-is the radius of the atom(smaller of the two that are involved) , t- is the
thickness of the layer which corresponds to the light photon (3.3 x
10^{-11} cm), θ - is the angle of contact. For the example chosen, we
have the angle of contact (the angle subtended by the adsorbate
atom at the center of the adsorber atom) is about 53°. Thus we have
the contact volume as $(4\pi)(0.78x10^{-8})$ $x(3.3x10^{-11})$ x(53/360) =
$4.05x10^{-26}$. The density in this elemental volume will correspond
to the geometric mean of the atom's internal density of 10^{45} p/cc
and the external radiation density of $1.24x10^{42}$ p/cc giving =
$\sqrt{10^{45} x1.24x10^{42}}$ = $3.5x10^{43}$ p/cc. Hence the number of particles
contained in the small volume will be $4.05 x10^{-26}$ x $3.5x10^{43}$ ÷ $1.42x$
10^{18} .If now we assume the energy of condensation as 8 kcal./mole
(which can also be deduced otherwise by the GET-refer (21)) , the
energy particles per atom will be $(8x10^3 x4.2x10^7 x6.67x10^{30})/$
$(6.025x10^{23})$ = $3.7x10^{18}$. Hence the time required to build up the
buffer or lose it otherwise during the adsorption , will be$(3.7x10^{18})/$
$(1.42x10^{18})$ = 2.7 seconds. This may be compared with the experimental value of 1-2 seconds(for 90 to 95 % adsorption.).

    Chemical adsorption will occur on the above lines at a much
longer times owing to the large buffer transfer unless the same
had been effected at an elevated temperature(particle density). On
this score, the retarded physical adsorption brought in by simultaneous chemisorption , presence of air(which tends to reduce the
rate of energy imparted to the adsorption zone) etc., will all become apparent.

    The variation of the adsorption of a gas in relation to its
critical temperature (or with the boiling .point) can be easily
understood. The significance and the properties of the critical

temperature, the boiling point and their inter relationships(the two-thirds law) etc., have all been brought in great detail already by the GET to which reference should be made(21). In the present context, we will observe that for any given gas, as its critical temperature and volume are related to the standard temperature and volume as

$$(\; T_c \; / \; T_s \;) = (4\pi/3) \; R_{xT}^{\;3} / \; (4\pi/3) \; R_a^{\;3} = R_{xT}^{\;3} \; / \; R_a^{\;3}$$

where R_{XT} and R_a are the radii of atoms at T_c and T_s respectively Hence for a given atom, the greater the critical temperature, the larger the value of R_{XT}. From the above equation , we can find for most gases of higher T_c , the nature of R_{XT}-variation from R_a will be marginal and $R_{XT} < R_a$ so that when such an atom adsorbs on to a surface it will only marginally part away with its energy clouds. On this basis, the adsorber atoms can entertain a larger number of host atoms without getting themselves highly saturated with energy particles. Thus the adsorption of the gases with higher T_c will be very high . In chemisorption such relations are not exhibited,since, the engagement at zone between the adsorber and adsorbate atoms will be over larger volumes and this would also tend to disturb the internal particle density of the two atoms. Thus the quantities of energy particles injected or abstracted will be large and can bear no simple relations to the properties like T_c etc., .

In the studies involving the liquid – gas adsorption phenomena, the property of surface tension will be mainly used. From the GET point of view, the surface tension, (being a force),is the manifestation of the density gradation law. Thus when atoms are relatively free as in a liquid, they tend to be attached to each other so that accretion can proceed to improve and fix the density gradation law.But, in this regard, any atom in the liquid can act as a nucleating center and therefore the accretion does not proceed smoothly to solidification. But when atoms are forced to come closer (as for example, by reducing the temperature)then the physical touch between the atoms is forced to exist which will lead to the buffer coalescence and hence solidification. The internal cohesive forces which is otherwise identified as the compaction factor (CF) by the GET plays the key role in this aspect. Its maximum value is 15,200 for a phase transformation, for example, for water of relative density equal to 1.0 . For liquid water, the experimental value for the internal pressure will be 14800 atmospheres signifying that it can solidify by taking a marginal compression(to raise the value to 15200 atmosphere -equivalent). On this basis, it can be seen that the formation of ice becomes much easier by bringing in this extra compaction of water molecules through recourse to temperature reduction.

When two liquids are in contact, the relative CF value of the liquids will tend to decide the behaviour of the interface. The genetical aspects of the various atoms will show that the internal nuclear structures of the atoms are so made that they can entertain only a specific buffer at a given temperature and pressure. Hence these parameters are the main criterion in deciding the inter

relationships of the molecular constitution and further aggregation.
On this score, when an atom A is introduced into the liquid made
of atoms B , it will find itself differentiated from the surroundings through the buffer considerations only. If the buffers are
compatible(oppositely charged), the atom A may even lose part of
its buffer by way of neutralisation and adjust its density gradation effect. In this case the atom A goes truly into the solution.
On the other hand, if the atom A had an incompatible buffer or even
larger than as can be tolerated by the system of atoms B, it is
merely pushed out of the surroundings. In this case, the atom A
comes out and floats on the liquid (when contained in a jar).When
the atom A is replaced by a molecule A which has a compatible buffer
only over part of its structure, then the location of the molecule
will be part into the liquid and part outside(this will be a typical
molecule with polar and non-polar parts as per the conventional
parlance). Thus the entire phenomena of liquid-gas adsorption and
the various observations made in this connection must be answerable
by the above general mechanism.

The behaviour of the surfactants will be easy to visualise.
In as much as its equivalent CF-value would be less than that for
the liquid (the surface tension being lower), the surfactant molecules will tend to float on the surface . In this regard, the diphilic nature of the molecules (with p+ and p- buffer portions)
will effectively prevent their entry into the bulk of the liquid.
The polar portion of the molecule will be highly p+ and hence the
corresponding buffer will be p- . On the same score, the water
molecules are highly p- and hence their buffer will be p+ .
Therefore, the polar ends exhibiting affinity to the water molecules
will be logical.

A little consideration will show that the hydrocarbon end of
the surfactant molcule will also will bear a net p+ charge(2.0 p+
for every CH_2 unit)so that the hydrophobic character of this portion
of the molecule will appear to be in contradiction to the above
deduction. Actually, this property comes in because of the geometrical shape of the molecule . The polar group is usually clumped
and hence it will easily build up its spherical symmetry in 3-D
space . On this account, it can attract water molecules (especially
because of the compatibility of the buffers) and thus build up the
stable density gradation law. However, the long hydrocarbon chain
can be shown to build up a density of 1.24×10^{42} p/cc along its
entire length . This value is close to the density of the surrounding which is 2×10^{40} p/cc. Therefore along the planes perpendicular
to the long axis of the chain,no further addition of atoms will be
possible • This will be the general property of the long chain hydrocarbons. However, small atoms (like N,O etc.,) of either buffer
character can be attsched to the chain at regular intervals which
will simulate local 3-D density law provided each such zone is well
displaced from its neighbour. Substitution of the atoms other than
H in the hydrocarbon chain will then become apparent. However, with
the water molecules , the substitution is not possible since the H
atom is already having a p+ buffer (in the CH_2 -CH_2 .. chain ,
the H are located sideways) and hence it will effectively prevent

the approach of a similar buffer(water) leading to the characteristic hydrophobic property.

The surface inactive substances must necessarily have greater CF(larger surface tension) so that they do not stay at the surface. They are therefore pulled into the solution , where for reasons stated earlier they could stay stably. In this regard, the absence of the hydrophobic moiety will be noticed. For example, the molecule of H_2SO_4 (which is a typical surface inactive substance) has 18.2p- and hence its buffer will be mainly p+. It can well go into the zone of water molecules since by itself it has a high spherical symmetry and so it can build its 3-D density law much against the buffer repulsion. In fact the surrounding water molecules (3-D distribution)would keep the H_2SO_4 molecule, by the mutual buffer repulsion stably in space and this complex set up is what is otherwise referred to as the hydration in the conventional parlance. Finally, we note that substances having the same CF would have no preferential behaviour with that of the bulk liquid and hence can be located at any place-from the surface proper and into the bulk of the liquid.

The Duclaux-Traube rule can now be investigated. At the aqueous solution-air interface, we will find that the water molecules have each 8.5p- and this would go in conformity with the basic tetrahed of the air complex (4N + O)as found in the atmosphere which will have 17.2p+ . Thus two water molecules will engage each complex-unit of air. In this respect, the H of the water molecules will bond preferentially with the O-atoms of the air or water as shown. Thus into this environment, if the surfactant molecule is brought in, it will provide for every CH_2 unit a net 2p+ (8-2x3). When the buffer system formed between the O-atoms of air and water are provided with this extra 2p+ , it is equivalent to reducing the net charge on each water molecule into 8.5-2.0 = 6.5p- . Since the share is for two O-atoms, then the effectiveness of each CH_2 is 6.5/2 =3.25 which is the observed value. On this basis, when the temperature is raised from the ambient , the heat photons pump extra p+ into the system(depending on the temperature) and hence we find that a CH_2 unit will appear to have its effectiveness reduced(since now the share is 1/2(8.5-2-x) where x tends to increase with the temperature). In the limit, the value of x can be increased until the net charge value drops to 2.5p- which corresponds to that of the O-atom alone so that the effectiveness is 1/2(2.5) =1.25 which is found in practice. It must be noticed , in this connection, as to how the buffer effect of the H of the water molecule is gradually reduced by the similar buffer of the H of the CH_2 radical, with the assistance of increasing temperature. When the non-polar solvent is involved, it is easy to see that by virtue of the high p+ character of both the solvent and the air molecules the net interfacial field becomes highly p+ . Into this if a CH_2 is introduced , it will experience a net repulsion that can easily push it into the solvent. We find that the solubility of the surfactant increases and hence as the size of the hydrocarbon chain increases this effect will also increase. In otherwords, the Duclaux-Traube rule must be reversed.

Depending upon the CF-values of substances coming into contact, a variety of interfaces can be formed as explained earlier. Thus purely water soluble substances will tend to collect partially into a spherical object. The coalescence cannot be complete due to some of the molecule of the substance tending to get into the water and hence a lens shaped object results. When the water molecules are able to attract better, the buffer becomes compatible and thus the monomolecular layer of the substance on water is formed. With non-polar substances the buffer compatibility does not take place and hence the molecules of the substance, on water, tend to seggregate forming a spherical droplet.

The formation of the gaseous film over water in the presence of surfactants of diphilic character is logical. The water being p- , will engage directly the polar(p+) end of the surfactant. The non polar hydrocarbon chain will easily allow the physical adsorption of the gas atoms(which are usually small) so that the density gradation effect is not violated. In this respect, it can be seen that the hydrocarbon end of the surfactant must be protruding on the water surface. This happens because the moving gas atoms on the water surface (which have usually the same type of buffer as the hydrocarbon chain)will tend to repel the surfactant (hydrocarbon) end. In this endeavour, the end so deflected will find an easy neutral zone on the surface of the water(zero force location and hence equilibrium). This will also be clear from figure-5a. As the water molecules are uniformly spread out , we notice that the symmetry would give the least configuration of 3 water molecules triangularly placed(partially filled tetrahedral structure) when they would have a combined cloud of 8.5x3= 25.5p- . In as much as each CH_2 has a net of 2p+ , the minimum number of units of CH_2 to balance this water clouds will be 25.5/2 = 12.75(quoted minimum is 12). Thus the zero field(no p+ or p- effect)is existing for this configuration and the h.c end of the surfactant can bend itself to find a null zone where the buffer effect of neighbouring gas atoms (of the film) will not be realised. The gas atoms by themselves would find this neutral zone highly conducive for their adsorption (or location in the gas film). This implies that the surfactant molecule is essential for this gaseous film. The limiting for this process occurs when symmetry is shown extended by 4 and maximum of 5 (fully filled up tetrahedral structure)of water molecules . Correspondingly,the h.c. equivalent will be 8.5x5/2 = 21.5 which is the observed result. For h.c. units less than 12, the water effect is considerable so that its buffer will tend to push out the gas molecules (destroying the film). The same occurs when the CH_2 units are larger than 22 when the self-field of the h.c. unit will repel the gas atoms.

When the condensed film is considered, the stronger p+ of the liquid(and solid) molecules will tend to push the h.c. end of the surfactant from all sides . Since the polar end of it is well anch-ored into the water, the molecule obviously tends to attain equili-brium only in the vertical direction. On the same basis, a large number of surfactant molecules can be crowded together when they form a coalesced buffer so as to generate the typical islands .Each such group of units will act collectively (because of their

coalesced buffer) and will tend to push similar units away and so
floating islands are formed in the base water surface. The palisade
is formed because of the buffer coalescence . The limiting condition for this can be easily deduced. Referring to the figure-5b, we
find that the basic unit of 3 water molecules (of the tetrahed)will
have an overall radius of influence of about 4A°.Adjacent such units
will tend to be away from one another by 16 A° so that the effective
zone of control of each unit is 8A°. This will hold as many as
$(8/2.9)^2$ = 9 units of the surfactant(the overall radius of the CH_2
is 1.45 A° and so with its zone of influence considered, the effective radius is 1.45x2 = 2.9A°). Then consider the h.c. chain to be
made of about 24 units of each 2p+ . The density for the cylinder
of base radius 8A° and height of 36A°(for 24 units of CH_2) will be
$(9 \times 24 \times 2 \times 10^{-12} \times 6.67 \times 10^{30})/(\pi \times 8^2 \times 10^{-16} \times 36 \times 10^{-8}) = 3.32 \times 10^{42}$ p/cc.
Thus each island builds up its buffer that tends to radiate freely.
Hence adjacent units at the contact point like X(see figure)will
develop about 6.64×10^{42} p/cc which will promote buffer coalescence
and the islands thus get locked up with each other. Clearly the
molecules contained in each island are arrested in their place simulating the "solidness" for the condensed film. As shown above, this
occurs for all CH_2 units exceeding 24 in number.

7. PROPERTIES OF THE AEROSOLS :

Aerosols have a high rarifaction and low viscosity and exhibit
very intense Brownian motion. The particles of aerosols often sediment heavily and they do not exhibit the typical electrical double
layer(owing to the gaseous medium). The optical properties of aerosols are the same as for the lyosols but often intense in character.
The Stoke and Cunningham equations would give the frictional force
for the acrosol systems . Coupled with these, the details of the
Brownian motion would enable to identify the optimum size of the
stable aerosol particles . In this attempt, the diffusion and the
sedimentation times would be reckoned to be the same. Aerosols
often exhibit the properties of thermophoresis, photophoresis and
thermoprecipitation. The first describes the ability of the aerosol
particles to drift from higher to lower temperature zones. The second involves the motion of the particles(opaque)either towards the
direction(+ve photophoresis) or the movement of the particles(trans-
luscent)along the direction(-ve photophoresis)of light. This is
attributed to the unequal heating of the particles in the front or
back with respect to the incident light. The last effect describes
the ability of the particles to precipitate when the aerosol comes
into contact with a cold surface. Although the aerosol particles
do not have the electrical double layer in them, they often show
charges due to the adsorption of ions or movement of the particles
(Brownian)etc., . Aerosols of metals and their oxides like Fe_2O_3,
MgO, Zn, ZnO and flour (starch) bear -ve charges. Non metal particles and their oxides such as SiO_2, P_2O_5, NaCl, coal, starch etc.,
are charged +vely. However, owing to the great complexity of the
aerosol dynamics, these charges cannot be easily related to their
dispersity. Experiments show that the charge on the aerosol particles are usually low(of the order of a few electronic charges). The

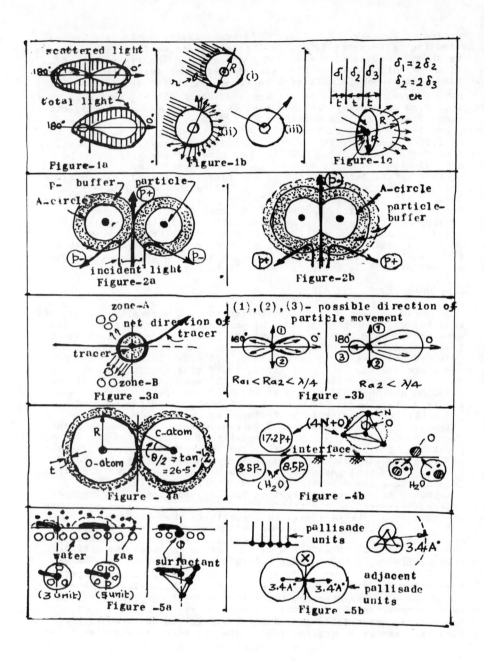

Figure-1a

scattered light
180°

total light
180°

Figure-1b

(i)

(ii)

(iii)

Figure-1c

$\delta_1 = 2\delta_2$
$\delta_2 = 2\delta_3$
etc

Figure-2a

p- buffer
A-circle
particle
P+
P~
incident light

Figure-2b

A-circle
particle-buffer
P+

Figure-3a

zone-A
net direction of tracer
tracer
zone-B

Figure-3b

(1),(2),(3)- possible direction of particle movement

180°

$Ra_1 < Ra_2 < \lambda/4$

$Ra_2 < \lambda/4$

Figure-4a

R
C-atom
$\theta/2 = \tan^{-1}\frac{1}{2} = 26.5$
O-atom
t

Figure-4b

$(4N+O)$
N
$17.2P+$ interface
$8.5P-$ $8.5P-$
(H_2O)
O
H_2O

Figure-5a

water
gas
surfactant
(3 unit) (5 unit)

Figure-5b

pallisade units
3.4A°
adjacent pallisade units
3.4A° 3.4A°

origination of this charge is difficult to comprehend. Frumkin showed that a +ve electric potential of about 250 mV exists at the interface of the aerosol of water. This is presumed as due to the dipole moment of the water. On this basis, a typical rain cloud (drop size of about $10^{-3} - 10^{-2}$ cm) can build a voltage of about 100 volts per centimeter.

Earlier work centered around the studies of aggregative stability of aerosols assumed a gas shell around the surface of the particles. The great stability of the powders in storage, their fluidic flow properties, and non wettability were attributed to this gas shell. It is known that smoke can easily pass through water . Some investigations showed that a great amount of air can be adsorbed by the aerosol particles. Later investigations had discredited the adsorption of the air and the existence of the gas shell. Aerosol particles can have liquid shells on them. However, the stability of such films are doubtful. Orr etal., studied the effect of humidity on the coagulation of smoke and found that the rate of sedimentation of MgO and specially NH_4Cl increased in a humid atmosphere. The microscopic examination show compact structure for these particles . Thus , although the aerosol particles possess sufficient stability and dispersity, they suffer from aggregative instability and rapid coagulations. Coagulation occurs much faster in aerosols when compared with hydrosols. (gold sol can contain 10^{18} p/cc). Coagulation is improved by polydispersity and anisodiametric shape of particles. Aerosols are rapidly destroyed when the particles are oppositely charged but retain stability (scatter) for long if charges are of the same polarity. Aerosol particles of a liquid with high vapour pressure rebounce resiliently with one another. Coaglation rate is altered by convection currents, stirring, ultrasonic vibration etc., . Coagulation by aggregation can occur also by isothermal distillation. (bigger particles growing still bigger by evaporation of the smaller ones). In aerosols, the resistance of the medium(evaporation rate and cooling rate) is found proportional to r(radius)of the particles of sizes from 10^{-2} to 10^{-5} cm. It varies as r^2 for particles from 10^{-5} to 10^{-7} cm. The coagulation rate is almost constant for the first region and rapidly rises and falls in the second region. Light scattering for the above regions will be proportional to r^2 and r^6 respectively. While sedimentation is predominent in the first region, diffusion predominates in the second region. The vapour pressure(of droplets of water)is almost constant and increases(rapidly) respectively in the two regions.

8. APPLICATION OF THE GET TO THE STUDIES OF AEROSOLS :

In the earlier sections, the optical properties and Brownian motion of the fine particles have been brought out by the GET and these will directly apply to the aerosols. While the aerosol particles have sizes ranging from 10^{-2} to 10^{-7} cm, only particles of intermediate sizes (10^{-6} cm) tend to generate stable aerosols. The cause of sedimentation for bigger particles and diffusion for small particles as advocated will be found to be untenable. In obtaining the size of the particles ,if the time of diffusion is set equal to that of sedimentation(as is done), then it implies that the

41

direction of motion imparted by the two effects must be opposite
and distance moved as equal. Otherwise a given particle will contin-
ously move in space in one way or the other, which is not the condition for stability. However, if the diffusion is say horizontally
to the wall of the container while the sedimentation is vertical
(gravitational effect)then clearly the particle will not maintain
its stability . In general, a given particle can have both the components of motion in different direction and speed so that stability cannot be easily related to the particle dimension as is
advocated. The GET traces this aspect to the relative sizes of the
buffers involved. In this regard, it has already been shown that
the free Avagadro radius of an atom (at STP) will correspond to
20.7×10^{-8} cm and hence as the size of the aerosol particle(10^{-6} cm)
approaches this value , its buffer size just is sufficient to make
a physical touch with its neighbour. In this state, its movement
(Brownian) is restricted and orderly since if any one particle tended to move, the clearence being restricte, the movement of the neighbours must also occur in unison. Bigger particles will have larger
buffers and their consequent coalescence will lead to aggregation
of particles and thus rapid sedimentation. When the particle size
is smaller, their free movement is made easier and their aggregation by coming into contact with one another and building common
buffer becomes much easier, leading to coagulation. The particle
size so fixed up as above can then lead to the determination of the
mass of the particle(with its density being known) directly.

The electrophoretic effect observed in aerosol particles had
already been explained in detail by the GET to which reference
should be made (21). Photophoresis effect(which is difficult to be
explained by the conventional sciences)can be easily visualised.An
opaque particle has a stronger A-circle and hence no light beam can
pass through it. However, the light photons are deflected at the
front end along the A- circle contour and will be discharged on the
opposite end and the trajectory corresponds to that of the air flow
around a fast moving projectile(with a shock cone in front and eddies at the tail). This flow pattern will generate a thrust only at
the back end because of which the particle moves in the forward direction (towards the light source). In this regard, in the translus-
cent particle, since the A-circle is not strong , the light photons
will penetrate it and move forward. This action permits the light
photons to push the A-circle at the front and consequently the body
moves in the same direction of light. These are identified as the
+ve and -ve photophoresis respectively. In some substances the bigger particles with larger A-circles will obviously simulate higher
strength and so will exhibit +ve photophoresis. When the size of
the particle is smaller, the A-circle becomes weaker permitting the
push by the light photons leading to -ve photophoresis.

Electrophoresis occurs when the aerosol particles of the given
buffer sizes (which depends on the temperature)comes into contact
with a colder surface. The heat photons transferred to the cold
atoms will imply that the buffer of the hot (aerosol)particle has
reduced permitting crowding of similar such reduced (bufferwise)
particles to coalesce to form condensation.(A liquid is formed when

the atoms(or molecules) come closer than is possible for a gas,as a result of buffer reduction). The origin of charge on the aerosol particle is very difficult to understand from the conventional concepts. However, this is the manifestation of the spectral absorption and the consequent buffer build up as will be seen through the GET. For the substances showing -ve polarity we get (from the spectral data) Fe_2O_3 is 2.3p+ ; MgO is 2.8p+ ; ZnO is 7p+ ; Zn is 9.5p+ ; (starch as a complex structure of amylose)flour is net p+. We see in all cases, the structures are made as tetrahedral,dumb-bell,dumbbell, single atom and complex(spherical)structures respectively. Hence the p+ effect of spectral absorption will establish a high p-buffer which is indicated by the -ve charge, experimentally, On the same basis, the +vely charged substances are: SiO_2 is 3p+; P_2O_5 is 3.5p+ ; NaCl is 2.6p- ; . Structurally SiO_2 is linear and hence its outer O-atoms dominate. In P_2O_5, the tetrahedral is made with dominating outer O-atoms . Hence these buffers will be +ve. (oxygen is 2.5p- , spectral). In NaCl the Na atom has a +ve buffer. In coal, the large number of hydrocarbon chains each of which is p+,but all their H atoms are at the sides and generate the net p- effect so that the overall buffer tends to show p+ . In starch(amylopectin), a net p- exists, so as to have a p+ buffer. Thus all observations are correctly accounted by the GET.

The computation of the voltage in a water droplet can be made when we observe that the droplet of 10^{-3} cm in size will be made in turn from finer drops of about 0.207×10^{-6} cm in radius(Avagadro radius). The charge that can be accumulated by this tiny drop will be confined to the surface sheath of 6.6×10^{-11} cm(electronic diameter) The volume of this sheath will be 1.41×10^{-22} cc.(in using the value for the radius , a factor of 2 has been included to account for the zone of influence of the particle). It should be seen that at this radius aggregation of the particles will be initiated as already seen. The density of packing of the shell will correspond to the geometric mean of the interstellar value(2×10^{40} p/cc). Therefore the particles involved will be $(1.41 \times 10^{-22})(2 \times 10^{40}) = 2.82 \times 10^{18}$. In as much as one volt is 1.1×10^{19} particles, the sheath voltage will be $(2.82 \times 10^{18})/(1.1 \times 10^{19}) = 0.25$ v = 250 mV. As the drops coalesce the bigger particle that develops typically to 10^{-3} cm, radius will show up the same potential that is observed experimentally.(for the equipotential surface, the voltage remains the same).Thus in a cloud, about 500 such drops will be in series giving $0.25 \times 500 = 125$ volts which is quoted above.

Resistance to flow is manifested in the form of a force opposing motion. From the force-energy relation(21), it can be seen for smaller particles, the buffers are small enough to demand $\delta p = (\delta 1 v /2)$ and v to become constant ($f = 2\pi r^2 v \, \delta 1 v/2$)so that $f \propto r^2$. For bigger particles, the Avagadro radius sets up the physical limit for the buffer and so the area(πr^2) remains constant , while the moving particle is squeezed so as to be elongated in that the l is altered to give $f \propto r$. The light scattering will be dependent as $f_1 \propto r^4$(from the Stefen's law for radiating body and the definition of temperature in the GET- cf:21)and area of particle $f_2 = \pi r^2$. Hence for scattering in small particles, the energy is proportional to $f_1 f_2 \propto r^4 \times r^2 \propto r^6$. When th size of the particles becomes larger,scattering tends to change into reflection and hence f_1 becomes a constant leading to scattering as proportional to r^2 .

REFERENCES :

Being common, references are listed in paper on Sedimentation Theory.

Nuclear Aerosol Transport Correlations: Comparisons with Data, Shape Factors, and Accident Sequences

M. EPSTEIN and P. G. ELLISON*

Fauske & Associates, Inc.
16W070 West 83rd Street
Burr Ridge, Illinois 60521, USA

Abstract

The universal curve already derived in [1] for the settling of aerosol clouds undergoing Brownian and gravitational agglomeration is compared with the measured decay of the aerosol concentration for a number of different aerosol materials. It is shown that the universal curve provides a simple route to the inference of particle shape factors. Also, a brief review of the correlational procedure used in the MAAP nuclear accident analysis code is presented, together with an assessment of its adequacy by comparing the results of the MAAP code with a new, discretized version of MAAP in which a direct numerical attack of the integro-differential equation of aerosol agglomeration and deposition is made.

1.   INTRODUCTION

A major portion of the fission product mass released from the fuel is transported as an aerosol in severe nuclear reactor core melt accidents. Thus, accurate modeling of aerosol behavior is an important aspect of predicting the consequences of a core melt accident. The modeling of fission product transport as an aerosol has evolved along two paths. Detailed separate effects computer calculations (i.e., TRAPMELT, NAUA, CONTAIN, QUICK computer codes) which directly solve the integro-differential equation of aerosol agglomeration and deposition. Some integral analysis severe accident codes [2,3] in development are using the direct solution approach. The second approach to the solution of aerosol physics problems in severe accident analysis is the aerosol correlation scheme reported in Refs. [1,4,5] and incorporated into the MAAP accident analysis code [6]. The use of the correlation method results in an efficient procedure for coupling aerosol behavior with the "hydraulics" of an accident sequence, and one that apparently has a numerical accuracy that is within the spread in predictions by the various direct numerical ("discretized" or "sectional") methods [7,8].

In Refs. [1,4,5] a coagulating and depositing aerosol was shown to obey a principle of similarity, in the sense that the solution of the equation of agglomeration for the distribution function can be made universal by introducing suitable scale factors in the dependent and independent variables. The similarity principle allowed universal curves (or correlations) to be obtained for the scaled aerosol mass concentrations versus time. In the present paper we wish to briefly review the correlation method and report on

---

*Present address: E. I. DuPont de Nemours & Company, Inc., Atomic Energy Division, Savannah River Laboratory, Aiken, South Carolina 29808, USA

further developments and implications of the method, with emphasis on (i) as yet unpublished comparisons with experiment data, and (ii) a comparison between a direct numerical attack of the aerosol transport equations and the correlation method for aerosol behavior in the primary system of a nuclear power plant under severe accident conditions. As by-products of Item (i), we obtain shape factors for several nuclear aerosol materials of interest in severe accident analysis and discuss the sensitivity of the shape factors to experimental uncertainties.

2. AEROSOL SIMILITUDE

A review of the aerosol literature indicates that many experiments on aerosol coagulation and decay have been carried out in large and small vessels with different materials and initial aerosol cloud densities. The majority of the experiments have been conducted with dry aerosols (as opposed to hygroscopic aerosols in a wet environment) and considerable data has been accumulated under conditions where the principle means of removal is sedimentation with agglomeration due to Brownian and gravitational motion. Figure 1 shows measured sodium oxide aerosol suspended mass concentrations as a function of time for several experiments performed in the same vessel but beginning with different initial mass concentrations [9,10]. It is seen that the decay of the aerosol is represented by a single asymptotic curve that is independent of the initial conditions of the aerosols. This asymptotic condition of aerosol similitude is achieved because sufficient particle coagulation takes place so that the distributions become independent of the initial distributions of particle sizes.

As a consequence of this asymptotic behavior, the significant dimensionless variables that characterize aerosol behavior have been identified by reference to the basic integro-differential equation of aerosol coagulation [1,4,5]. In particular, it is shown in Refs. [1,5] that the solutions of the aerosol equation for the aerosol cloud density m as a function of time t for an aerosol undergoing Brownian and gravitational coagulation and losing mass by sedimentation must be of the form

$$M = \left[\frac{\gamma^9 \, gh^4 \, \epsilon_o^5}{\alpha^3 \, K_o \, \rho^3 \, \mu} \right]^{1/4} m = M(\tau) = M \left[\left[\frac{\alpha \, g \, \rho \, K_o}{h^2 \, \chi^2 \, \mu \, \gamma \, \epsilon_o} \right]^{1/2} t \right] \tag{1}$$

The symbols are defined in the nomenclature section. The quantities γ and χ are constant coefficients (or shape factors) which correct for nonspherical particles and α is a constant which corrects for particle porosity. If γ and χ are used then α must be set equal to unity [10]. The quantity ϵ_o is a particle capture efficiency constant that appears in the expression for the gravitational coagulation rate [5]; it is defined such that the identifications $\epsilon_o = 1$, 1/3 represent Fuchs [11] and Pruppacher and Kletts' [12] models of gravitational coagulation, respectively. In the next section values for γ and χ for a variety of aerosol materials will be inferred from available experimental data.

Figure 2, which first appeared in Ref. [1], shows the dimensionless suspended aerosol mass concentration M as a function of dimensionless time τ. These curves have been established by a great many numerical solutions

46

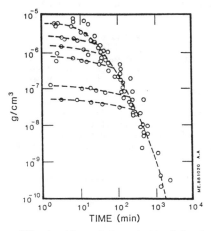

Fig. 1. - Illustration of Asymptotic Behavior of Aging
Sodium Oxide Aerosols [9, 10].

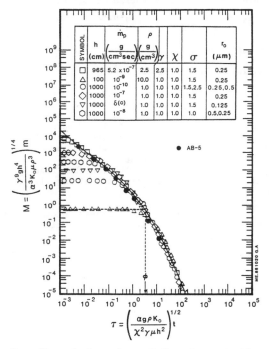

Fig. 2. - Dimensionless Aerosol Mass Concentration as a
Function of Dimensionless Time [1] .

of the aerosol coagulation equation. Several representative solutions are shown in Figure 2 for log-normal source distributions having standard deviations σ and initial geometric mean particle radii r_o. In most of these numerical "experiments" the aerosol is introduced at a steady rate into an initially empty compartment for a period of time that is sufficiently long so that the particle size distribution achieves an equilibrium or steady-state distribution. After the steady-state is reached, the source is turned off and the aerosol is allowed to decay. Aerosol behavior after source termination consists of a transition from the steady-state to the decaying aerosol curve. It is important to note that, much like the behavior of real coagulating and sedimenting aerosols, the predicted decay of aerosols is governed by a single asymptotic or similitude curve (compare Figure 1 with Figure 2). The similitude decay curve is also applicable to instantaneously generated aerosols. One of the sample numerical solution curves shown in Figure 2 is for an instantaneous release and is identified in the figure legend by the Dirac δ function.

The asymptotic curve in Figure 2 is well-represented by the algebraic fit

$$M(\tau) = 74.24 \ \tau^{-3.05} \left[1 + 3.74 \ \tau^{-1.12}\right]^{-1.70} \tag{2}$$

If we ignore the relatively small decrease in aerosol concentration required to make the transition from the flat portions of the curves onto the similitude curve, then the time at which the aerosol concentration begins to fall, ϕ, is given approximately by the inverse of Equation (2), viz.,

$$\phi = 4.11 \ M(o)^{-0.33} \left[1 + 0.623 \ M(o)^{0.66}\right]^{-0.82} \tag{3}$$

where $M(o)$ is the aerosol concentration at time zero.

3. METHOD FOR DETERMINING SHAPE FACTORS

As already mentioned, the four adjustable parameters used in aerosol analyses are α, γ, χ, and ϵ_o. Typically the particle porosity parameter α is set equal to unity and the shape factors γ and χ are employed to match the experimental data. This recipe will be followed here. The value of the remaining parameter ϵ_o is dictated by ones choice of gravitational coagulation model. Here we will assume the Fuchs expression for the particle capture coefficient to be appropriate and set $\epsilon_o = 1$ [5]. It turns out that uncertainties in ϵ_o are not too serious. It can be readily shown from Equation (2) that $m \sim \epsilon_o^{.275}$ for small time and ultimately goes as $\epsilon_o^{-.675}$ for $\tau \gg 1$.

Equation (2) can be transformed into a linear function from which the values of χ and γ can be found from a least-squares fit of experimental data. This linear form is given by

$$y = A + Bx \tag{4}$$

where

$$y = \left(mt^{3.05}\right)^{-\frac{1}{1.70}} \qquad (5)$$

$$x = t^{-1.12} \qquad (6)$$

and γ and χ are related to A and B through the inverted forms of the definitions of A and B:

$$\gamma = \left[\frac{K_o \, \rho^3 \, \mu}{g \, h^4}\right]^{1/9} \left[\frac{2.053 \, B^{2.72}}{A^{1.02}}\right]^{4/9} \qquad (7)$$

$$\chi = \left[\frac{K_o \, g \, \rho}{\gamma \, \mu \, h^2}\right]^{1/2} \left(\frac{B}{3.74 \, A}\right)^{\frac{1}{1.12}} \qquad (8)$$

The procedure for determining γ and χ for an aging, sedimenting aerosol is to select m versus t data from experiments involving a period of aerosol decay and substitute the m,t pairs taken from the measured, asymptotic decay curve into Equations (5) and (6) to generate the corresponding pairs x,y. This is followed by a least-squares fit of Equation (4) to the x,y data to determine A and B. The values of γ and χ can then be computed from Equations (7) and (8). Finally, Equation (3) and the inferred values of γ and χ are used to check that the selected data points lie along the asymptotic curve. This procedure circumvents the difficulties associated with the trial and error method used in fitting numerical solutions of the aerosol agglomeration equation to experimental data in order to infer γ and χ.

4.   COMPARISON WITH EXPERIMENT

The theory is compared with experimental data obtained from sodium fire tests. These tests were characterized by a buildup in aerosol concentration by sodium burning followed by the termination of the fire source and a period of aerosol decay. Sodium fire tests were conducted at laboratories in the U.S.A., in Japan, and in France at CEA. In all of these experiments the investigators noted that the aerosol mass-time histories were asymptotic to a common decay curve when the tests were conducted in the same vessel, but with different initial aerosol concentrations. Again, this observation was used to develop the aerosol similarity theory described previously and in Refs. [1,4,5]. The Atomics International (AI) LTV Test 3 (Koontz and Baurmash [13]) and the Hanford Engineering Development Laboratory (HEDL) Test AB-1 (Hilliard et al. [14]) sodium oxide data are plotted in the dimensionless M-τ coordinate system in Figure 3. The corresponding shape factors γ and χ were obtained using the procedure outlined in the previous section and are listed in Table I.

The shape factors inferred from the two sodium fire tests are seen to differ substantially, despite the fact that the tests were quite similar in most respects. Test AB-1 was a sodium pool fire that lasted for 3600 s in a vessel of effective height h = 9.65 m. The LTV-3 sodium fire lasted 3800 s and the test vessel has an effective height of 9.0 m. The AB-1 aerosol

49

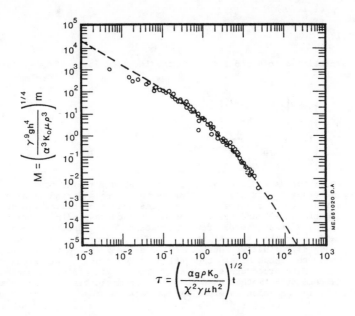

Fig. 3. - Similitude of Aerosol Experimental Data; See
Table I for a List of Data Included in Figure.

Test	Substance	γ	χ	ρ (g/cm^3)	h (cm)	Refs.
HEDL AB-1	Na_2O	1.6	0.7	2.3	965	14
AI-LTV-3	Na_2O	8.8	2.8	2.3	910	13
BNL-PuO$_2$	PuO_2	18.2	9.8	11.46	100	15
BNL-UO$_2$	UO_2	21.7	9.4	10.96	100	15
ORNL NSPP-511	Fe_2O_3	11.9	6.6	5.5	525	16
ORNL NSPP-631	Fe_2O_3, U_3O_8	7.3	4.0	7.02	525	17

TABLE I: SHAPE FACTORS AND REFERENCES FOR DATA PLOTTED IN FIGURE 6

suspended mass concentration versus time data are compared with that of LTV-3 in Figure 4. The plot shows that the rate of sodium oxide aerosol decay in Test LTV-3 was somewhat greater than in Test AB-1. Comparison at 10^5 seconds shows more than a factor of two disagreement. By this time, however, decay has reduced the aerosol concentrations by more than three decades. Given the current state of aerosol measurement techniques and other uncertainties (such as departures from the spatially uniform aerosol assumption and the presence of removal mechanisms besides sedimentation), the contrast between the test data results should not be regarded as significant. What is significant is the sensitivity of the shape factors to such experimental differences. Moreover, the shape factor $\chi = 0.7$ determined for Test AB-1 is a nonphysical result. Deviations from Stoke's law by irregular particle shapes are corrected by the dynamic shape factor χ; it has a lowest possible value $\chi = 1$. Thus achievement of even physically reasonable shape factors is difficult when accurate experimental results are not available.

The experimental data on the aging of plutonium oxide and uranium oxide aerosols reported by Castleman et al. [15] at Brookhaven National Laboratory (BNL) and on the aging of iron oxide aerosol by Adams and Tobias [16] at ORNL are also included as part of the data set displayed in similitude coordinates in Figure 3. The shape factors inferred from the data are given in Table I. The PuO_2 and UO_2 tests were performed in the same vessel at BNL and reveal similar shape factors. The last entry in Table I refers to the ORNL Test NSPP-631 [17] in which the aerosol particles were a mixture of U_3O_8 and Fe_2O_3. As might be expected, these components were observed to fallout at the same rate. The suspended mass-time history of the mixture aerosol is very well correlated by the universal aging curve in Figure 3 for $\gamma = 7.3$ and $\chi = 4$.

5. REVIEW OF CORRELATION METHOD

    In Refs. [1,4] and in the previous sections special cases involving either prolonged sources and mass loss by sedimentation (steady-state) or pure aerosol decay by sedimentation in closed geometry were considered in some detail. It was demonstrated that if one knows the source strength or the initial aerosol mass concentration then the desired suspended aerosol mass concentration [1] or particle size distribution [4] can be immediately obtained from simple algebraic correlations or by reference to universal plots (e.g., see Figure 2). This is possible because the integro-differential equation of Brownian and gravitational coagulation and sedimentation can be made universal for these two limiting cases by introducing suitable scale factors in the dependent and independent variables. Because more complex aerosol problems do not obey this condition of "aerosol similitude", we are led to seek another way of correlating the solutions to the integro-differential equation for problems of engineering interest. This is accomplished by constructing correlations of the removal rate λ for the two special cases of aging or steady-state aerosols [5]. Aerosol behavior in more complex situations is then estimated by using the λ's for these special cases in combination with an interpolation procedure.

    The removal rate λ formally appears by taking the first moment of the governing integro-differential equation. That is, if the integro-differential equation is multiplied through by particle volume v and

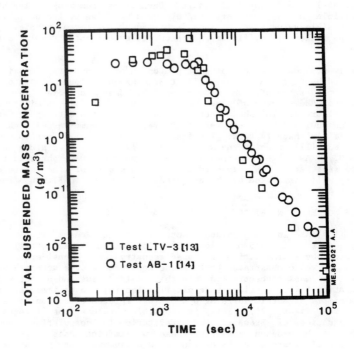

Fig. 4. – Measured Aerosol Concentration Histories
for Tests LTV-3 [13] and AB-1 [14].

integrated over all particle sizes it becomes the ordinary differential equation for the density of the suspended aerosol mass m:

$$\frac{dm(t)}{dt} = -\lambda(t)\, m(t) + \dot{m}_p \tag{9}$$

In the above equation $\dot{m}_p$ is the constant mass rate of production of aerosol particles per unit volume, t is time, and λ is defined by

$$\lambda(t) = \frac{\displaystyle\int_0^\infty vn(v,t)u(v)dv}{h\displaystyle\int_0^\infty vn(v,t)dv} \tag{10}$$

The variable $n(v,t)$ is the particle size distribution function and $u(v)$ is the particle deposition velocity.

To utilize the simple macroscopic description of aerosol behavior suggested by Equation (9), rational correlations for $\lambda(m)$ have been developed for the two special situations for which universal solutions of the integro-differential equation of coagulation are known to exist [5]. First the particle source term in this equation was set equal to zero and a correlation for the λ of an initially specified aging or decaying aerosol was developed and represented by the symbol λ_D. Then the transient term describing the rate of change of the particle distribution was ignored and a steady-state λ (denoted by λ_{SS}) was derived for an aerosol in which the source rate of input of particle mass is balanced by the loss rate of particle mass by deposition.

During the stages of aerosol evolution from the initial condition of zero aerosol concentration to steady-state, or between steady-state and decay, the true relation between λ and m is unknown. However, Equation (9) is accurate far from the steady-state during the early stage of aerosol evolution, even though there is the absence of a correlation between λ and m during this period. Early in time λ is a function of the properties specified for the source. However, m is small, the sedimentation term λm is negligible compared with $\dot{m}_p$, and the early behavior is insensitive to λ. The correlation procedure then is based on the notion that the solution of Equation (9) will not depart from the true solution in a significant way during the transition from aerosol buildup with no particle removal and steady-state, since the solution to Equation (9) is constrained by these known limits. With regard to the transition between steady-state and pure aerosol decay, the two curves $\lambda_D(m)$ and $\lambda_{SS}(m)$ are close to each other in λ-m space. Thus the actual solution path in λ-m space cannot stray too far from the similitude curves and is constrained to remain between them. This fact plus the reasonably rapid transition from steady-state to pure decay, especially at high values of m, suggests the use of either λ_{SS} or λ_D within the transition region. For the situation of aerosol behavior following source termination, λ_D is the natural choice since the solution is asymptotic to the similitude curve for pure aerosol decay. The decay along this curve occurs in accord with the source-free dimensionless form of Equation (9) with $\lambda = \lambda_D$. The resumption of a source triggers a return from the decay curve to the steady-state curve. In this case Equation (9) with λ

- λ_{ss} is recommended since its accuracy increases during the transition from the decay to the steady-state curve.

The correlation technique is limited to power law representations of the deposition velocity, $u(v) \sim v^n$. However, this does not present a major limitation to the application of the method, as a number of deposition mechanisms besides sedimentation exhibit a power law relationship between deposition velocity and particle volume. In fact, λ versus m correlations have been developed for deposition from turbulent flow, inertial deposition in convection around blunt bodies, and processes with size-independent removal rates. The latter include leakage, thermophoresis and particle deposition by steam condensation. Also, the range of these correlations has been extended to include several cases in which two deposition processes are operating simultaneously. This was accomplished with interpolation formulae which express a "combined λ" in terms of the λ correlations for the two processes acting independently. Space limitations prevent us from reviewing the development of the individual correlations or their algebraic combinations. The correlations, together with practical example problems that illustrate their use, may be found in Ref. [5].

6. COMPARISON WITH DISCRETE-BASED AEROSOL CODE

The coupling of aerosol behavior to integrated computer-code-calculated nuclear reactor accident sequences has been accomplished with the incorporation of the present correlation technique into the MAAP accident analysis code [6]. Naturally it is of interest to compare the correlations as used in MAAP to a state-of-the-art numerical technique. Accordingly, a new version of MAAP was constructed. The MAAP models for the primary system thermal-hydraulics and fission product release were left intact, but extensive changes were made to the MAAP fission product transport modules to implement the sectional solution method [18] of the aerosol coagulation and transport equation. Briefly, the method involves dividing the continuous particle size spectrum into sections within which all particles are assumed to be of the same size or the particle size distribution function is assumed to be known. An ordinary differential equation is derived for the net rate at which particle mass is added to a section due to coagulation of particles within the section with particles both inside and outside the section. The number of ordinary differential equations required is equal to the number of sections that are selected.

Two LWR major accident sequences were used to evaluate suspended concentration results from the MAAP correlational technique by comparing them to the sectional version of MAAP, namely a station blackout and a small break loss of cooling accident (LOCA) in a PWR. The station blackout accident assumes a loss of on- and off-site power, reactor trip, pump coastdown, and loss of auxiliary feedwater. In addition all containment safeguards are unavailable due to the loss of all power. After the steam generators dry out, the accident commences when the coolant inventory is lost through the PORV. When the water level drops to the top of the core, fuel heatup commences and accelerates as the Zircaloy cladding oxidizes. During this core heatup phase, the majority of the volatile fission products are released to the primary system. The core support plate fails when a significant fraction of the core has melted. The debris resulting from the core support plate failure enters the lower plenum and causes the vessel to fail at the lower head penetrations.

The above sequence of events is modeled in detail by the severe accident code MAAP. The comparison of the suspended aerosol mass history as predicted with the two versions of MAAP for the station blackout accident is shown in Figure 5. The "sectional MAAP" prediction is slightly higher than that of the correlational version throughout the transient due to a different source rate of fission products and structural materials being released from the core. This difference was traced to a different core flow rate of steam and hydrogen between the sectional version and the correlational version of the MAAP code. Even with these different source rates the models agree within about 30% of each other in total suspended mass concentration. Thus, even with the question of the source rates still to be resolved (see below), the correlation model can adequately replace the sectional model in severe accident analysis.

The small-break LOCAs are characterized by a short residence time of the aerosolized fission products and structural materials in the primary system. Thus, the LOCA sequence is a good choice to determine if the scheme developed to handle aerosol leakage from individual volumes of the MAAP correlational code is adequate.

The small-break LOCA sequence chosen for this analysis was a break in the intermediate cold leg in a Zion-like PWR. The break size was chosen to be 2.18×10^{-3} m^2. This area gives a short residence time in the primary system for the aerosol and vapor. The accident commences with a break in the cold leg followed by a failure of the high and low pressure injection systems. The reactor is scrammed at time zero and the letdown and charging pumps are assumed to be inoperable. At about 60 seconds into the accident the main coolant pumps trip off due to vibration. The core begins to uncover at about 3050 seconds into the accident, with vessel failure occurring at about 6200 seconds. Again, this accident was simulated with the MAAP-PWR code using both the sectional and correlational models of aerosol transport.

The results of the two MAAP calculations showing the suspended fission product and structural material aerosol mass are given in Figure 6. The curves agree quite well up to the point of vessel failure. After vessel failure the aerosol cloud density drops, since the aerosol cloud is ejected into the lower cavity at vessel failure. Figure 6 shows that the two MAAP results indicate a somewhat different amount of aerosol material ejected into the containment. The differences after vessel failure are caused by different estimates of the vessel failure time and concomitant differences in gas mass flow rate from the vessel immediately following vessel failure. It is felt that the discrepancy has more to do with the sensitivity of the core melt progression model to time step selection in the numerical integration than to limitations associated with the aerosol correlation technique.

The general result of these comparisons is that the sectional and correlational versions of MAAP produce results that are well within the uncertainty bounds required for severe accident analysis and that the correlations are an efficient substitute for discrete distribution codes.

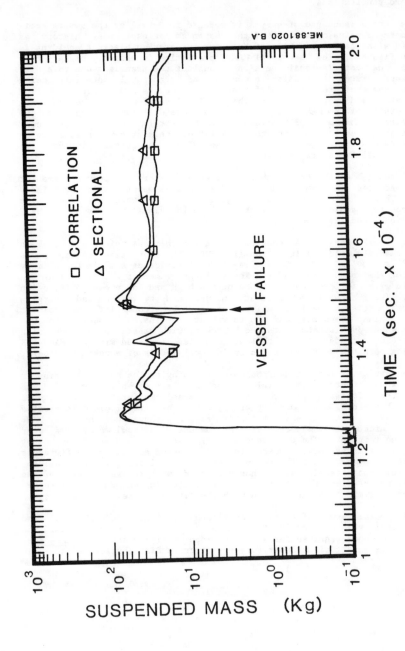

Fig. 5. – Comparison of Predicted Suspended Mass in the Primary System as Obtained with the Sectional and Correlational Versions of MAAP for the Station Blackout Accident.

56

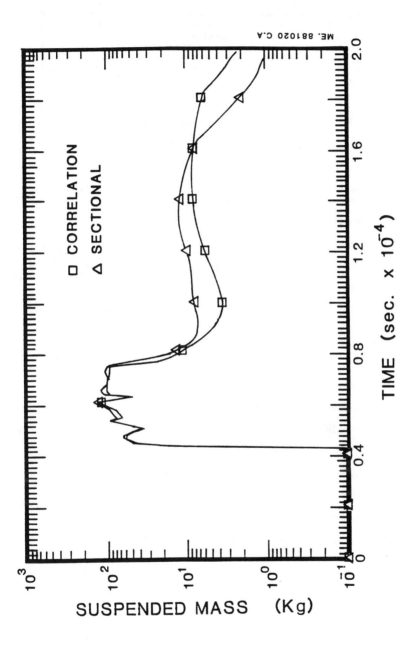

Fig. 6. – Comparison of Predicted Suspended Mass in the Primary System as Obtained with the Sectional and Correlational Versions of MAAP for the Station Blackout Accident.

57

7. CONCLUSIONS

Direct evidence of the existence of aerosol decay similitude has been provided by a comparison of the theoretical universal decay curve with experimental data on the aging of several aerosol materials. Using the universal decay curve, a simple method for the inference of aerosol particle shape factors from these data was presented. The attainment of accurate shape factors, however, requires more extensive and precise data. Finally, the comparisons presented herein of the MAAP correlational accident analysis code with its discretized version indicate that the radiological aerosol concentration-time history is fairly well represented by the correlation technique.

8. NOMENCLATURE

A   Coefficient for least-squares fit, Equation (4).

B   Coefficient for least-squares fit, Equation (4).

g   Gravitational constant.

h   Effective height of the particle cloud; volume of particle cloud (compartment) divided by deposition area.

k   Boltzmann constant.

K_o Normalized Brownian collision coefficient, $= 4kT_\infty/(3\mu)$.

m   Density of particle cloud; mass per unit volume of particle cloud.

M   Dimensionless density of particle cloud.

$\dot{m}_p$ Mass rate of production of particles per unit volume of particle cloud.

n(v,t)   Particle size distribution function.

t   Time.

T_∞ Gas temperature.

u(v)   Deposition velocity for particles of volume v.

v   Particle volume.

x   Transformed time for least-squares fit, Equation (6).

y   Transformed mass and time for least-squares fit, Equation (5).

Greek Letters

α Density correction factor.

γ Collision shape factor.

ϕ Approximate dimensionless time at which the aerosol cloud density of an aging aerosol begins to decrease.

$\delta(o)$ Delta Dirac function, indicates instantaneously generated aerosol at time zero.

ϵ_o Adjustable particle capture efficiency constant.

λ Removal rate constant.

μ Gas viscosity.

ρ Density of particle material.

τ Dimensionless time.

χ Particle settling shape factor.

Subscripts

D   Refers to a pure decaying aerosol (no source).

SS   Refers to a source-reinforced aerosol in steady-state.

REFERENCES

1. Epstein, M., Ellison, P. G., and Henry, R. E., "Correlation of Aerosol Sedimentation", J. Colloid and Interface Sci., Vol. 113, 342-355 (1986).

2. Camp, W. J., et al., "MELPROG-POW/MODO: A Mechanistic Code for Analysis of Reactor Core Melt Progression and Vessel Attack Under Severe Accident Conditions", NUREG/CR-4909 (April, 1987).

3. Leigh, C. D., (editor), "MELCOR Validation and Verification", NUREG/CR-4830, SAND86-2689 (March, 1987).

4. Epstein, M. and Ellison, P. G., "Similitude of Aerosol Size Distributions", Journal of Colloid and Interface Sci., Vol. 119, 168-173 (1987).

5. Epstein, M. and Ellison, P. G., "Correlations of the Removal Rate of Coagulating and Depositing Aerosols for Application to Reactor Safety Problems", Nucl. Eng. and Design, Vol. 107, 327-344 (1988).

6. MAAP User's Manual Volumes I and II, IDCOR Technical Task Reports 16.2 and 16.3, AIF (May, 1987).

7. Otter, J. M. and Vaughan, E. U., "Evaluation of Empirical Aerosol Correlations", EPRI NP-4974 (December, 1986).

8. Vaughan, E. V. and von Arx, A. U., "Evaluation of Aerosol Correlations", EPRI NP-5602 (January, 1988).

9. Kitani, S., Matsu, A., Uno, S., Murata, M., and Takada, J., "Behavior of Sodium Oxide Aerosol in a Closed Chamber", Journal of Nuclear Science and Technology, 10, 9, 566-573 (September, 1973).

10. Silberberg, M., (editor), "Nuclear Aerosols in Reactor Safety", Nuclear Energy Agency, OECD, Paris (1979).

11. Fuchs, N. A., "The Mechanics of Aerosols", Pergamon, Oxford (1964).

12. Pruppacher, H. R. and Klett, J. D., "Microphysics of Clouds and Precipitation", Reidel, Dordrecht, Holland (1980).

13. Koontz, R. L. and Baurmash, L., "Analysis of Aerosol Behavior for FFTF Hypothetical Accidents", AI-69-MEMO-95, Atomic International (October 28, 1969).

14. Hilliard, R. K., McCormack, J. D., and Postma, A. K., "Aerosol Behavior During Sodium Pool Fires in a Large Vessel - CSTF Tests AB-1 and AB-2", HEDL-TME 79-28, Hanford Engineering Development Laboratory Report (June, 1979).

15. Castleman, A. W., Jr., Horn, F. L., and Lindauer, G. C., "On the Behavior of Aerosols Under Fast Reactor Accident Conditions", BNL-14070, Battelle Memorial Institute, Columbus, Ohio (1970).

16. Adams, R. E. and Tobias, M. L., "Aerosol Release and Transport Program Quarterly Progress Report for April-June 1983", NUREG/CR-3422, Vol. 2 (February, 1984).

17. Adams, R. E. and Tobias, M. L., "Aerosol Release and Transport Program Semi-Annual Progress Report for October 1983 to March 1984", Draft.

18. Gelbard, F., Tambour, Y., and Seinfeld, J. H., "Sectional Representations for Simulating Aerosol Dynamics", J. Colloid and Interface Sci., Vol. 76, 541-556 (1980).

PARTICULATES

Experimental Study of Ash Erosion of Boiler

J. Q. SUN, J. Q. WANG, and N. ZHUO
Shanghai Institute of Mechanical Engineering
516 Jun Gong Road
Shanghai 200093, PRC

Abstract

A specific rig for test of ash erosion was constructed. Its system and the experimental technique are detailed in this paper.

The results of velocity exponential have been obtained by test of impacting the carbon steel specimen using Al_2O_3 and S_iO_2 particles under different working conditions.

The test results of four types of ash from different power plants of China and the respective formulas of calculating the ash erosion rate are also shown.

1. INTRODUCTION

A large amount of low grade coal with high ash content has been used in power boilers in China in order to use the fuel source rationally. Therefore, the problem of ash erosion on the convection heating surface of boilers have become more important to ensure the economy and safety operation of boilers. For example, tube failure caused by ash erosion occured to the reheater of a boiler with capacity of 400 T/H after 5000 hours of operation.

Since 1962, numerous studies, concerning the rule and the protection measures of ash erosion, have been conducted in China [1,2,3,4,5,6]. However, the studies of the basic behavior could not provide a practical solution to technical problems about ash erosion, such as researches on main factors influencing ash erosion and calculation of the erosion rate. We performed a series of experimental researches on the property of ash erosion using AL_2O_3 and S_iO_2 particles and four types of coal-ash, and obtained some technical information for the design and economy, safety operation of coal fired boilers in power station.

2. TEST APPARATUS AND MEASUREMENT METHOD

Tests had been conducted in a specific rig (Fig.1) using compress air as gas source.

Gas coming from gas tank 1 flows through the path with bypass, in which the temperature, the pressure and the discharge of gas are measured by the thermometer 2, the pressure gauge and the rotameter 3. Then it divides into two way. One of them, as complement gas, inters ash funnel. The others, as main efflux inters jet 4, where it introduces grains from ash funnel forming phase flow, through acceleration section, then impacts the specimen fixed on test section 5. After impacting specimen, it passes through the grains remover 6. Finally, passes micro-grains remover 7 into atmosphric. The test section and the specimen are shown in Fig.2.

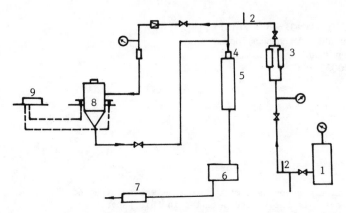

Fig.1.-Schematic Drawing of The Experimental System

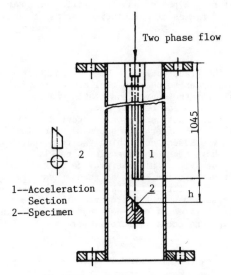

Two phase flow

1045

2

1

2   h

1—Acceleration
   Section
2—Specimen

Fig.2-Test Section And Specimen

The weight of ash was continually weighed by the electronic-weighing system (precision 0.5%) with printer. The lost weight of the funnel before and after test was checked by a standard scale with 1/5000 precision. The lost weight of the specimen was weighed by the balance with 1/1000 g precision.

3. THE TEST RESULTS OF VELOCITY EXPONENTIAL ABOUT AL_2O_3 AND S_iO_2 PARTICLES

The velocity of particles is a main factor of affecting ash erosion. The following equation is used to determine the velocity exponential:

$$I = C.W^n \qquad (1)$$

where: I-- erosion rate, mg/g;
         $I = g/G$
         g--absolute erosion quantity of the specimen, mg;
         G--weight of impacting particles, g;
         C--erosion constant;
         W--velocity of particles, m/s;
         n--velocity exponential.

The experiments of impacting carbon steel specimen with impact angle $40°$ were conducted by AL_2O_3 particles (40-42 um diam.) at a velocity range from 40 to 90 m/s under fourteen working conditions. Fig.1 shows the experimental result. The formula of calculating the erosion rate is

$$I = 2.375 \times 10^{-6} \cdot W^{2.54} \qquad (2)$$

The velocity exponential n is 2.54. Linear relative coefficient r equals 0.98.

The experimental results of impacting carbon steel specimen (angle $40°$) were obtained by using S_iO_2 particles (36.40 um diam.) at the velocity range from 73 to 90 m/s under five working conditions. That is shown in Fig.3. The relative equation used to determine the erosion rate is:

$$I = 8.6115 \times 10^{-6} W^{2.234} \qquad (3)$$

The velocity exponential n is 2.234. Linear relative coefficient r equals 0.997.

It was shown by experiments that the velocity exponential of AL_2O_3 particles was over that of S_iO_2 particles.

4. THE TEST RESULTS OF COAL-ASH EROSION

The following mathematic model of erosion rate was introduced on the basis of the theoretical study and analysis of some test information [7,8,9],

$$I = a \cdot d^m \cdot W^n \cdot f(\theta) \qquad (4)$$

where: I--erosion rate, mg/g;
         a--erosion behavior coefficient of particle, $S^2 \cdot mg/g.um.m^2$;
         d--average size of grain, um;
         W--velocity of grain, m/s;
         $f(\theta)$--a function of impact angle;

65

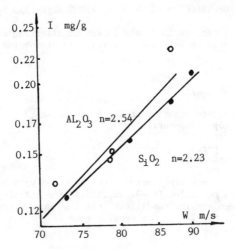

Fig.3.-The Velocity Exponential
About AL_2O_3 And SiO_2 Particles

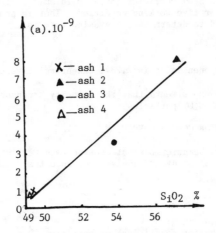

Fig.4.-The Relation Between Erosion
Behavior Coefficient   and S_iO_2

m,n--exponential of size and velocity of particle,respectively.

It is convenient using the mathematic model above for data processing and determination of formula of erosion rate calculation.

The formula of calculating average size may be given by

$$d=[\sum_{i=1}^{n}(x_i \cdot d_{si}^m)]^{\frac{1}{m}} \tag{5}$$

Where x_i is percentage of particle whose size is d_{si} % and d_{si} is size of i kind particle, um.

A series of experiments had been conducted by using four types coal-ash from different power plants. Table I and II show their physico-chemical properties respectively. Table III gives the erosion behavior coefficient (a) of four type coal-ash.

The Fig.4 shows the relationship between erosion behavior coefficient of four types of coal-ash and S_iO_2 content.

It may be seen by Fig.4 that S_iO_2 content of coal-ash has an effect on the erosion property.(a)will increase with S_iO_2 content in coal-ash.

The Fig.5 shows the experimental results of four coal-ash samples by impacting carbon steel specimen with 45° under different velocity conditions.

Table IV shows the formulas of calculating erosion rate of four type ash.

The size exponential m (m=1.876) in Table IV was obtained by experiment[8].

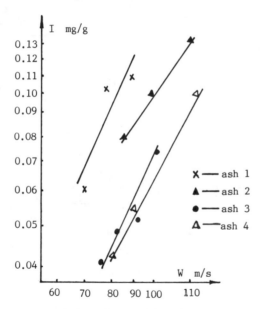

Fig.5.-Test Curve of The Property of Ash Erosion

TABLE I: MAIN CHEMICAL COMPOSITION ABOUT FOUR TYPE ASH

COAL-ASH COMPOSITION	ASH 1	ASH 2	ASH 3	ASH 4
S_iO_2	49.36	57.20	53.80	49.22
Fe_2O_3	5.20	6.20	4.60	16.80
AL_2O_3	36.64	29,29	34.72	22.66
C_aO	4.64	2.52	2.52	2.32
M_gO	1.40	2.80	0.20	2.30

TABLE II: PHYSICAL PROPERTIES OF FOUR TYPE COAL-ASH

COAL-ASH SIZE um	ASH 1	ASH 2	ASH 3	ASH 4
280--125	2.65	2.32		16.50
125--90	2.50	3,50		6.55
90----76	1.85	2.65	8.70	3.96
76---61	5.30	8.35	10.10	8.05
61---55	1.05	1.15	1.15	0.75
55---50	3.75	6.77	5.30	3.70
50---48	0.40	0,20	0.40	0.30
48---40	11.50	29,70	9.55	11.70
40---38.5	7.50	9.50	6.75	1.75
< 38.5	63.50	35.86	58.05	46.74
AVERAGE diam.um	33.08	38.60	34.52	46.90

TABLE III: THE EROSION BEHAVIOR COEFFICIENT (a)

COAL-ASH	ASH 1	ASH 2	ASH 3	ASH 4
$(a) \times 10^{-9}$	3.42	7.84	0.72	0.79

TABLE Ⅳ : THE FORMULAS OF CALCULATING EROSION RATE

COAL-ASH	EQUATIONS
ASH 1	$I = a.d^m.W^n = 3.42 \times 10^{-9} \times 48.9^{1.876} \times W^{2.07}$
ASH 2	$I = a.d^m.W^n = 7.84 \times 10^{-9} \times 53.1^{1.876} \times W^{1.95}$
ASH 3	$I = a.d^m.W^n = 0.79 \times 10^{-9} \times 86.23^{1.876} \times W^{2.33}$
ASH 4	$I = a.d^m.W^n = 0.72 \times 10^{-9} \times 44.95^{1.876} \times W^{2.44}$

REFERENCES

1. Chen,X, J.,Xi'an Jiao-Tong University,"Study of Gas Velocity in the Convection Heating Surface of the Boiler", Journal of Mechanical Engineering, Vol. 10,No. 5, June 1962.

2. Chou,A.,"Test Study of the Coal Erosion Behavior", Heating Power, No.5, 1979.

3. Xian Thermal Engineering Research Institute,"Wear of the Boiler Heating Surface", Boiler Boiler Technique, No. 9, 1972.

4. Sun, J. Q., Shanghai Institute of Mechanical Engineering "Study of Ash Erosion on Power Boiler Convection Heating Surface", Boiler Technique, On. 7, 1980.

5. Finnie, I.,"Erosion of Metals by Solid Particles", Journal of Materials, Vol. 2, No. 3, 1967.

6. Raask, E.,"Tube Erosion by Ash Impaction", Wear 13, 1969.

7. Tabakoff, w., Kotwal, R. and Hamed, A.," Erosion Study of Different Materials Affected by Ash Particles," Wear 52, 1979.

8. Sun, J. Q. and Wang, J. Q.,"Two-Phase Flow and Heat Transfer" China-U.S. Progress. P.577-585. Hemisphere Publishing Corporation, 1984.

9. Wang, J. Q. and Sun J. Q.,"An Experiment Study of the Property of Ash Cutting", Journal of Shanghai Institute of Mechnical Engineering, Vol.9,

    No.2, June 1987.

Flow Visualization of a High Concentration Particulate Flow

C. S. K. CHO
Department of Mechanical Engineering
Western Michigan University
Kalamazoo, Michigan 49008, USA

ABSTRACT

A highly loaded two-phase flow field cannot be visualized due to the opaqueness caused by a random light scattering from the large number of particles. A technique that measures a highly loaded particle-fluid flow is investigated by matching the refractive index of carrier fluid and that of particle. The proper matching of refractive indices can provide a transparency of two-phase flow while simulating an actual behavior of two-phase flow. The flow visualization of very high particle concentration field can be measured by mixing colored particles as tracers.

INTRODUCTION

The Laser Doppler Velocimeter can measure velocity profiles of fluid phase and particulate phase simultaneously as both phases pass through an optical volume created by the crossing of laser beams. The Laser Doppler Velocimeter can provide velocity and size information from an instantaneous local measurement, but reducing the locally collected data into two dimensional information introduces ambiguities. The flow visualization technique of a flow field is getting more attention because the advancements in computer technology enables digitization of a large image and reconstruction of the image into a new multi-dimensional image which is useful in flow modelling. The optical measurement technique of a two-phase flow has been studied since 1970's by many researchers [1,2,3]. Kobayashi [4] utilized an image digitizing technique to trace a motion of particle and obtained a path line and velocity vectors. Herman [5] investigated a fuel spray characteristic by processing a photographic image generated by a pulse laser.

As the particle concentration becomes higher, the corresponding increase of opaqueness through a particle field limits the accessibility of the view field and reduce the signal to noise ratio in the LDV measurement. The opaqueness of flow field can be bypassed if transparent particles and a fluid with matched refractive index are used in a

71

simulated flow. Yannekis and Whitelaw [6] measured a particle-fluid flow using a Laser Doppler Velocimetry. They reported that the rate of data collection was severely influenced by the particle concentration, and the measurable particle concentration was limited to 10% by volume because of drastic decrease of laser beam intensity. In order to provide a better optical view of a flow field, various ways of matching particle-liquid refractive indices were reported. [7,8,9]. Edward [10] described a method of matching refractive indices and reported the measurement results of a velocity profile near a simulated heat exchanger surface.

EXPERIMENTAL APPARATUS

The reservoir was made of 6 inches ID glass chamber as shown in Fig. 1 to insure a chemical compatibility with a carrier fluid, and the reservoir was tested to a pressure of 30 psi. The flow velocity in a test section was controlled by regulating the compressed air pressure in the reservoir. The density difference between particles and fluid caused a settling of particles, therefore a stirrer was installed to maintain an uniform mixing of particles and fluids.

An identical fluid with the fluid in the test section was filled into the square channel to match the refractive indices. The test section consisted of a tube and an outer square channel so that a curvature effect of the tube surface could be eliminated. For a test with glass particles a 0.55 inch ID glass tube was chosen for a test section, and 1 inch ID acrylic tube was chosen for a test with 1/8 inch diameter acrylic particles.

The laser beam was delivered by two mirrors to a cylindrical lens where a laser beam diverged into a two dimensional sheet of laser beam that had a thickness of 1 mm. The sheet of laser beam was aligned parallel to the vertical channel. The laser sheet was moved across the radial direction of the test section by moving the reflection mirrors, and the flow visualization tests were performed at every radial location.

FLOW VISUALIZATION WITH GLASS PARTICLES

It was found that Tetralin had some degree of long term incompatibility such as softening and discoloring of plastic. Glass particles of average diameters of 30 microns and 1.3 mm were purchased from the Potters Industry. In the process of selecting carrier fluids, the environmental hazard and the compatibility of fluid with the test section materials were considered. Methyl Benzoate had a refractive index of 1.516 which was very close to that of glass, but it reacted strongly with plastic and rubber products like a solvent. Dibutyl-Phthalate had a relatively good compatibility with plastic and was chemically stable. After some deliberation a mixture of Tetralin and Diabytyl-Phthalate was selected as a carrier fluid.

Tetralin had a refractive index of 1.546 and Dibutyl-Phthalate had a refractive index of 1.496. The glass particle had a refractive index of 1.51. The refractive index of fluid could be matched with that of a glass particle if a proper ratio of mixing of two fluids were found. The

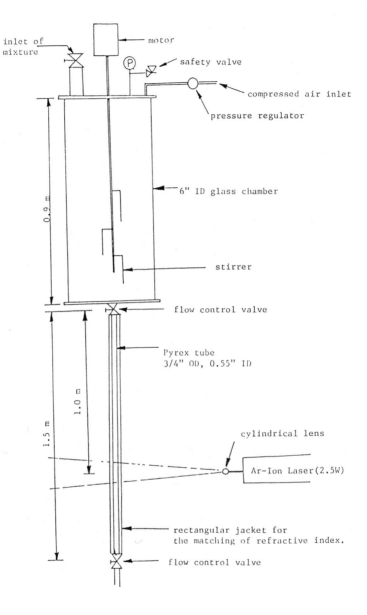

Fig. 1 Schematic of Flow system.

change of refractive index of carrier fluid with respect to the mixing ratio of two fluids by volume was tested as shown in Fig 2. The Fig 2 shows that the proper mixing ratio was 66% of Dibutyl-Phthalate and 34% of Tetralin by volume at 27 °C. The corresponding refractive index measured by a refractor meter was 1.508. The best matching of the refractive indices were checked by measuring a laser beam intensity after the passage of the laser beam through the test section.

The density ratio between glass particle and the mixture of Tetralin and Dibutyl-Phthalate was 2.45. The densities of glass, Tetralin and Dibutyl-Phthalate were 2.50, 0.973 and 1.043, respectively. The large density difference between glass particle and fluids created an nonuniform particle concentration at the entry section of the test tube. Individual glass particle of 30 micron diameter was clear and transparent under a microscope, but the thick layers of particles across the radial direction of test section created a random scattering of laser light and blocked the passage of laser beam. The glass particles of 30 microns diameter in a test section had severe opaqueness problem, and it was almost impossible to pass a laser beam across the test channel. In this regard 1.3 mm diameter glass particle was chosen for the experiments. The sheet of laser beam was sent parallel to the vertical test section and the high speed video images were taken by the Spin Physics high speed video system at a frame rate of 200 per second. The higher frame rate was not tried, because the light intensity was too low to capture an image. A 28 mm extension tube was added to a lens to capture a magnified image of the particle. The obtained video images showed that the outer boundary of the particles which were located in the sheet of laser beam could be recognized as circles. The capability of image recognition suggested that this flow visualization technique could be utilized even with the higher particle concentration. But the impurity in a glass particle gained from a manufacturing process reduced the light transmissibility in the test section. In order to overcome these short comings, acrylic particles were tested with a fluid which had matched refractive index with acrylic particles.

FLOW VISUALIZATION WITH ACRYLIC PARTICLES.

Dibutyl-Phthalate was selected as a carrier fluid which had a refractive index of 1.493 at 25 °C. Acrylic particle of 1/8 inch diameter were purchased from the Precision Plastic Co. Since the acrylic particle had a refractive index of 1.491, the refractive indices of both phases could be matched within 1% deviation without an elaborate adjustment. A density of acrylic was 1.19 and Dibutyl-Phthalate had a density of 1.043. The density ratio between two phases was 1.141 which was far less than the density ratio between the glass particle and the fluid mixture of Tetralin and Dibutyl-Phthalate. The problem encountered with glass particles, such as the nonuniform particle concentration and the gravity settling problem, could be reduced by using acrylic particles. The traveling speed of acrylic particles within Dibutyl-Phthalate was slow enough, so that the gravity induced free fall of particles was used in this experiments, and photographs were taken with a camera.

The Fig 3 shows a photograph taken with acrylic particles of 7% by

74

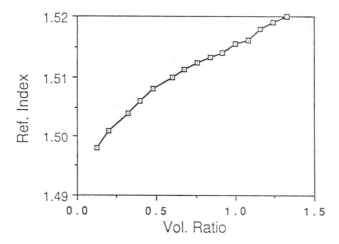

Fig. 2 Change of Refractive index with respect to the volume
ratio of Tetralin and Dibutyl-phthalate.

Fig. 3 Falling motion of 1/16" dia. acrylic particles
in water. (particle volume concentration: 7%)

volume in water. It shows that the refractive indices between two phases were not matched properly. The light scattering at the interface between the fluid and particle surface was intense due to the reflection and deflection of laser light. In order to compare the light scattering characteristics between a case with matched refractive indices and a case with unmatched refractive indices, water was selected as an unmatched carrier fluid. Water has a refractive index of 1.332 at 27 °C, and the refractive indices between two phases were deviated by 12%. The particles out side the laser sheet were also illuminated as a result of rescattering of the scattered light from particles within a laser sheet. The Fig 4 shows a case of matched refractive indices between acrylic particles and Dibutyl-Phthalate as a carrier liquid. The particle concentration was 7 % by volume. It shows that only the particles within the laser sheet were illuminated and the rest of particles outside the laser sheet were barely noticeable.

The Fig 5 shows a test result with Dibutyl-Phthalate and the acrylic particles of 22 % by volume. Again, only the particles within the laser beam were shown with spherical images while the particles located radially before or after the laser sheet were not noticeable. In a very high particle concentration flow, colored acrylic particles, which had the same geometry and properties with transparent particles, were mixed with transparent acrylic particles. The colored particles were used as tracers to visualized the flow field. Red acrylic particles of 1 % by volume were mixed with transparent acrylic particles of 22 % by volume, and the tests were run with Dibutyl-Phthalate at the same flow conditions with the previous experiments. During the experiment only the particles within the laser sheet was illuminated, especially the red particles within the laser sheet were illuminated as an orange color with high intensity as shown in Fig 6. If an intensity discrimination and color matching technique were applied properly, only the image of the colored particles will be shown. The mixing of colored particles as tracers with transparent particles can be a promising technique to measure a motion of very high concentration particle flow. The motion of whole particles within the laser beam can be represented by the motion of colored particles. If colored particles is mixed uniformly with the transparent particles at the upstream of test tube, the particle concentration at the down stream can be calculated from the radial number density of the colored particles.

CONCLUSION

A plane of Ar-Ion laser beam was created by a cylindrical lens, and a high speed video and photographic technique were used to captured the two-phase flow field. In order to measure a highly-loaded particle-fluid flow, the matching of refractive indices were investigated. Glass and acrylic particles were tested with the liquids of properly matched refractive indices. A particle concentration of 22 % by volume was tested and demonstrated that the measurement of two-phase flow field was feasible. The mixing of colored particles with transparent particles can increase the applicability of the refractive index matching technique, because the general flow characteristics can be represented by the behavior of colored particle. It indicates that the transparency of particle is important to apply the refractive indices matching scheme.

Fig. 4   Falling motion of 1/16" dia. acrylic particles
in Dibutyl-phthalate.   (Particle volume
concentration : 7%)

Fig 5. Falling motion of 1/16" acrylic particles in
        Dibutyl-phthalate. (particle volume
        concentration : 22%)

Fig 6. Falling motion of a mixture of clear and red
        acrylic particles. (1/16" dia, clear particle
        volume concenrtation : 22%, Red particle volume
        concentration : 1%)

REFERENCE

1. Bachalo, W.D., "Method for measuring the size and velocity of spheres by dual-beam light-scatter interferometry," Applied Optics, Vol. 19, No.3, 1980, pp 363-370.

2. Chigier, N.A., Ungut, A. and Yule, A.J., "Particle size and velocity measurement in flame by laser anemometer," 17th International Symposium on Combustion, The Combustion Institute, 1979, pp 315-323.

3. Lee, S.L. and Cho, S. K., "Simultaneous measurement of size and two-velocity components of large droplets in a two-phase flow by Laser-Doppler Anemometry," IUTAM symposium on Measuring Techniques in Gas-Liquid Two-Phase Flows, Nancy, France, 1983, pp 149-164.

4. Kobayashi, T., Ishihara, T. and Saaki, N., "Automatic analysis of photographs of trace particles by microcomputer system," Proc. of 3rd Int'l Symp. on Flow Visualization, Ann Arbor, MI, 1983, p 231.

5. Herman, M.A., Parikh, P. and Sarohia, V., "Digital image processing application to spray and flammability studies," Proc. of 3rd Int'l Symp. on flow visualization, Ann Arbor, MI, 1983, p 269.

6. Yanneskis, M. and Whitelaw, J.M., "An investigation of refractive index matching of continuous and discontinuous phase," Third Int'l Symp. on Application of Laser Anemometry to Fluid Mechanics, 1986, Lisbon, Portugal, pp 4.4.

7. Bichen, A.F., "refraction correction for LDA measurement in flow with curved optical boundaries,' Imperial College report, FS/81/17.

8. Durst, F. and Zare, M., "Laser doppler measurement in two-phase flow," Proc. on the Accuracy of Flow Measurement by Laser Doppler Method, Copenhagen, 1975.

9. Horvay, M. and Leuckel, W., "LDA measurement of liquid swirl flow in converging swirl chamber with tangential inlets," Proc. of 2nd Int'l Symp. on Application of Laser Anemometry to Fluid Mechanics, 1984, Lisbon, Portugal.

10. Edward, R.V. and Dybbs, A., "An index matched flow system for measurement of flow in complex geometries," Proc. of 2nd Int'l Symp. on Application of Laser Anemometry to Fluid Mechanics, Lisbon, Portugal.

ACKNOWLEDGEMENT

I would like to acknowledge that this paper content is based upon the work supported by the WMU FACULTY RESEARCH SUPPORT FUND under Grant No. 22-221102.

Notice on a Method for Measuring Size Range of Particles in Gases from Stationary Sources

JOAO FERNANDO P. GOMES and MANUEL A. E. P. SILVA
Instituto de Soldadura e Qualidade
Department R & D R. Tomas de Figueiredo, 16-A
1500 Lisbon, Portugal

Abstract

The size range of particles is an important parameter for
subsequent dispersion estimates and sometimes is quite useful for
analysing the performance of combustion equipment as well as
dedusting equipment efficiency.
So far no standard method has been developed but several
laboratories use laser-beam techniques which are obviously very
expensive ones.
In this work a method is presented where the particles are
collected isokinetically in a filter from gases circulating in a
stack as described in EPA method 5. Afterwards particles are
observed at an optical microscope and measured. In spite of some
limitations this appears to be and inexpensive method which has
been given reproducible results on applications described on this
work.

1. INTRODUCTION

Instituto de Soldadura e Qualidade is currently involved in a
full emission inventory estimate of pollutants content in flue
gases from stationary sources all over Portugal.
The aim of this work is to establish a coerent data base so that
emission limits from stationary sources could be fixed.
This work is being made in co-operation with the portuguese
government agency for environment, who recommended that EPA
method should be followed as these will probably be the standard
methods in Portugal.
Frequently we come to situations where EPA methods are not
applicable or do not even exist for certain determinations.
One of these situations is the determination of size range of
particles as emmited from stationary sources.
The size of a particle is an important parameter in environmental
studies, as it affects plume opacity and is also an important
factor for the selection and design of air pollution control
equipment, namely cyclones, bag-filters and electrostatic
precipitators.

2. SCOPE OF THE METHOD

Since the individual particles in any particulate sample are of many different sizes, statistical methods offer an excellent way to present particle size data. Particle size data are usually presented as frequency or cumulative distributions in terms of number or mass. The distributions are generated by grouping observed size data in classes or intervals of size and determining the number of particles or amount of mass in each interval. The results of such a data analysis are then displayed using a histogram or a curve, plotting the number of particles or mass of each interval.
If cumulative frequency distributions are desired the number of particles or mass of particles in each interval can be successively added and plotted against the interval numbers.
Reference [1] discusses the statistics of particle size data in considerably more detail.
Determining the size distribution of particulate air pollution emissions is a complex problem. Even if an analytical method exists that will yield accurate size and size-mass distribution data, it is virtually impossible to collect a particulate sample without altering its size distribution.
Furthermore in any finite duct or stack (a non point source) the particle size distributions at individual points within the dust are unique. Therefore, a distribution of individual point particle size distributions exists.
Since it is usually impossible to sample the full flow of a source, an integration of many point particle size distributions must be used. Approximate integrations can be accomplished by traversing the cross-sectional area and sampling at several points. This approximate integration method works reasonably well in determining unbiased particulate concentartions, but the possible errors encountered are of even greater concern when trying to determine a size distribution.
Ideally to reduce the sampling error incurred, the ultimate collection device should itself be directly in the gas flow as close to the location of interest.
However, this is not usually possible or even desirable.
The usual "in situ" stack or duct environment of air pollution sources is too extreme to allow direct insertion of the ultimate collection device. Furthermore, the physical size of most collection devices is usually sufficient to disturb the local velocity profile and thus the local particle size distribution.
Therefore, some sort of a sampling probe is usually used. This probe can be fabricated of glass or stainless steel and are as small as possible to reduce local flow disturbances without being so small as to discriminate against the larger particles in the gas sample.
Properly designed and used probes could deliver reliable samples to a collection device. However, when using any probe to extract a particulate sample for subsequent collection and sizing analysis, the following limitations should be recognized :

a)   If nonisokinetic conditions exist at the probe nozzle
     entrance, the particulate size distribution may be altered.

b)   Particulate impaction may occur at all bends in nozzle and
     probe lines, resulting in particulate losses.

c)   Particulate may settle out in horizontal probes.

d)   The gas velocity in vertical probes may be insufficient to
     support larger particles.

e)   Water and other condensible fractions in the original gas
     stream may condense or chemical rections may occur that
     will change the nature of the sample particulate.

f)   Small particulates may diffuse to the probe and sample
     collecting device walls due to thermal or electrostatic
     potential gradients and brownian diffusion.

g)   Agglomeration or fracturing of particles in the probe lines
     or collection device may alter the size distribution.

If a representative sample of an aerosol can be reliably
collected, several analytical methods exist that can be used to
quantify the size distribution.
However, the size analysis of a given particulate sample as
determined by a specific method of analysis is usually unique and
cannot be duplicated using another analysis method.
Due to the size of the particles of interest, microscopy is the
only method commercially available for direct counting and or
sizing of individual particles.
Generally, a particle size distribution can be determined in only
three ways :

1) Collecting an integrated sample of particulate with subsequent
microscopic counting and sizing of the individual particles.

2) Collecting a particulate sample that has been separated into
size intervals by some indirect physical means with subsequent
garvimetric weighing of the intervals.

3) Collecting a particulate sample that has been separated into
size interval by some indirect physical means with subsequent
microscopic counting of particles within the intervals.

Optical microscopes can be used to measure particle size
distribution. However since the images are two-dimensional, only
linear or area measurements of size can be made. Linear distances
can be measured very accurately with a microscope and the lower
sensitivity limit of light microscopes is about 0.5μ m [2].

Sample preparation is the most critical operation of the whole process. This is because optical methods require a very dilute dispersion, and dilution of the sample can affect the nature of the particle. In fact the results of a particle size analysis are more dependent of the sample itself than any other step of the method [3].
The sample is collected according directions of EPA method 5 using a stainless steel probe, designed so that flow perturbations are as negligenciable as possible. The filtering medium is a simple Whatman filter.
During sampling an appropriate flow control must be attained in order to obtain isokinetic conditions. Otherwise the sampling is rejected.
After weight gain is determined, the particles are mounted on glass lamela, using a suitable non-dissolving liquid, and viewed on the the optical microscope.
The particles are usually photographed as shown in figure 1 and are then counted and measured until a number that is considered representative, which means that the size distribution range does not changes significantly when more particles are counted and measured.
In several cases where this method has been followed it has been noticed that a counting of about 500 particles is enough for these purposes.
The actual size of particles is measured by direct counting over transparent graticules that are superimposed on photographs.

3.   PRESENTATION OF SOME CASE STUDIES

Tests were made on a set of two electrostatic precipitators from a kraft pulp mill in order to determine particle collection efficiency and improve its performance. Among other determinations size range of particles was determined following the method described above.
Tests were made in three different conditons of gases flow rate : 220,000 m3/h (tests 1 and 2), 250,000 m3/h (test 3) and 290,000 m3/h (test 4); before and after the electrostatic precipitator system. The obtained results are shown in table I.
It should be noted that isokinetism was obtained in all tests within the limits prescibed in EPA method 5 (100 ± 10%). The mean dimension presented is simply a weighted aritmetic mean of diameters.
As expected a significant reduction of mean dimension of the particles occurs from before to after the system, that is, there is a bigger occurrence of small particles in the outlet gas stream than in the inlet.
This fact is in accordance with the principles of operation of electrostatic precipitators which collect mainly relatively bigger particles as noted by Duprey [4] who conduced tests over standard silica dust with a particle density of 2.7 g/cm3 and observed the collection efficiency shown in table II.

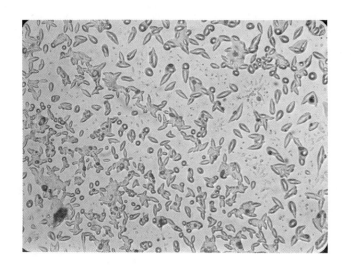

Figure 1 - Particles as collected from
a gas stream (500x)

TABLE I

TESTS OVER AN ELECTROSTATIC PRECIPITATOR

Particle Diameter Interval (μm)	Q = 220,000 m3/h				Q=250,000 m3/h		Q=290,000 m3/h	
	test 1		test 2		b	a	b	a
	b	a	b	a				
< 1	22.1	99	20.1	95	33.9	33.0	25.7	73.7
1 - 2	15.1	1	6.7	5	18.1	40.8	45.2	18.3
2 - 4	28.1	–	9.5	–	31.0	25.7	20.7	7.1
4 - 8	29.8	–	42.2	–	14.1	0.4	4.4	0.7
8 - 16	4.9	–	18.9	–	2.7	–	3.9	–
16 - 30	–	–	–	–	0.2	–	–	–
> 30	–	–	2.6	–	–	–	–	–
Mean dim. (μm)	3.6	0.5	5.3	0.5	2.6	1.6	2.2	0.9
Isokine-tism (%)	90.0	100.3	98.7	101.7	93.7	95.0	98.0	100.2

Note : b = before electrostatic preecipitators
a = after electrostatic precipitators

TABLE II

COLLECTION EFFICIENCY OF PARTICLE EQUIPMENT [4]

Collector equipment type	Efficiency (%)					
	net	0-5	5-10	10-20	20-44	>44
Electrostatic precip.	97.0	72.0	94.5	97.0	99.5	100
Baghouse-filter	99.7	99.5	100	100	100	100

It should also be noted that there is a reasonable agreement between the observed size range of particles in tests 1 and 2 which is an indication of reprodutibility of the method.
The obtained results were found to be quite useful in the optimization of the system.

Similar tests were made over a bag-house filter from a foundry plant and the obtained results are presented on table III.

TABLE III

TESTS OVER A BAG-HOUSE FILTER

Particle diameter interval (μm)	Before	After
< 1	82.5	90.3
1 a 2	17.1	9.3
2 a 4	0.3	0.4
> 4	0.0	0.0
Mean dimension (μm)	0.68	0.60
Isokinetism (%)	91.8	103.3

It can be noted that the majority of particles has a diameter that is less than 1μm.
Also, the percentage of particles in the interval 2/4 μm is somewhat bigger after the filter which should be regarded with precaution because this small percentage is near the experimental error of the measurements.
Apart from that the tendency noted by Duprey [4] is again observed.

5. FINAL REMARKS

Many methods have been developed for measuring particle size because of an almost infinite variety of needs. Only a few are generally applicable and no one of them satisfies all requirements.
When air pollution problems are concerned one can use several sophisticated methods which are, in some cases, extremely complex and subject to misinterpretation [5]. However the equipment involved is far too expensive and, for the majority of applications this extra cost on equipment does not pay as is does not bring an extra accuracy on the results.
Microscope examination is one of the most obvious methods in spite of being somewhat slow and tedious and was used in some applications described in this work with a certain degree of reliabillity.
As noted before, the most critical step of the method is the sampling procedure. However if certain precautions are taken the results are, at least, consistent.
Other techniques such as the use of a series of dry screens could be used for sampling the particles before the microscope observation.
It should be noted that the cost of a commercially available equipment for these purposes is sometimes similar to the one of a stack sampler, and the size intervals obtained for particles need frequently to be narrowed.
Tests are under way to compare results obtained by using these two sampling procedures.
Apart from that a digital processing image system is being developed for sizing automatically particles on the microscope.

REFERENCES

[1] "Air Sampling Instruments", American Conference of Government Industrial Hygenists, Cincinnati, Ohio, 1972

[2] McCrone, W.C. ; Draftz,R.G. ; Delly, J.G. , "The Particle Atlas, Ann Arbor Science Pubs, Ann Arbor, Michigan, 1967

[3] "Particle-size analysers : Out of the lab, into the plant", Chemical Engineering, November 7, 1988, 95-96

[4] "Compilation of Air Pollutant Emission Factors", National Air Pollution Contro Administartion, PHS Publication, 999-AP-42, 1968

[5] Brenchley, D.L. ; Turley, C.D. ; Yarmac, R.F. , "Industrial Source Sampling", Ann Arbor Science Pubs., Ann Arbor, Michigan, 1982

The Electro-Suspension of Powders—Basic Principles and Application

S. G. SZIRMAI
Commonwealth Scientific and Industrial Research Organization
P.O. Box 136, North Ryde
Sydney, Australia

Abstract

Electro-suspension (ES), an electro-kinetic effect for the electromagnetic suspension of particulates, has been the subject of attention at the CSIRO for over a decade. In light of present understanding, closed systems can be brought into dynamic equilibrium with a static bed, where macroscopic parameters, such as concentration and flux, coexist with microscopic ones relating to particle size, momentum and charge. It can be shown, that for the electrostatic case the minimum dispersing field exceeds the field strenght needed for levitation, so that suspended particles are in an oscillatory mode. An important modifying influence on air-suspensions is a partial electric breakdown of the gas, ascribed to corona points intermittently formed by particles attaching themselves to the electrode. It was observed, that suspensions of metal dust emit x-rays in vacuum, a fact used for the development of powder x-ray tubes. Various other applications of ES include electro-suspension filters and the coating of suspended powders.

1. INTRODUCTION

Electro-suspension has first been detected by the author in 1974, independently of others who may have been working along the same lines at this time [1] resulting from work on measuring the tensile properties of consolidated powder-columns under high voltage conditions [2].

Studies have indicated, that ES is a complex effect, one resulting from a number of competing processes. Some are due to forces experienced by the electrically charged particles within the externally applied field (Lorentz force), while others, such as ponderomotive forces and the mechanical force of ionic wind, are independent of charging. As well, there are electrical image forces, gravitational attraction and the cohesion between particulates which all have a role to play, while high-density suspensions are further complicated by mechanical and electrical interactions between the particles. – In contrast, when produced within closed insulating vessels, the suspensions show a remarkable degree of macroscopic stability and are easy to generate.

This paper will describe the main results of studies

89

conducted at the Commonwealth Scientific and Industrial Research
Organization (CSIRO) initially aimed at understanding the main
processes responsible for the effect. Increasingly, there are
practical reasons for carrying on with the work, as a number of
industrial-type applications began to appear. Some of these will
be explained in detail.

2.   ELECTRO-SUSPENSION FUNDAMENTALS

Definition

        Electro-suspension is a process for the electromagnetic
dispersion and suspension of particulates, leading to a dispersed
system characterised by steady conditions on a macroscopic scale.

Experimental systems

        ES is generated by subjecting a bed of powder to intense
electric forces through electrodes set inside the insulating
container, which also contains the powder. In practice, this
usually means applying a potential difference in excess of 10 kV
at the right frequency, depending on the particular dust and the
type of suspension required. Often, the frequency is set to zero,
i.e. an electrostatic field is applied.

        Using the electrostatic mode for suspending dust, one can
demonstrate some of the fundamental aspects of the system. With
a closed dispersing vessel containing a bed of alumina powder
(fig.1), a suspension is generated by simply applying a potential
gradient of 0.23 kV mm^{-1} to the powder, where the surface of the
bed is regarded as an electrode and is kept at earth potential.
Using the same applied voltage on the electrodes, but reducing
their distance, results in increased dust-concentration, clearly
indicating that field-strength (rather than voltage) is the main
determining factor. In other basic experiments, one could keep
the same field-strength while raising inter-electrode distance
and the potential applied to the electrodes. This leads to an
overall increase in the quantity of suspended matter, while dust-
concentration remains the same.

        These and other simple experiments performed on closed
suspension-systems can lead to analogies with evaporating fluids.
When generated within a closed insulating vessel, the suspended
cloud of particulates reaches equilibrium with the powder bed,
rather like vapours over some liquid in closed space, with
electrical energy substituting for the thermal energies usually
associated with vapour molecules (fig.2). By externally heating
the liquid, this raises the concentration of evaporated molecules
(as well as their energy), leading to an increase in the
equilibrium vapour pressure and, by increasing the volume above
the liquid, this leads to the evaporation of more liquid in
direct analogy to powder suspensions, where energy is added by
raising the applied voltage rather than heating. While useful in
helping to visualize the behaviour of ES-systems, particularily

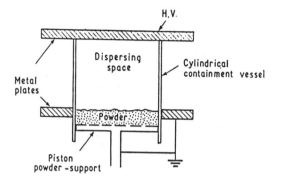

Fig.1. - Simple ES-generator.

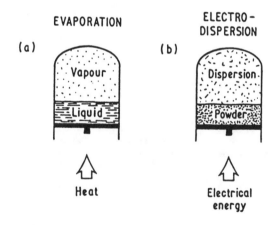

Fig.2. - Analogies between ES and evaporation.

its macroscopic behaviour, the analogy can not be carried into the microscopic events which underly the effect, as these are basically different from the behavior of liquids on a molecular level.

ES may be generated in a number of dispersing media and within a wide range of pressures. In vacuum, the effect can not be sustained in the range 10^{-1} to 10^{-3} torr, where corona-discharges and cathode rays are prevalent, effectively preventing powder-dispersion by shorting the applied electric field. However, it is possible to suspend powders in high vacuum, particularily well below the pressure of 10^{-4} torr. In fact, this is the only known technique for continuously suspending dust in vacuum. As well, ES can be produced inside dielectric liquids e.g.: oils, hydrocarbon liquids and non-aqueous solvents. While these type of suspensions have not been studied in adequate detail, current-voltage curves obtained for alumina dispersed in C_2Cl_4 reveal sharp differences compared with air-dispersed alumina, using identical electrode geometries. Particularily noticeable is the large hysteresis loop in case of liquid suspensions (fig.3), a feature totally absent from air-dispersions. It was earlier identified as due to viscous drag, affecting the settling velocity of the particles [3].

The dispersing mechanism

As shown in earlier, more comprehensive treatments of the effect [3,4], ES is the result of a number of competing forces and is further influenced by the electrical and mechanical properties of the powder, by its shape and size distribution, as well as some environmental factors, such as humidity.

From the point of view of generating a suspension from an initial static bed, the relevant forces may be classified as either dispersive, or restraining forces. (The mechanical action of ionic wind may be counted towards the first cathegory). *For the electrostatic case, this may be expressed as*

$$F_{total} = F_{dis} - F_{res} \qquad \cdots \cdots \cdots (1)$$

The dispersive forces are $F_{dis} = qE + m\,dv/dt + k\,dE^2/dy$, where the 2nd and 3rd terms on the right are due to ionic wind- and ponderomotive forces respectively.
The restraining forces are $F_{res} = F_{image} + F_{pol} + mg + H_0 + F_c$ where the image- and polarization forces are determined by the field E, while particle weight and interparticle cohesion (H_0) are not. Capillary forces (F_c) are due to the surface tension of a physically adsorbed liquid film.

For obtaining dispersion, the following condition must hold

$$F_{dis} > F_{res} \qquad \cdots \cdots \cdots (2)$$

92

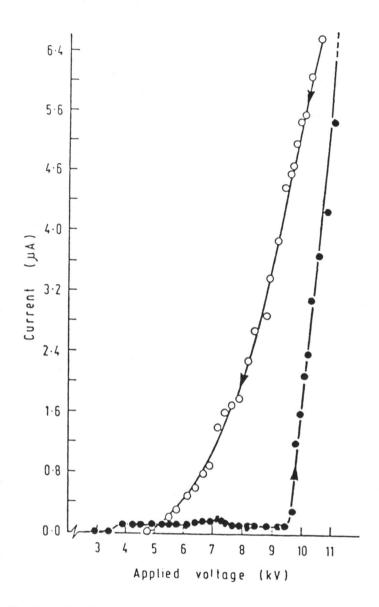

Fig.3. – Conductometric curves for electro-suspended alumina in a
dielectric liquid (C_2Cl_4).

In a large number of instances, condition (2) can be satisfied, while in some cases, such as for polyvinyl chloride, it can not.

Refering to the previous expression for the dispersing force, it was experimentally shown, that the relative size of the term qE is larger by at least one order of magnitude than the term dE^2/dy and that forces due to ionic wind (m dv/dt) can also be neglected in their supposed role of initiating suspensions, after all, ES can be generated in high vacuum, where such forces are absent [3]. Therefore, the dispersing force F_{dis} is largely determined by the term qE, actually part of the wider expression referred to as the Lorentz force :

$$F = q \{ E + \nu \times B \} \qquad \ldots\ldots\ldots\ldots\ldots \quad (3)$$

where q is the particle charge and vectors ν and B refer to charge velocity and to the magnetic induction field respectively. Since $\nu = 0$ during the initial stage of the process, the main dispersing mechanism is due to the force qE, where the particle charge q for spherical particles is the derived expression [4]

$$q = 4 \pi \varepsilon_0 k E r^2 \qquad \ldots\ldots\ldots\ldots\ldots \quad (4)$$

while 'charging constant' k is within the range

$$1 < k < 3.43 \qquad \ldots\ldots\ldots\ldots\ldots (5)$$

a result, which was theoretically derived and later verified against published values [4].

It is an experimental fact, that the applied field must exceed a certain threshold value before dispersion can take place. This is also obvious from equation (2), which is the basis for deriving the value of this threshold field (E_0) for special instances. For the case of spherical metal particles, it was shown that

$$E_0 = \sqrt{\frac{r \rho g}{\varepsilon_0 k (3 - 0.473k)}} \qquad \ldots\ldots \quad (6)$$

Finally, it is relevant to determine the analytical form of the 'levitating field' (ξ), which must be applied to float particles above the bed. Clearly, one must have the condition

$$m g = q \xi \qquad \ldots\ldots\ldots \quad (7)$$

therefore the levitating field ξ is expressible as

$$\xi = mg/q = \frac{4/3 \ r^3 \pi \ \rho \ g}{4 \ \pi \ \varepsilon_0 k \ \xi \ r^2} \quad , \text{ so that}$$

$$\xi = \sqrt{\frac{r\rho g}{3\varepsilon_0 k}} \qquad \dots\dots\dots (8)$$

Comparing equations (5) and (7), it is clear, that in every case

$$\xi < E_0 \qquad \dots\dots\dots (9)$$

which can not be realized in practice, since field value ξ needed for particle-levitation is exceeded by the threshold value of the field E_0 which must be applied to start the process. It follows, that suspensions within an electrostatic field will consist of oscillating particles.

Disperse phase processes

As previously indicated, suspended particles do not levitate, but are in a continuous oscillatory mode. The physical mechanism which ensures this, is the repeated process of charge-exchange at the electrodes, as particles change polarity in order to take on the potential of the electrode with which they make contact. Since the field-vector always points in the same direction when dc-potential is applied, the periodic polarity-change of the particles ensure their reciprocating motion, unless particles become permanently attached on making contact. This simple mechanism also accounts for the flow of current through the system, at least in the initial stages of the process, when dust-concentration is low (typically below 400 particles per cm³, or about 0.33 mg/cm³ for the case of 60 μm copper spheres). Current densities are also low, rarely going beyond the figure 4.2×10^{-8} amp/cm², even for the very dense suspensions.

In additon, there is evidence for a second mechanism of charge-transfer, gaining rapidly ground as dust-loading is increased to the point where the mean free path of the particles is less than their individual path-lengths between electrodes. This consists of 'charge hopping' between particles, for which visual evidence has been gained by observing the motion of suspended particles in viscous liquids, such as paraffin oil [5]. Observing through a cathetometer, it was found, that in dense suspensions approaching opacity to transmitted light, increasing fraction of particles perform rapid oscillations between neighbouring granules, as well as taking part in the overall shuttle-motion between electrodes. This has been interpreted as evidence for charge-hopping and ties up with considerations involving mean free path, as well as a general increase in the observed current density. Such events are too rapid for allowing them to be observed in low-viscous media, such as air. However, a number of experimental measurements convinced the author, that this mode of charge transfer is not unique to liquids and forms part of the normal pattern of behaviour exhibited by high-density suspensions.

Much of the evidence for the behavior of electro-suspended dust had been gained through controlled experiments conducted by the author [5], using a specially constructed apparatus refered to as the 'dispersion analyser'(DA). The equipment was designed to study the properties of confined suspensions within uniform field conditions and employing a non-intrusive measuring probe (fig.4). Measurements were carried out of such parameters as the momentum and flux of particles, their size and velocity-distribution as functions of the applied field. The technique of measurement was based on 'sampling' the cloud by allowing an insignificant portion of the particles to escape through a small opening at the center of the upper plate. Information carried by these escaping particles was analysed by capturing them on the sticky underside of an external probe, positioned above the opening at pre-detemined altitudes above the plate. By exposing the probe to the particle-beam for a set time and repeatedly counting the captured particles under a microscope, this also allowed the calculation of the flux by extrapolating to zero elevation. Using the reverse process, it is possible to determine the maximum height of rise, yielding some added information on the momentum and velocity distribution of the species as function of applied field. Some of these results are displayed in figs. 5,6 and 7.

The role of ionic discharge in suspensions

While not all powders disperse equally well and can even remain completely inactive on applying a dispersing field, a few substances earlier classified as non-dispersive are sometimes found to disperse, following appropriate adjustment of relevant parameters. A point in question is a published classification of coal-dust, as one unable to form suspensions [6]. After careful anlysis of the problem, the present author has been able to suspend a range of finely ground coals (below 30 µm), including black coals, brown coals and lignite. Part of the solution is to sufficiently modify electrode-geometry for avoiding the build-up of ionization points to which coal dust is susceptible by its tendency to form short filaments and needles which attach themselves to electrodes. The author's own list of seemingly non-dispersive dielectric crystalline dusts (e.g.: ascorbic acid) has also been reviewed in light of later findings [7].

This brings up the subject of gas-discharges within the ES-system. These are mostly in the form of either intermittent discharges, or continuous discharges which may lead to sudden spark-breakdown and/or arcing. With low voltage-gradients present (typically below 0.1 kv/mm), discharges are a drift-velocity based mechanism, where conduction is due to the drift of existing

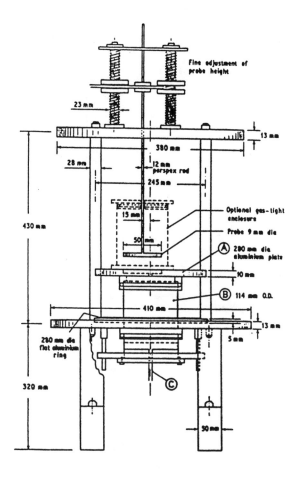

Fig.4. - The dispersion analyser (DA).

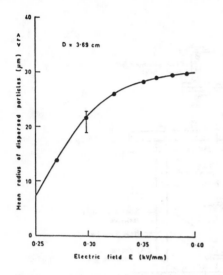

Fig.5. - Mean size of suspended
      copper spheres.

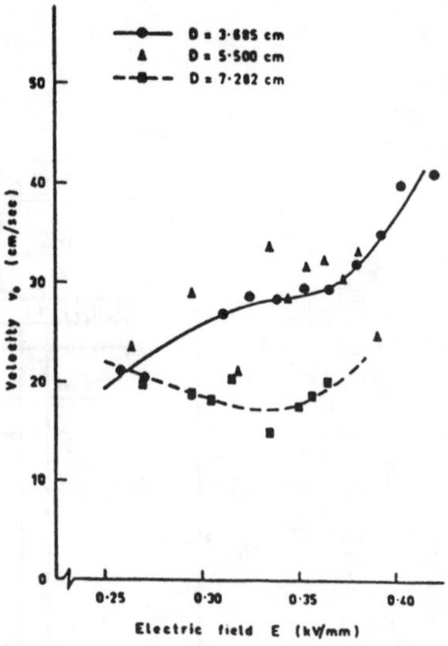

Fig.6.- Velocity of suspended
      copper spheres.

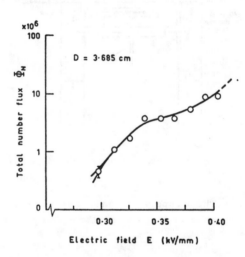

Fig.7 - Total particle flux in a closed system (Copper spheres).

ions in the direction of the field. Appropriately, current is proportional to the applied field until saturation is reached, where the whole mechanism depends on ion-pairs maintained through external factors [8] :

$$J = e_0 \, n_0 (\, \nu_+ + \nu_- \,)E \qquad \ldots\ldots\ldots \quad (10)$$

where J is current density,
n$_0$ denotes the number of ion-pairs
ν_+ and ν_- are the ionic drift-velocities
e_0 is the electronic charge.

For higher applied field, the current-avalanche region appears, characterized by a current which is no longer linearily dependent on E and various secondary mechanisms become evident, markedly increasing current-gain. They are mostly due to collision-ionization by energized primary electrons and by secondary elctrons emitted by the cathode under positive ion bombardment. This region, which is still not self-maintaining, is usually described by the Townsend equation for the steady state current

$$I = \frac{I_0 \, e^{\alpha d}}{1 - \gamma(e^{\alpha d} - 1)} \qquad \ldots\ldots\ldots \quad (11)$$

where α and γ are the first and second Townsend coefficients and d is the electrode gap. It therefore follows, that a complete breakdown of the gas occurs when

$$\gamma(e^{\alpha d} - 1) = 1 \qquad \ldots\ldots\ldots \quad (12)$$

which can also be expressed in the form

$$V_0 = f(pd) \qquad \ldots\ldots\ldots \quad (13)$$

after appropriate substitutions for α and γ. Specific forms of equ.(13) are known, e.g.: for STP-conditions it may be written as [9]

$$E = 24.22 + \frac{6.08}{\sqrt{d}} \qquad \ldots\ldots\ldots \quad (14)$$

where E is the strength of the uniform field (kV/cm) at the break-down potential V_0 and d is the spacing between the electrodes (cm). The strength of the break-down field E for 1-cm long gaps is usually quoted as 30 kV/cm.

Most DC electro-suspensions are generated by electrostatic fields in the region of 0.25 to 0.55 kV/mm strength, so that clearly E ≪ E and one does not expect breakdown conditions to occur. In spite of this, the evidence obtained from suspensions of alumina in a parallel-plate system appears to contradict this, as seen from the conductometric curves of Fig.8. The curves are plots of current vs. field, measured in air and SO_2 -suspensions respectively, demonstrating that at the relatively low field strenght of less than 0.13 kV/mm, there is already an appreciable portion of the current ($\approx$ 30%) which is carried by air-ions. (This can be seen by comparing the current in air with that measured in pure SO_2, known to be efficient for suppressing ionic currents [5]. Some other gases which can also act this way include SF_6 and C_5F_8, they have relative electric strenghts of 2.5 and 5.5 times the electric strenght of air respectively [9]). Part of the conductometric curves in Fig.8, below the field-strength of 0.5 kV/mm, thus indicate pre-breakdown ionic currents in agreement with equations (10) and (11) above. However, at the threshold value of 0.51 kV/mm, there is a sudden sharp increase of the total current typical of the breakdown conditions in air, but at a much lower value than could be expected on the basis of equ.(14), which predicts a value of 2.71 kV/mm for the particular electrode-spacing involved. This apparent contradiction can be removed by postulating a corona-ionization type partial breakdown of air, typical of strongly non-uniform field geometries, the field in the corona region being in excess of 5 x E_0, where E_0 is the overall field obtained by dividing the applied potential difference by electrode spacing. While this is usually achieved through a point-plane or wire-plane electrode system, the present geometry (Fig 1.) is clearly not designed for this, as also shown by the unsuccessful attempts to produce breakdown conditions by applying high potential difference across the clean system.

On the above basis, it follows that the apparent electric breakdown of air-suspensions is caused by the corona ionization of the gas, where corona-points are provided by the particles temporarily attaching themselves to the dispersing electrodes. This conclusion is also supported by recent work on the suspension of crystalline substances [7].

In the experience of the author gained by working with ES-systems over some years, gas discharges are one of the major influences on the behaviour of the system. Their effect is to disrupt the orderly progress of dispersion, causing irregular surface eruptions and cloud-turbulance. On a more significant level, e.g.: when continuous corona-ionization takes place, interference can even extend to a complete suppression of the process. The problems caused by the presence of gas-ions can be traced to the large discrepancy between the ionic and particle mobilities, which can result in an ion-current at the expense of the flux of charged particles. Since angularity of the particles and their tendency to line up and form filaments and/or needles have close bearing on the gas-discharge level within the system, the matter must be regarded as an important factor in determining disperseability. A practical example of this is the electro-suspension of lycopodium powder, which was found to be completely inactive in air, irrespective of applied voltage.

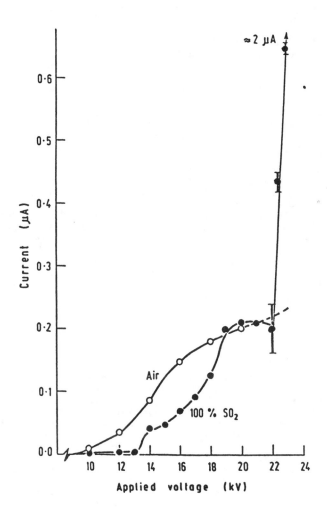

Fig.8 - The electrical conductivity of a suspended cloud of
alumina in air and dry SO_2.

When examined under the microscope, the particles were found to be covered by sharp needle-like protrusions. Based on the expectation, that the needles act as field-concentrators, thus giving rise to gas-discharge, air was displaced from the system by dry SO_2. The test was done in the previously mentioned DA-apparatus, which incorporates means for this type of work. It was found, that the previously inactive powder dispersed well, forming dense clouds of suspended particles on applying a field strenght of about 0.3 kV/mm [5].

Further matters likely to influence dispersion behaviour include the various restraining forces peculiar to a particular dust, including dipole forces, electrostatic forces arising from currents in a resistive body and inter-particle cohesion. These matters have been thoroughly reviewed in the literature [6]. It is also widely accepted, that particles must have a finite surface-conductivity in order to disperse, as shown by the ES-behaviour of highly resistive dusts such as sulphur, which can not be suspended under any known conditions. The role of surface conductivity is to produce a conducting path through the static bed, by which the surface-particles are able to acquire their charge needed for overcoming the various restraining forces and for lift-off as per equation (3). Experimental verification of this mechanism is by evacuating an appropriately constructed dispersion cell to a pressure of approx. 10^{-5} torr. If the cell contains some alumina powder, which disperses well in air, then evacuation stops the dispersion process, re-gained over some minutes after opening the system to air. This can be interpreted, as due to loss of a conducting surface-layer acquired in air, but lost by boiling off in vacuum. Taken with other evidence, it is clear that surface-conductivity is the electrical link between the in-bed electrode and the particles situated at the upper surface of the bed. Actual figures for the critical surface-conductivities are not available at present.

3.   APPLICATIONS

Introduction

It is increasingly realized, that ES has a strong potential for industrial applications. The reasons are often economical, as the electrical energy needed to produce suspensions are minute compared to the more conventional techniques, e.g.: fluidization. There are cases where an application, or even whole classes of application appear possible which could not be obtained by other known techniques. A case in point is the suspension of powders in high vacuum. Clearly, there are no other means for producing such an effect on a continuous basis, with the additional benfits of control over the concentration of the suspended cloud. Already, there are some vacuum applications which exploit this feature, such as the vapour-coating of powders and powder x-ray tubes.

The following section is a short survey of the various

102

application opportunities, also containing details of selected topics.

Powder x-ray tubes

These novel sources are based on the author's observation (1981) of continuous x-ray emissions from a system of electro-suspended metal dust in vacuum. The reasons for the emission have been subsequently traced to the following mechanism : on applying a dispersing voltage of 25-45 kV DC in vacuum (pressure range of 5×10^{-8} to 5×10^{-7} torr), an electro-suspension is generated, which induces the emission of electrons from the cathode. These are emitted in bursts whenever a charged particle approaches the cathode and the local field reaches about 10 kV/mm. The electrons are subsequently accelerated towards their target, mostly the individual dispersed particles, when x-rays are generated on impact. Since each suspended particle is a potential target for the electrons, the rays are generated throughout the dispersion, so that the device is a volume source of radiation. This is particularily so for the dispersing voltages which can generate a dense suspension, with the mean free path of electrons less than the separation distance between the cathode and the upper surface of the bed (refer to figs. 9 and 10).

Quantitatively, the x-ray mechanism depends on the emission of electrons from the cathode, governed by the Fowler-Nordheim law [10]

$$I = B E^2 e^{-\beta/E} \qquad \ldots\ldots\ldots\ldots\ldots(15)$$

where I is the emission current, B and β are constants containing the work function and E is the electric field. It may be shown, that emissions from the cathode can start well before impact and, to a first approximation, continue increasing for approx. 1-10 μsec as function of $1/d^4$ until impact, where d is the momentary separation of the particle from the cathode.

Overall, the factor which finally determines the current in the tube, as well as the intensity of x-rays which are finally emitted, is clearly the number of particle-induced emissions from the cathode per unit time, numerically equal to the particle flux multiplied by the avarage current density of a single emission. For any one emission of electrons from the cathode, the resulting x-ray intensity after impact with a target is given by equ.(16), due to Beatty [11], but modified by the author for the powder source :

$$\omega_\tau = k Z V^2 i_\tau \qquad \ldots\ldots\ldots\ldots \quad (16)$$

where ω_τ is the x-ray intensity of each single emission (watts), Z is the atomic number of the target, V the potential difference

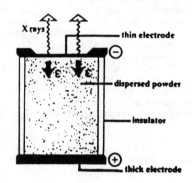

Fig.9. - Principle of the powder x-ray tube.

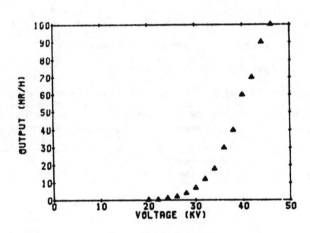

Fig.10. - Output of an 11-mm powder x-ray tube.

(volts) between the electron-emitter and target, while i_τ is the current (amps) induced by the proximity of a single particle to the cathode and k a constant, usually around 1.3×10^{-9} per volt. Microscopically, each emission of electrons and x-rays are on a time scale of about 10 μsec or less (as already indicated), where $\tau = t_1$ denotes the start of emission and $\tau = t_2$ the end of emission, sharply defined by the impact of the particle with the cathode, when emission ceases.

Overall, tube current I may be approximated by the expression

$$I = \psi_p \, A_c \, i \qquad \dots\dots\dots\dots\dots \quad (17)$$

where ψ_p is the particle flux (number/cm^2 sec)
A_c is the surface area of the cathode (cm^2)
i is the mean emission current induced by a single particle , i.e.:

$$i = 1/\tau \int_{t_1}^{t_2} i_\tau \, d\tau \qquad \dots\dots\dots\dots \quad (18)$$

Finally, the overall intensity of x-ray emissions can be expressed as

$$\Omega = k \, Z \, V^2 \, \psi_p \, A_c \, i \qquad \dots\dots\dots\dots \quad (19)$$

Since the product ($V \, \psi_p A_c \, i$) represents the electrical power-input, the overall conversion efficiency is

$$\phi = k \, Z \, V \qquad \dots\dots\dots\dots\dots\dots \quad (20)$$

which is the same as in conventional x-ray tubes (usually less than 0.4 %). However, since ψ is a function of the applied field and was found to linearily depend on voltage (true only as first approximation, since the exponent of V appears somewhat greater than unity), the intensity of x-rays are proportional to the cube of V instead of the usual V^2 :

$$\Omega \, \propto \, V^3 \qquad \dots\dots\dots\dots\dots \quad (21)$$

so that one can expect a very sharp increase in the emitted x-rays with applied voltage, while conversion efficiencies are no greater than usual.

In practice, powder x-ray sources consist of evacuated 40-80 mm long 10-200 mm wide cylinders, made of silica-glass or a suitable ceramic, also containing a small quantity of a suitable metal powder and are sealed at both ends by metal electrodes. While these are generally representative of the dimensions currently in use, sources up to 500 mm width appear feasible, the dispersing volume being important in determining x-ray output. In one type of design, the upper electrode (cathode) is the radiation window itself, which can be made of a 300 μm Be-foil for near-optimal transmission, or a 100 μm thick aluminium foil, which is nearly as good and much more cost-effective (with small 20 mm sources, the cost of using a berillium window is estimated as 98% of the cost of the whole unit). The type of powders which can be used are only restricted to those which disperse in vacuum. Some of the best target materials (high value of Z) which can not be used in conventional tubes on account of their low melting-point, are suitable for use inside powder tubes, as there is much less tendency for heating of the target due to efficient heat transfer from within. Output of present tubes is a modest 2000 R/h at the window of the 10 mm wide powder-sorce, but likely to increase as a result of present research. Another feature of the tubes is that electrons are generated by cold-emission from the cathode, as previously mentioned, thus eliminating the need for fragile heated filaments that limit the lifespan of present x-ray tubes to around 1500 hours.

Finally, powder-tubes use dispersed targets (the suspended dust) and are panoramic sources, emitting broad-beam radiation, unless restrained by appropriate windows and/or collimators. At present, this represents a limitation for their possible use, as imaging applications such as needed in the medical field, depend on focussed beams for producing sharp images. However, a modification of the basic geometry is under cosideration, which could remove this problem.

Coating applications

Electro-suspension coating (ESC) represents a range of applications, some with industrial potential, where a cloud of suspended particulates are treated by the vapours or mist of some other substance. In general, this result in powders, where the individual granules aquire a thin coating of the substance, while remaining un-agglomerated in the process.

A particular form of the process is one, where the coating takes place in vacuum, the chamber housing both the suspended dust and the vapour generating unit. This usually consists of an electron-gun and vapour source, the latter preferably machined from a block of high grade carbon-graphite housing the evaporant (in the form of a wire or foil), while internally heated by the electron-beam. The vapours are directionally emitted from the source through perforations and are allowed to diffuse through the cloud of suspended dust, when coating takes place by conden-sation on the surface of particles kept significantly below the temperature of the vapour [12,13]. The vacuum coating unit based on these principles, was successfully used for producing small quantities of coated metal powders, e.g.: copper, tin and nickel-alloy powders coated with zinc, silver, tin, gold and palladium. While it is possible to build up the coating thickness by repeated applications, this turned out to be a very lengthy process, involving repeated re-evacuations of the system coupled with the need to mechanically rotate the bed to ensure coating uniformity. On a single application, coating can proceed rapidly, producing particles with a few atomic layers of deposit. The technique appears therefore particularily suited for ultra-thin vapour depositions and is able to produce near-continuous films. As well, various exotic coated substances can be made by using the technique, such as thin alloy films prepared by co-depositing two or more simultaniously generated vapours.

While the vapour-coating of suspended powders in vacuum is a promising technique for preparing some unusual, and even unique substances, the author feels that a lot more development effort is needed before it can be fully utilized. As well, it is a batch process, unlikely to become useful in the mass-production of coated powders. At present, there is strong R & D effort for the development of a continuous ESC-technique at atmospheric pressure aimed at high-volume industrial production and uses are expected in such areas as the manufacture of controlled-release powders in the pharmaceutical industry and in modifying surface properties of powders used in powder-metallurgical applications.

Other applications

New potential applications continue to appear with increase in awareness of ES-techniques. These include possible use in dry papermaking to replace the present costly technique of dispersing paper fibres in water, for keeping them apart until a deposit is formed on a suitable substrate. The technique requires the use of extremely large quantities of water, typically 0.25 tons for each pound of paper. Apart from representing considerable handling costs, it also requires facilities for the large-scale treatment of the water for environmental reasons. It was demonstrated, that a suspension of paper fibres may be produced by an appropriate modification of the ES-system and the resulting 'fibre-gas' deposited on a moistened substrate, so as to ensure the hydrogen bonding of the fibres while also drastically reducing the mass of water normally required for the process.

There are further applications in the areas of filtering dusty gases, particularily where back pressure from clogged-up filters can not be tolarated (e.g.: diesel engine exhaust), or in cases where electrostatic precipitators are not generally effective. Further uses are envisaged in the combustion of coal dust, both in reactors and engines, as well as in the fields of medical device technology.

Finally, new applications may arise in areas which are incompletely understood at present, such as suspensions in a high pressure environment and by high frequency AC-fields.

4.   FUTURE DIRECTIONS

        While currently there is noticeable upsurge of interest for
applying ES-technology in diverse areas, it would be a mistake to
lose sight of the many unresolved questions surrounding the whole
subject. Accordingly, a lot more work should be devoted to basic
investigations for understanding the processes which govern the
ES-behavior of individual powders, with particular reference to
minimum specific surface-conductivities that are essential pre-
requisites for initating and maintaining the effect.

        There is further need to extend the study of ES to embrace
conditions known to support the effect, such as dielectric liquid
suspensions and suspensions under high atmospheric pressure. Much
basic information could be obtained from studying the effect
under zero gravity, as gravity distorts the velocity-field of the
particles by reducing their speed in one direction and increasing
it in the other, as well as imposing directional constraints.

        There is a clear distinction to be made between open and
closed systems, the latter enabling the study of macroscopic
equilibrium conditions, which are the conditions for much of the
work by the author [3,4,5,7,12], while open systems of the kind
studied by Colver are typically non-equilibrium systems, though
they can be brought to steady condition [1,6]. Clearly, there is
need to review the whole area of study in order to unify, and
possibly simplify theories as well as perform experiments to
enable comparisions.

        An area which has yet to receive detailed attention is the
suspension by AC-voltages at various set frequencies. While the
application of such fields did not reportedly elicit any unusual
influence on open systems, the author found that closed systems
respond with noticeably changed appearance, already evident when
AC-voltage is merely superimposed on a dispersing electrostatic
field. As well, the use of 50 c/s dispersing voltages appear
to increase emission from the earlier discussed powder x-ray
tubes, as well as responding to changes in frequency. Clearly,
the whole area deserves to be studied and understood.

        Finally, safety matters connected with electro-suspensions
is one requiring urgent attention. It is known, that many serious
accidents have occured by the explosion of suspended combustible
dust, where the term 'combustible' includes a wide range of
substances, even some metal dusts such as aluminium. With respect
to the present electro-suspension technology, the question
occurs, whether the presence of high dispersing voltages could in
fact add to the dangers of explosion already present in finely
suspended matter. On the basis of a number of relevant studies by
Bartknecht [14], it is known, that spark-ignition of dust clouds
can only take place above a certain threshold spark-energy, which
have been tabulated [15]. On the basis of calculations by the

author, spark-energies encountered in ES-systems are a long way below this threshold, however, further detailed studies must be made of the problem if questions on the safe use of such systems are to be answered.

Acknowledgements

The above work is receiving continued support by the CSIRO, which is gratefully acknowledged.

REFERENCES

1. G.M. Colver, J.Appl.Phys.,47;11 (1976), 4839.

2. S.G. Szirmai and E.C. Potter, J.of Phys.E: Sci.Inst.,9 (1976),985-989.

3. S.G. Szirmai, D.H. Morton and E.C. Potter, J.Appl.Phys., 51;10 (1980),5215-5221.

4. S.G. Szirmai, D.H. Morton and E.C. Potter, J.Appl.Phys., 51;10 (1980),5223-5227.

5. S.G. Szirmai, Ph.D.Thesis (1981),Univ.of N.S.W.,Sydney,Aus.

6. G.M. Colver, J.Powder&Bulk Solids Tech.,4;2/3 (1980),21-31.

7. S.G. Szirmai, *The suspension of dielectric crystals by an electrostatic field*. J.Appl.Phys.,(1989),submitted.

8. B.Yavorsky and A.Detlaf, *Handbook of Physics*. MIR publ., (1980),3rd ed.

9. L.L. Alston (ed.), *High-voltage technology*. Oxf.Univ.Press (1968), 48-49.

10. R.H. Fowler and L. Nordheim, Proc.Roy.Soc.(London), 119A; 173 (1928).

11. R.T. Beatty, Proc.Roy.Soc.(London), 89;314 (1913).

12. S.G.Szirmai, J.Appl.Phys., 55;11 (1984) 4088-4094.

13. S.G.Szirmai, *An electron-beam operated graphite source for the emission of directional vapours in high temperature vacuum coating applications*. J.Appl.Phys.(1989),submitted.

14. W. Bartknecht, *Staub explosionen*. Springer-Verlag, (1987), Berlin, New York.

15. P. Field, *Dust explosions*. Elsevier,(1982),Amsterdam, 205-227.

Effect of Joule and Viscous Dissipations on Temperature Distributions through Electrically Conducting Dusty Fluid

A. A. MEGAHED, A. L. ABUL-HASSAN, and H. SHARAF EL-DIN
Department of Engineering Physics and Mathematics
Faculty of Engineering
Cairo University, Egypt

The present work is devoted to the study of unsteady temperature distributions   through a viscous incompressible dusty fluid. The fluid is electrically conducting and is allowed to flow between two parallel fixed infinite porous plates, subjected to a constant pressure gradient under the action of a transverse uniform magnetic field.

The energy equations describing the heat flow within the system are formulated in a non-dimensional form. By using the results of author's previous work concerning the velocity distributions of fluid and dust particles, expressions for Joule heat and viscous dissipations are evaluated.

A numerical solution is applied to solve the system of energy equations.

The effect of magnetic field intensity, the porosity of the plates and density of dusty particles on temperature distributions are represented graphically. The effect of Joule and viscous dissipations on heat flow is examined and it is deduced that viscous dissipations has a negligible effect on heat transfer, while Joule heat has a comparatively large effect.

1.   INTRODUCTION

The study of fluid or gas having uniform distribution of solid particles has engaged the attention of several scientists and engineers for a long lime and has been enhanced recently due to its wide applications. This field has a great interest in areas of technical importance, including fluidization (flow through packed beds), combustion (use of dust in a gas cooling system) and in centrifugal separation of particulate matter from fluid. Other industrial applications include purification of curde oil, polymers technology and fluid droplets sprays. Ather forms of dusty fluids are observed in natural applicatications including environmental pollution, sediment transportation by water and air, soil solvation by natural wind and movement of dust Laden air.

The magnetic field affecting the motion and heat transfer of an electrically conducting dust fluid may be introduced artificially in industrial applications or may be presented naturally due to earth's magnetic field.

Review of Previous Work

In 1962, Saffman [1] while studing the stability of
laminar flow, introduced a mathematioal model for a dusty gas.
This model is based on the fact that a mutual force between the
fluid and solid particles which is proportional to the relative
velocity between them appears and acts on each of them in opposite directions.

Recently, the study of flow of dusty fluids has been considered by a large number of research workers and scientists,
with no account of magnetic field. In 1980, Nag [2] studied
the flow of dusty fluid past a wavy wall. The solution is obtained by using perturbation methods at moderate amplitudes of
the motion of the wall and for large wave length.

The steady flow of a dusty fluid subjected to a transverse
magnetic field was studied by Singh S.S. et al. (1981), [3],
by assuming the velocity of dusty particles to be parallel to
that of the fluid. While, in (1981), Mitra et al. [4] studied
the flow of a dusty fluid between two parallel plates in the
presence of transverse magnetic field, caused by the oscillations
of one of the plates.

The problem of heat transfer through a dusty viscous
fluid has been studied by Datta and Mishra, [5] in 1983. The
dusty fluid is allowed to flow between two parallel plates kept
at arbitrary temperatures. The flow is due to the motion of one
of the plates. In addition to solving the momentum equations,
the energy equations have been solved neglecting viscous dissipations, with no account of external magnetic field.

2. BASIC EQUATIONS

The present work is mainly concerned with the unsteady
flow of a dusty conducting viscous incompressible fluid between
two parallel porous horizontal in finite plates under the influence of a uniform external magnetic field and a time dependent pressure gradient. The two plates are kept at two different constant temperatures. The fluid motion is subjected to
uniform suction and injection at the upper and lower plates respectively. The direction of main flow is the positive x-direction while the external applied mognetic field is assumed in
the y-direction. The magnetic field is undisturbed assuming
low magnetic Reynold's number. The bounding plates coincide
with the planes y = +_h.

The dust particles obey the following descriptions :

i.   Spherical in shape and uniformly distributed.

ii.   Small in size such that Buoyancy force is neglected.

iii.   Concentration N is uniform and independent of time.

iv. Dust particles are not affected by the uniform suction and injection of the clean fluid.

The fluid velocity has the components $\underline{u} = (u, u_w, 0)$, where $u = u(x,t)$ is the x-component of the fluid, while u_w is the uniform transverse component due to suction and injection.

The dust particles are assumed to have one velocity component in the x-direction, so $\underline{v} = (v, 0, 0)$. The magnetic field is undisturbed and $\underline{B} = (0, B_0, 0)$

Following the above assumptions, together with the usual MHD approximations, the system of equations desciping the problem is reduced to [6,7] ;

Continuity Equations

$$\frac{\partial u}{\partial x} = 0 \qquad\qquad \text{For the fluid,} \qquad (1)$$

and

$$\frac{\partial v}{\partial x} = 0 \qquad \text{for the dust particles.} \qquad (2)$$

Equations of Motion:

For the fluid in x-direction ,

$$\rho\left[\frac{\partial u}{\partial t} + u_w \frac{\partial u}{\partial t} \right] = - \frac{\partial p}{\partial x} + \mu \frac{\partial^2 u}{\partial y^2} + KN(v-u) - \sigma B_0^2 u , \qquad (3)$$

and in y-direction :

$$0 = - \frac{\partial P}{\partial y} + KN(0 - u_w) .$$

While for the dust particles, we have :

$$m\left(\frac{\partial v}{\partial t}\right) = K(u-v)$$

The initial and boundary conditions are

$$u=0 \quad , \quad v=0 \quad \text{at} \quad t<0$$

$$u=0 \qquad v=0 \quad \text{at} \quad t>0 \quad \text{and} \quad y = \pm h \,] \qquad (6)$$

Energy Equations:

For the fluid particles :

$$\rho c\left[\frac{\partial T}{\partial t} + u_w \frac{\partial T}{\partial y}\right] = k \frac{\partial^2 T}{\partial y^2} + \mu\left(\frac{\partial u}{\partial y}\right)^2 + \sigma B_0^2 u^2 + \frac{2KNk}{3} (T_p - T). \qquad (7)$$

113

and for the dust particles ;

$$\frac{\partial T_p}{\partial t} = - \frac{T_p - T}{\tau_T} \tag{8}$$

with the initial and boundary conditions :

$$T = T_p = T_p = 0 \qquad \text{for} \qquad t < 0$$
$$T = T_p = T_1 = \text{constant for } t > o, \qquad y = -h \tag{9}$$
$$T = T_p = T_2 = \text{constant for } t > o, \qquad y = +h$$

Introducing the following non-dimensional variables :

$$\tilde{y} = \frac{y}{h} , \quad \tilde{t} = \frac{\nu t}{h^2} , \quad \tilde{u} = \frac{uh}{\nu} ,$$

$$\tilde{v} = \frac{vh}{\nu}, \quad \tilde{p} = \frac{ph^2}{\rho \nu^2} , \quad \tilde{\theta} = \frac{T - T_1}{T_2 - T_1} , \quad \tilde{\theta}_p = \frac{T_p - T_1}{T_2 - T_1} \tag{10}$$

equations of momentum and energy are re-written in the form (with ~ dropped)

$$\frac{\partial u}{\partial t} + M \frac{\partial u}{\partial y} = - \frac{\partial P}{\partial x} + \frac{\partial^2 u}{\partial y^2} - Ha^2 u + R(v - u) \tag{11}$$

$$G \quad \frac{\partial v}{\partial t} = u - v \tag{12}$$

with initial and boundary conditions :

$$u = o , \quad v = o \qquad \text{at} \quad t < o , \quad -1 < y < 1$$
$$u = o , \quad v = o \qquad \text{at} \quad t > o , \quad y = \pm 1 \tag{13}$$

and

$$\frac{\partial \theta}{\partial t} + M \frac{\partial \theta}{\partial y} = \frac{1}{P_r} \frac{\partial^2 \theta}{\partial y^2} + E(\frac{\partial u}{y})^2 + Ha^2 Eu^2 + \frac{2R}{3P_r} (\theta - \theta_p) \tag{14}$$

$$\frac{\partial \theta_p}{\partial t} = -L (\theta_p - \theta) \tag{15}$$

with initial and boundary conditions :

$$\theta = \theta_p = 0 \qquad \text{for} \quad t<0 \ , \quad -1<y<1$$

$$\theta = 1 \ , \qquad \theta_p = 1 \qquad \text{for} \quad t>0 \ , \quad y=1 \qquad (16)$$

$$\theta = 0 \ , \qquad \theta_p = 0 \qquad \text{for} \quad t>0 \ , \quad y=-1$$

where

M = suction parameter = $\dfrac{u_w h}{\nu}$

Ha = Hartmann number = $B_o h \sqrt{\dfrac{\sigma}{\nu}}$

R = Particle concentration parameter = $\dfrac{KNh^2}{\rho\nu}$

G = Particle mass parameter = $\dfrac{m\,\nu}{Kh^2}$

P_r = Prandtl number = $\dfrac{\mu C}{k}$

E = Eckert number = $\dfrac{\nu^2}{h^2 C (T_2 - T_1)}$

L = $\dfrac{h^2}{\nu\tau_T}$

The analytical solution of the equations of motion (11) and (12) subjected to the conditions (13) when the pressure gradient is represented by any arbitrary function of time, is obtained by the authors in a previous work [8], using the laplace transform techniques. The solution for the fluid velocity distribution corresponding to the case of constant pressure gradient has been deduced in the closed form :

$$u(y_1 t) = \frac{I_1}{z_1}(e^{z_1 t} - 1) + \frac{I_2}{z_2}(e^{z_2 t} - 1)$$

$$+ \sum_{n=1}^{\infty} \left[\frac{I_{n1}}{z_{n1}}(e^{z_{n1} t} - 1) + \frac{I_{n2}}{z_{n2}}(e^{z_{n2} t} - 1) \right.$$

$$\left. - \frac{I_{n3}}{z_{n3}}(e^{z_{n3} t} - 1) - \frac{I_{n4}}{z_{n4}}(e^{z_{n4} t} - 1) \right] \qquad (17)$$

where

$$z_{1,2} = \left[-(GHa^2 + RG + 1) \pm ((GHa^2 + RG + 1)^2 - 4GHa^2)^{1/2} \right]/2G \ ,$$

$$z_n = \left[(NG - B) \pm ((NG - B)^2 - 4(Ha^2 - N)G)^{1/2} \right]/2G \ ,$$

$$N = -\pi^2(n-1)^2 - \left(\frac{M}{2}\right)^2 \quad \text{when calculating } Z_{n1,2},$$

$$N = -\left(\frac{\pi}{2}\right)^2(n-0.5)^2 - \left(\frac{M}{2}\right)^2 \quad \text{when calculating } Z_{n3,4},$$

$$I_{1,2} = \frac{1+GZ_1}{G(Z_1-Z_2)}\left[1 + e^{\frac{My}{2}}\left\{Sh\,\frac{M}{2}\,\frac{Sh\,my}{Sh\,m} - Ch\,\frac{M}{2}\,\frac{Ch\,my}{Ch\,m}\right\}\right]$$

$$m = \sqrt{\left(\frac{M}{2}\right)^2 + w},$$

$$w = \frac{G\,S^2 + (G\,H_a^2 + RG + 1)\,S + H_a^2}{1 + G\,S}$$

$$I_{n1,n2} = \frac{1 + G\,Z_n}{G(Z_n - Z_1)(Z_n - Z_2)}\,e^{\frac{My}{2}}\,Sh\,\frac{M}{2}\,\frac{Sh\,my}{Ch\,my}\frac{dm}{ds}$$

In I_{n1} $P_n = P_{n1}$, m and (dm/ds) are evaluated at $S = P_{n1}$ with a similar argument for P_{n2}

$$I_{n3,n4} = \frac{1 + G\,Z_n}{G(Z_n - Z_1)(Z_n - Z_2)}\,e^{\frac{My}{2}}\,Ch\,\frac{M}{2}\,\frac{Ch\,my}{Sh\,m}\frac{dm}{ds}$$

In I_{n3} $P_n = P_{n3}$, m and (dm/ds) are evaluated at $S = P_{n3}$ with a similar argument for P_{n4}.

Numerical Solution Of The Energy Equations

The finite difference method is introduced to solve equations (14) and (15) subjected to conditions (16). Two approaches have been used; the explicit method and Crank Niclson method [9,10]. It is found that the explicit method leads to more accurate results when compared to those obtained by solving the equations of energy analytically neglecting viscous & Joule dissipations. So, it has been chosen to apply the explicit method to solve the problem numerically, taking into account viscous dissipation and Joule heat.

A computational grid is presented in space-time (y-t) domain. The variable y ranges from -1 to +1 and the y-domain is devided into I = 20 divisions. The width of each division is $\Delta y = 0.1$ The t-domain is devided into J equal intervals with $\Delta t = 0.001$.

The numerical values for the terms including velocity u and its partial derivative w.r.t. y is calculated at a relatively small number of mesh points and an interpolation method is applied to calculate u and $\partial u/\partial y$ at each mesh point of the grid.

Consider the difference forms :

$$\frac{\partial \theta}{\partial t} = \frac{\theta_{i,j+1} - \theta_{i,j}}{\Delta t} \quad ,$$

$$\frac{\partial \theta}{\partial y} = \frac{\theta_{i+1,j} - \theta_{i,j}}{\Delta y} \quad ,$$

and

$$\frac{\partial^2 \theta}{\partial y^2} = \frac{\theta_{i+1,j} - 2\theta_{i,j} + \theta_{i-1,j}}{(\Delta y)^2} \quad ,$$

so equations (14) and (15) become :

$$\frac{\theta_{i,j+1} - \theta_{i,j}}{\Delta} + M \frac{\theta_{i+1,j} - \theta_{i,j}}{\Delta y} =$$

$$= \frac{1}{P_r} \frac{\theta_{i+1,j} - 2\theta_{i,j} + \theta_{i-1,j}}{(\Delta y)^2}$$

$$+ Ec \left(\frac{\partial u}{\partial y}\right)^2 + Ha^2 E u^2 + \frac{2R}{3P_r} (\theta_{P_{i,j}} - \theta_{i,j}) \tag{18}$$

and

$$\frac{\theta_{P_{i,j+1}} - \theta_{P_{i,j}}}{\Delta t} = -L(\theta_{P_{i,j}} - \theta_{i,j}) \tag{19}$$

where u and $\frac{\partial u}{\partial y}$ are calculated for every mesh point. Starting by the given initial and boundary conditions, equations (18) and (19) are solved progressively to obtain the temperature distributions within the fluid and for dust particles.

The proposed numerical solution is applied to equation (18) in the following cases :

(1) Including both viscous and Joule Dissipations.

(2) Neglecting Viscous Dissipation.

(3) Neglecting Joule Dissipation.

The results obtained are tabulated and drawn versus different flow parameters.

3. RESULTS AND CONCLUSIONS

The temperature distribution of both fluid and dust particles are calculated numerically from the equations (14) & (15) which include viscous and Joule dissipations terms. The results are plotted in figures (1) to (13).

Figures (1) to (5) illustrate the temperature distribution of the fluid, while figures (6) to (11) illustrates the temperature distribution of the dust particles. These figures show the following :-

1) Figure (1) shows the advance of the solution toward the steady state solution as the time increases.

2) Figure (2) shows that the increase in particle concentration R to lead a decrease in the fluid temperature distribution.

3) Figure (3) shows the relation between the temperature distribution and Prandtl number. Decrease in Pr lead to increase in temperature distribution that is reasonable because Pr is inversely proportional to thermal conductivity.

4) Figure (4) shows that as suction increases the temperature distribution decrease which is expected since suction current is from cold plate to hat plate. An opposite case is expected if suction parameter has negative value.

5) The effect of Hartmann number is shown in figure (5). The increase of Hartmann number leads to an increase in the fluid temperature distribution.

6) L, T, Ec, G have a negligible effect on the fluid temperature distribution.

7)   Figures (6) to (11) correspond to dust particles tempe-
     rature distribution, the same behavior is expected to all
     parameter except for L parameter which is effective   in
     dust particles temperature distribution, increasing L re-
     sult in increase in temperature distribution.

     It should be noted that dust particles temperature dis-
tribution takes much more time to reach steady state than that
of the fluid.

     The viscous dissipation is found to be negligible and al-
most has no influence while the Joule dissipation is found to
have larger effect an temperature distribution of both fluid
and dust particles.

     The temperature distribution corresponding to the flow of
dusty fluid between two parallel plates at no suction and with
no account of the magnetic field (M=O and Ha=O) is shown in
figure (12). This special case was studied by Datta and Mishra
[5] who solved that problem neglecting viscous dissipation by
taking Laplace transform followed by its numerical inversion.
They found oscillations in the temperature distribution for both
the fluid and the dust particles, specially at low times. They
also found that at low times the temperature may become negative
within a certain range of y. They did not account for these
negative temperatures nor were their results in agreement with
ours. In the other hand, a complete agreement exist between
our numerical and analytical solution.

     The above results show that the existence of dust partic-
les in a fluid has a marked effect on its velocity and temperature distribution. It follows that the motion of a conducting fluid can be controlled by the addition of the dust particles. This process is already carried out in blades of turbines
and other engineering applications. Moreover, the Effect of
Joule heat is studied and the fluid temperature is calculated
and torbulated for different values of Eckert number, and
Hartmann number Fig.(5 ,13).

4. NOMENCLATURE

a	radius of solid particles
$\underline{B}$	magnetic flux density vector
$\bar{c}$	specific heat at constant pressure for the fluid
C_s	specific heat of fluid
$\underline{D}$	electric flux density vector
$\underline{E}$	electric field vector
$\bar{E}c$	Eckert number
Fe	electromagnetic force
G	particle mass parameter
$\underline{H}$	magnetic field intensity vector
Ha	Hartmann number
$\underline{J}$	electric current density vector
$\bar{k}$	thermal conductivity
K	Stokes drag coefficient
L	parameter inversely proportional to T
m	mass of individual particle
N	concentration of dust particles
p	pressure
Pr	prandtl number
R	particle concentration parameter
T	fluid temperature
Tp	dust particles temperature
$\underline{u}$	fluid velocity vector
$\underline{v}$	dust particles velocity vector
μ	dynamic coefficient
ν	kinematic viscosity
ρ	density of the clean fluid
ρ_f	electric charge density
ρ_p	mass of dust particles per unit volume
ρ_s	material density of dust particles
σ	electric conductivity
τ_T	temperature relaxation time
τ_p	velocity relaxation time
Φ	viscous dissipation function
Π	suction parameter

TABLE I: JOULE DISSIPATION EFFECT ON FLUID TEMPERATURE
(Ha=0.5, Pr ≲ 1, Ec = 0.2 , R = 0.25 , L = 0.2)

y	T with Joule Dissipation	T without Joule Dissipation
-1.0	0.000000E+0	0.000000E+0
-0.9	2.281529E-3	1.778005E-3
-0.8	4.975480E-3	3.942629E-3
-0.7	8.496790E-3	6.924156E-3
-0.6	1.333111E-2	1.123839E-2
-0.5	2.008900E-2	1.752426E-2
-0.4	2.954045E-2	2.657373E-2
-0.3	4.263509E-2	3.934923E-2
-0.2	6.049857E-2	5.698331E-2
-0.1	8.440778E-2	8.075463E-2
+0.0	1.157349E-1	1.120360E-1
+0.1	1.558659E-1	1.522127E-1
+0.2	2.060888E-1	2.025736E-1
+0.3	2.674674E-1	2.641816E-1
+0.4	3.407051E-1	3.377384E-1
+0.5	4.260226E-1	4.234578E-1
+0.6	5.230618E-1	5.209691E-1
+0.7	6.308395E-1	6.292669E-1
+0.8	7.477552E-1	7.467224E-1
+0.9	8.716641E-1	8.711607E-1
+1.0	1.000000E+0	1.000000E+0

TABLE II: JOULE DISSIPATION EFFECT ON DUST PARTICLES TEMPERATURE
(Ha=0.5, Pr=1, Ec=0.2, R =0.25, L =0.2)

y	T with Joule Dissipation	T without Joule Dissipation
-1.0	0.000000E+0	0.000000E+0
-0.9	1.731414E-5	1.152208E-5
-0.8	3.851635E-5	2.659622E-5
-0.7	6.761778E-5	4.954652E-5
-0.6	1.101253E-4	8.632817E-5
-0.5	1.743350E-4	1.455569E-4
-0.4	2.726118E-4	2.397843E-4
-0.3	4.229835E-4	3.870763E-4
-0.2	6.509707E-4	6.129332E-4
-0.1	9.918303E-4	9.525446E-4
+0.0	1.493026E-3	1.453335E-3
+0.1	2.216976E-3	2.177690E-3
+0.2	3.243748E-3	3.205711E-3
+0.3	4.673661E-3	4.637754E-3
+0.4	6.629356E-3	6.596529E-3
+0.5	9.257229E-3	9.228448E-3
+0.6	1.272780E-2	1.270400E-2
+0.7	1.723504E-2	1.721696E-2
+0.8	2.299419E-2	2.298227E-2
+0.9	3.023853E-2	3.023273E-2

TABLE III: EFFECT OF HARTMANN NUMBER ON TEMPERATURE DISTRIBUTIONS
OF FLUID
(Ec=0.2, Pr=1, R=0.25, L=0.2, M=0)
"Joule Heat is Included"

y	(Ha = 0.1)	(Ha = 1)	(Ha = 5)
-1	0.0	0.0	0.0
-0.8	0.00399	0.00770	0.01619
-0.6	0.01132	0.01881	0.03395
-0.4	0.02670	0.03728	0.05646
-0.2	0.05713	0.06964	0.09081
0.0	0.11219	0.12534	0.14710
0.2	0.20272	0.21523	0.23640
0.4	0.33786	0.34844	0.36762
0.6	0.52105	0.52855	0.54369
0.8	0.74676	0.75048	0.75897
+1.0	1.0000	1.000	1.000

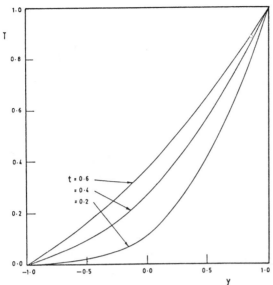

Figure (1) Fluid Temperature Distribution
t=0.2, 0.4, 0.6

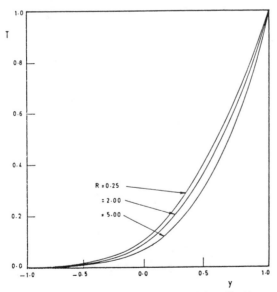

Figure (2) Fluid Temperature Distribution
R=0.25, 2, 5

123

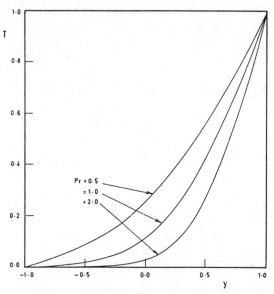

Figure (3) Fluid Temperature Distribution
Pr=0.5, 1, 2

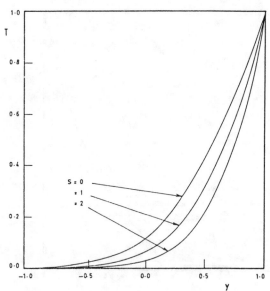

Figure (4) Fluid Temperature Distribution
S=0, 1, 2

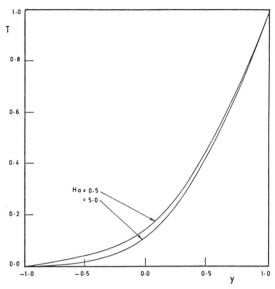

Figure (5) Fluid Temperature Distribution
Ha=0.5, 5

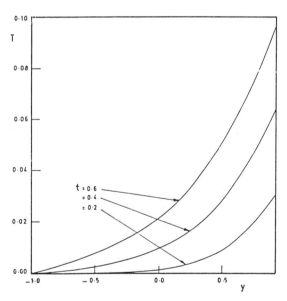

Figure (6) Dust Particles Temperature Distribution
t=0.2, 0.4, 0.6

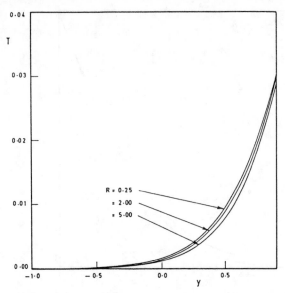

Figure (7) Dust Particles Temperature Distribution
R=0.25, 2, 5

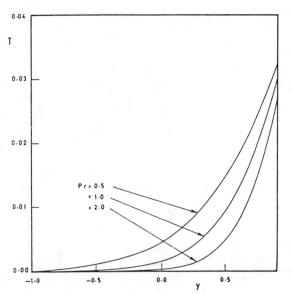

Figure (8) Dust Particles Temperature Distribution
Pr=0.5, 1, 2

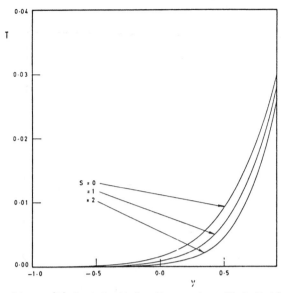

Figure (9) Dust Particles Temperature Distribution
S=0, 1, 2

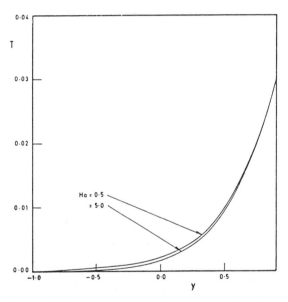

Figure (10) Dust Particles Temperature Distribution
Ha=0.5, 5

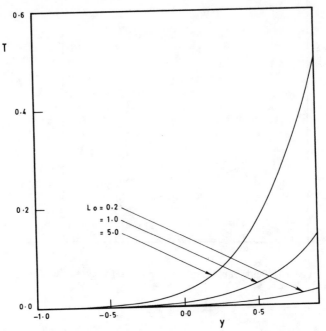

Figure (11) Dust Particles Temperature Distribution
Lo=0.2,1,5

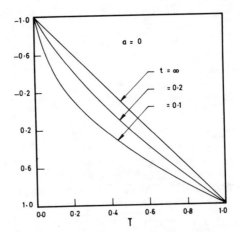

Figure (12)
Fluid temperature distribution in case of
no suction at different times

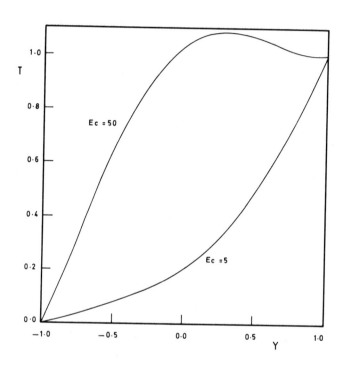

Fig. 13 Effect of Eckert Number
on Fluid Temperature (M=0; Ha=5)
Joule Heat is Included

REFERENCES

1. Saffman P.G., (1962), "On The Stability Of A Laminar Flow Of A Dusty Gas", Journal Of Fluid Mechanics, Vol.13, pp. 120.

2. Nag S.K., (1980), "Flow Of A Dusty Fluid Past A Wavy Moving Wall", Journal Of The Physical Society Of Japan, Vol. 45, No 1, pp. 391.

3. Singh S.V., Gangwar S.S., And Bobu R., (1981), "Steady Transverse MHD Flow Of A Dusty Fluid With Parallel Velocity Fields", Il Nuovo Cimento, Vol.65B, No 2, pp. 419.

4. Mitra P. Bhattacharyya P., (1981), "On The Hydromagnetic Flow Of A Dusty Fluid Between Two Parallel Plates, One Being Stationary And The Other Oscillating, Journal Of The Physical Society Of Japan, Vol.50, No 3, pp. 995.

5. Datta N., And Mishra S.K. (1983), "Unsteady Coutte Flow And Heat Transfer In A Dusty Gas", Int. Comm. Heat Transfer, Vol.10, pp. 153.

6. Cramer K.R., And Pai S.I. (1973), "Magnetohydrodynamics For Engineers And Applied Physicists", Scripta Pub. Company, Washington.

7. Eckert E.R., And Robert Drake, (1972), "Analysis Of Heat And Mass Transfer", McGraw Hill, Japan.

8. Megahed A.A., Abul.Hassan A.L. And H.Sharaf El-Din, (1988) "Unsteady Flow Of Dusty Conducting Fluid Between Two Porous Plates", Indian Journal Of Pure And Applied Mathematics, (Submitted For Publication).

9. Smith G.D.(1981), Partial Differential Equations, Finite Difference Method, Oxford University Press.

10. Ortega M., and Poole W.G., (1981), Numerical Methods for Differential Equations, Pitman Rublishing.

SLURRIES

Chevron Slurry Fuel Development and Handling Characterization

M. NIKANJAM and D. A. KOHLER
Chevron Research Company
576 Standard Avenue
Richmond, California 94802-0627, USA

Abstract

The typical bottoms product from a petroleum refinery solvent deasphalting process is solid to well above room temperature. Current practice requires cutting the viscosity with middle distillate petroleum fractions, which might otherwise bring higher value as diesel or jet fuels, to deliver a manageable fuel to the burner. Chevron Research Company has developed a petroleum-based solid/water slurry fuel for potential application as a replacement for these heavy residual fuels. Process development and testing are described for a slurry fuel containing up to 70 wt % of solvent deasphalting bottoms, which is produced by emulsifying this low value residue from petroleum refining at high temperature in water. The process parameters and additive packages for emulsification are described. Special handling aspects of this slurry fuel were also studied and requirements for pumps, strainers, fuel nozzles, etc., are also described.

I - INTRODUCTION

The bottoms from a solvent deasphalting process, SDA tar, is a very high softening point hydrocarbon mixture and is of very limited use in its solid and brittle form. It has a much higher viscosity than ordinary heavy fuels and crude oils. Some properties are summarized in Table I. Currently, this material is blended with equal volumes of more expensive petroleum fractions such as diesel or jet fuel to yield a liquid fuel of a practical viscosity to be used as high sulfur fuel oil. The use of the SDA tar in a water slurry will eliminate the need for the higher value fractions and will reduce the overall volume of the heavy fuel. The resulting product will have a lower viscosity than ordinary bunker fuel and does not require preheating in its application. The water in the fuel will also reduce the combustion-generated NO_x emissions considerably.

Another advantage of a slurry fuel compared to a solid fuel is that it can be transported through pipes, pumped through mostly conventional liquid fuel equipment, and burned in furnaces designed for a heavy oil. The development of such a fuel requires three stages: formulation, handling characterization, and combustion testing. Three methods were considered and tested to prepare a slurry fuel. These were grinding, atomizing, and emulsifying the SDA tar. The most practical and economical method was the emulsification process. This article will describe this process, the additive selection, and the handling characteristics of this fuel.

A separate article (Reference 1) will address the atomization and combustion tests.

II - PROCESS

An emulsification pilot plant was designed and constructed to handle temperature levels of well above 400°F and pressure values to 200 psi. Figure 1 is the schematic of this facility. Colloid mills have been used for many years to make asphalt emulsions. In this case, the mill consists of two circular plates, one stationary (stator), and the other, the rotor, rotating at about 3600 revolutions per minute (rpm). The clearance between the two plates is around 0.040 in. Each plate's surface has a series of teeth in a circular pattern to create a torturous path to apply high shear to the passing mixture of SDA tar and water.

TABLE I

TYPICAL SDA TAR PROPERTIES

Property	Range
Gravity, API at 60°F	0.90-1.1
Viscosity, cSt	
at 275°F	5,000-20,000
at 350°F	500-1,000
Softening Point, ASTM D 2398, °F	200-240
Heat of Combustion, Btu/Lb	17,600-17,800
Ramsbottom Carbon	29-31
Acid No., mg KOH/g	1-2
Proximate Analysis, %	
Moisture	0.03-0.06
Ash	0.2-0.3
Volatiles	82-84
Fixed Carbon	16-17
Elemental Analysis, %	
C	85-90
H	9-10
N	1.4-1.7
O (Difference)	0.5-2.0
Ash	0.1-0.15
S	1.6-2.5
Metal Analysis, ppm	
Al	6-14
Ca	128-148
Fe	108-171
K	2-8
Mg	12-26
Na	70-151
Ni	136-235
Si	4-26
V	154-196

FIGURE 1
SCHEMATIC OF COLLOID MILL PILOT PLANT

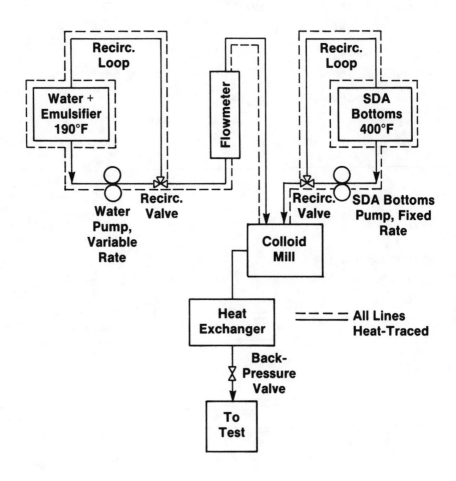

Water and additives are mixed in one reservoir, heated to
the required temperature, and circulated prior to emulsification. SDA tar is mixed and heated in another reservoir and circulated. Both streams are directed into the colloid mill to make
the slurry fuel. The fuel leaves the mill at a temperature well
above the boiling point of water. A backpressure valve maintains
sufficient pressure in the product line to prevent water vaporization and loss, until the fuel has passed through a heat
exchanger and has cooled to 140°F or lower.

III - ADDITIVE DEVELOPMENT

The production of a stable low viscosity oil in water slurry
fuel by emulsification requires specific additives as emulsifiers
and stabilizers. Table II includes the name and composition of a
few of many additives which were tested in the pilot plant.
Table III is a selected summary of the performance of these additives. Vinsol resin combined with sodium hydroxide has been used
in an economical formulation to produce asphalt emulsions. This
additive package was also capable of producing the SDA tar/water
slurry. The addition of sodium to a fuel is not desirable, however, since sodium may combine with the vanadium in the fuel to
form low melting eutectics, which can result in slagging of the
furnace tubes.

An additive package containing T-Mulz, a phosphate ester
manufactured by Thompson Hayward, was tested and selected as the
most effective formulation in this case. This additive package
is as follows:

T-Mulz Emulsifier, Wt %	0.5
Kelzan Stabilizer, Wt %	0.05
Ammonium Hydroxide, Wt %	0.4
Light Cycle Oil, Wt %	3
Formaldehyde, ppm	5

Kelzan, a xanthan gum, is used for long-term stability. Ammonium
hydroxide is used in place of sodium hydroxide to make the phosphate ester salt. Light cycle oil is used to adjust the softening point of the SDA tar to the 180-190°F range. The amount
used, therefore, varies with the initial softening point of the
material. Formaldehyde is used to preserve the Kelzan. The
slurry product with the above formulation has these typical
properties:

137

TABLE II

ADDITIVES FOR EMULSIFICATION

Trade Name	Manufacturer	Composition
Vinsol Resin + NaOH	Hercules	Rosin Acids
Witconate P10-59	Witco Chemical	Alkylbenzene Sulfonate
Dowfax	Dow Chemical Co.	Alkyl Aryl Disulfonate
CNA	Chevron Corporation	Naphthenic Acid + NaOH
Nopcosperse 092	Diamond Shamrock	Imidazoline Derivative
Nopcosperse RG1	Diamond Shamrock	Naphthalene Sulfonate
Vinsol Resin + TAAH	Hercules	Rosin Acids
Pluradyne	BASF Wyandotte	Polyether Polyol
TUSEV + Acetic Acid		Fatty Amine/Fatty Diamine
Orzan AE/NH$_4$$^+$	ITT Rainier	Lignosulfonate
Orzan AE/NH$_4$$^+$ + NH$_4$OH	ITT Ranier	Lignosulfonate
Naphthenic Acid 190	CPS Chemical	Naphthenic Acid + NH$_4$OH
T-Mulz 565 + NH$_4$OH	Thompson Hayward	Phosphate Ester
T-Mulz 598 + NH$_4$OH	Thompson Hayward	Phosphate Ester
T-Mulz 596 + NH$_4$OH	Thompson Hayward	Phosphate Ester
T-Mulz 734-2 + NH$_4$OH	Thompson Hayward	Phosphate Ester
Asphalt Emulsifier No. 3	Witco Chemical	Tall Oil Fatty Acids in Tar/Witconate + NH$_4$OH in H$_2$O

138

TABLE III

SELECTED PILOT PLANT RUN SUMMARY

Additive	SDA Tar			Product		
Name/Concentration, Wt %	Softening Point, °F	Temp., °F	Viscosity, cSt	Median Particle Size, Micron	Solids Content, %	Viscosity[1] (100 Sec.)
Vinsol Resin + NaOH/1	203	400	365	3	65	110
Witconate P1059/2	196	401	280	3	60	-
Dowfax/2	206	410	300	3	63	116
Chevron (NA) + NaOH/1	214	414	380	4	64	159
Nopcosperse 092/2	214	410	385	-	Failed	-
Nopcosperse RG1/2	214	410	385	-	Failed	-
Vinsol Resin + TAAH/2	216	410	440	7	66	44
Pluradyne/1	166	342	240	11	66	-
TUSEV + Acetic Acid/2	178	360	240	-	Failed	-
Orzan AE/NH$_4^+$/2	158	335	240	-	Failed	-
Orzan AE/NH$_4^+$ NH$_4$OH/2	163	335	245	-	Failed	-
Naph. Acid 190/2	165	335	246	4	67	281
T-Mulz 565 + NH$_4$OH/0.5	173	350	240	2.3	65.5	-
T-Mulz 598 + NH$_4$OH/1	160	335	240	1.8	63	-
T-Mulz 596 + NH$_4$/OH/1	160	335	240	1.8	64	-
T-Mulz 734-2 + NH$_4$OH/1	160	335	240	1.6	64	-
Asphalt Emulsifier No. 3	174	350	240	2.45	70.1	-

Median Particle Size, Micron	2-3
Solids Content, Wt %	65
Viscosity, cSt at 75°F	200
Heating Value, MMBtu/Bbl	4.2
Specific Gravity at 60°F	1.04

Table IV includes the test conditions and the product properties of a series of pilot plant productions. It should be noted that the combination of the emulsifier and the stabilizer is responsible for a good-quality slurry fuel. In some cases, when the Kelzan concentration was reduced, the product quality suffered. In one case, Kelzan was eliminated and the process failed.

One of Chevron's plants which produces asphalt emulsions commercially was modified to produce the SDA tar/water slurry on a trial basis. The modification included addition of a heat exchanger and a backpressure valve downstream of the colloid mill in order to handle the higher temperatures and pressures required for this material. The emulsification rate was around 80 gal./min., and the slurry fuel was loaded directly into tank trucks for use in the handling characterization pilot plant and extensive atomization and combustion tests (Reference 1).

IV - HANDLING CHARACTERIZATION

Although the fuel looks and acts like a liquid fuel, problems can occur if the proper equipment and methods are not applied. A pilot plant was designed and constructed to study the handling characteristics of slurry fuels. The flow diagram is presented in Figure 2. Fuel can be supplied from a 350-gal. tank equipped with water coils for moderate heating. A regular 55-gal. drum can also be connected to the system if the fuel quantity is limited. Temperature can be measured at several locations using Type J thermocouples. Data from the flowmeter, pressure transmitters, and the thermocouples are recorded on a YEW 30 channel recorder. This section summarizes the results of a series of tests to determine the handling characteristics of this fuel.

TABLE IV

PILOT PLANT PARAMETERS FOR T-MULZ

Tar Softening Point, °F	LCO in Tar, Wt %	Final Softening Point, °F	Product Temp., °F	T-Mulz Type	T-Mulz, Wt %	Kelzan, Wt %	Median Particle Size, μm	Solids Content, Wt %
186	4	163	148	598	1	0.05	1.8	63.2
186	4	164	153	596	1	0.05	1.8	64
186	4	162	150	565	1	0.05	1.9	64.5
186	4	160	156	734.2	1	0.05	1.6	63.8
186	2	174	150	565	0.5	0.05	2.2	64
186	2	173	–	565	0.5	0	Failed	
186	0	186	–	565	0.4	0.05	Poor	
186	0	186	–	565	0.5	0.03	Poor	
186	0	186		565	1	0	Poor	
186	0	186	157	565	0.5	0.05	3.6	63
203	0	203	158	565	1.0	0.05	9.9	67
173	0	173	147	598	0.5	0.05	2.8	64
184	0	184	113	565	0.5	0.05	2.7	64.9
184	0	184	120	565	0.4	0.05	5.4	61.4
201	2	189	128	565	0.5	0.05	2.3	64.4

FIGURE 2

FLOW DIAGRAM
SLURRY FUEL HANDLING FACILITY

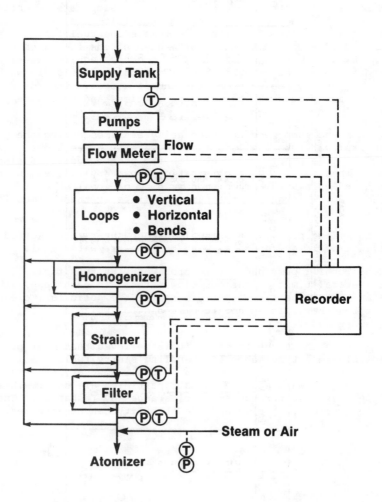

1. Pumps

Three positive displacement pumps were selected for testing. These included a Moyno progressing cavity pump, a Viking
gear pump, and a Tuthill impeller process pump. These pumps had
the flow rate range and the associated pressure required to provide fuel to a typical atomizer. There was very little flow rate
reduction with increasing fuel pressure.

The progressing cavity Moyno pump was by far the best of the
three pumps tested. This low shear pump provided the longest
operating time and the least damage to the product. In one test,
one drum of a 64% solids slurry fuel was circulated six times
through the pump at the rate of 25 lb/min. Samples were obtained
at the end of each cycle. The median particle size, 2.2 micron,
did not change. It is important to remove the fuel from the pump
at the end of each run to avoid breakdown of the fuel when mixed
with an excess amount of water. This process requires water flow
through the pump for several minutes. Regular liquid dishwashing
soap was added to the water at the end of a test. The rotor and
stator of the pump were checked several times and no damage was
observed.

The Moyno pump becomes quite noisy at times during the
switch from water to fuel and fuel to water. Water flows
smoothly in the pump. The fuel with the proper additives also
can be pumped easily. The combination, however, dilutes the surfactant and breaks the emulsion down to a material which is very
hard to pump. Increasing the pump flow rate at these times is
helpful and will clear the pump faster.

The Viking pump, with slightly higher shear rate, worked
relatively well in most cases with the SDA tar/water fuel.

The Tuthill pump had severe problems with this fuel. It was
tested several times at various flow rates. In each case, it
froze and had to be taken apart to remove the solid tar, cleaned,
and put together. It is not recommended for this use unless
changes are made.

Based on the study to date, it is believed that gear pumps
in general are not suitable for handling slurry fuels. Some gear
pumps are used on asphalt emulsion trucks. However, the emulsion
in this case is warm and the asphalt particles are soft. These
particles can pass through small clearances by reshaping. The
particle size distribution and the stability can be affected, but

143

since the emulsion is applied in its final step, there is no concern. To increase the chance of success, these gear pumps are
often run with abrasive material first to increase the clearances
to reduce down time. This will decrease the pump efficiency.
For tar/water use, even at warm fuel temperatures, the tar particles are solid and will not pass through small clearances in a
gear pump. This application is similar to the one with hard air-
blown asphalt. With this product, centrifugal pumps are used in
the plants since they handle much higher flow rate and do not
require a precise metering of the flow.

2. Flowmeter

Flow rates were measured by a Micro Motion Model D mass
flowmeter which provides direct mass flow measurements independent of properties such as pressure and temperature. It has a
nonintrusive sensor which is ideal for slurry fuels and has a
maximum operating temperature of 400°F. This was one of the most
reliable and convenient components of the system. This device
has been used for many slurry fuel applications such as coal/
water mixtures in recent years. The original model was sensitive
to vibrations, but the existing type has eliminated the problem. A larger unit was also installed at the commercial asphalt
emulsion plant to measure tar flow to the colloid mill at high
temperature and proved reliable.

3. Piping

Slurry fuels are generally transported under laminar flow
conditions. The shear rate encountered in power plant piping
systems is around 100 sec.$^{-1}$. The shear rate experienced in commercial pipeline application is in the range of 5 to 30 sec.$^{-1}$.
The power plant application has been the focus of this study.

The system consisted of two pipe loops, 1/2 in. and 1 in. in
diameter, to provide fuel circulation and flow data such as pressure drop. Throughout the testing period, no plugging or
restriction problems were identified. It is recommended to wet
the surface of the pipe first by water flow before fuel is introduced. It is required to flush the system with water at the end
of the test. Lines should generally be as short and as straight
as possible. Stagnation zones where material may accumulate and
remain for long periods of time should be avoided. These are
areas such as a "T" where one side is closed off by a valve. A
number of these locations in the pilot plant were plugged with
solid material and had to be opened up for cleaning. This

144

problem was eliminated once the procedure to flush with water was extended to these areas as well. No pipe flow problems were experienced during the combustion test.

4. Pressure Drop Measurement

The SDA tar/water slurry fuel is a non-Newtonian fluid. Its viscosity is not a constant and is a function of the shear rate. For non-Newtonian fluids, the design and specification of fuel handling equipment such as pumps and pipes requires knowledge of the rheological parameters and pressure drops associated with the fuel. The fuel handling pilot plant was used to characterize the rheological properties of SDA tar/water slurry fuel. The 1-in. and 1/2-in. pipe loops were equipped with four Taylor Instrument pressure transmitters. The horizontal test length between transmitters was 83 in. for the 1-in. pipe and 71 in. for the 1/2-in. pipe. A longer 1-in. pipe section would have been more desirable for better accuracy, especially for the lower viscosity (lower pressure drop) fuel. However, this was not possible due to space limitations in the area.

Pressure drop, as a function of shear rate in a pipe for this non-Newtonian fluid, was used to calculate the fluid viscosity. The Haake viscometer was used independently to measure the viscosity at these shear rates. These values were compared with the ones obtained from the pilot plant and were in good agreement.

Three fuel samples were used to provide a range of solids contents and, therefore, viscosity variation for the experiments. Sample information is included in Table V. Results are presented in Figures 3-5. Shear stress, viscosity from pressure measurements, and viscosity from the Haake viscometer are plotted as a function of shear rate for all three cases. In most cases, the agreement between the pilot plant data and the viscometer is excellent. The initial points in the low viscosity fuel case in the 1-in. line, Figure 3, are not accurate because the pressure drop values were too small to be measured precisely in the given length of the pipe. The first two values of the Haake viscometer for the high viscosity fuel case, Figure 5, are also not accurate. This is related to the measuring head used. The instrument has two torsional springs for torque measurement. The range of the more sensitive one was not large enough to cover the low shear rate conditions for the viscous fuel. The less sensitive head was used instead. The initial values were at the very low part of the torque range and are not reliable.

FIGURE 3
LOW VISCOSITY FUEL (H52)

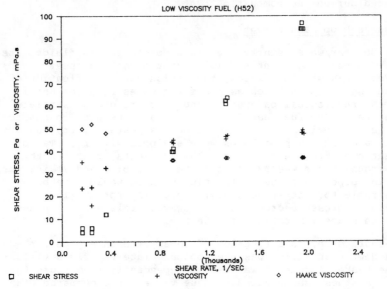

☐ SHEAR STRESS	+ VISCOSITY	◇ HAAKE VISCOSITY

FIGURE 4
MEDIUM VISCOSITY FUEL (H53)

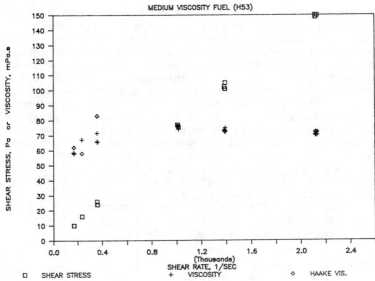

☐ SHEAR STRESS	+ VISCOSITY	◇ HAAKE VIS.

FIGURE 5

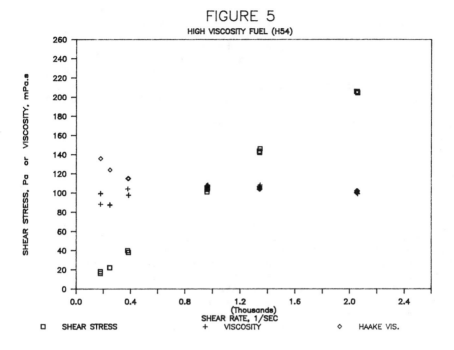

TABLE V

SLURRY FUEL SAMPLE PROPERTIES
PRESSURE DROP MEASUREMENTS

Median Particle Size, Micron	Solids Content, Wt %	Run I.D.
3.5	59	H52
2	63	H53
2.5	65	H54

147

Appendix I includes a more detailed statement of the relationship between pressure drop and viscosity, as well as the Haake viscometer description.

5. Homogenizer

The tar/water emulsion using the current additive package has very good storage stability and does not require the use of a high shear homogenizer. A three-stage in-line Tekmar homogenizer with a 5-hp motor to handle up to 10 gal./min. was used in one case earlier with an emulsion produced with BASF Pluradyne. This product had some degree of settling. The Tekmar was used to homogenize the fuel. It had no adverse effect on the product. The median particle size was 11.2 micron before and 11.1 micron after passing through Tekmar.

6. Strainer

The initial design of the fuel handling system included a 1-in. Simplex strainer Model 105 manufactured by Plenty Strainers. The basket has a 13 sq in. surface area . Three metal baskets, 20, 60, and 100 mesh, were included. None of these were able to handle the slurry flow and plugged shortly after the fuel was introduced. The critical factors were found to be the available surface area of the basket and the orifice size. The first is flow rate-related and the second has to do with the solid particle size. This is similar to the requirements for an atomizer nozzle. In the absence of the proper size, the fuel will dewater and the strainer will plug.

A larger single-bag filter manufactured by FSI Filter Specialists, Inc., was applied next. This FSP-35 model has a 1-sq ft surface area metal basket with 1/8-in. holes. Filter bags fit in this basket. The 20-mesh bag obtained with this system dewatered the fuel and failed at 25 lb/min. flow rate. There was a 1/4-in. thick cake on the basket with the pressure at 150 psig. The 1/8-in. metal basket without filter bags was used and was successful. This system was then used at flow rates as high as 110 lb/min. with excellent performance.

An AMF Cuno Auto-Klean filter was available and was tested. This system is motorized and is built to remove solids continuously. The cartridge has wheel-shaped discs, spacers, and cleaner blades, stacked on a rotatable shaft. The particular model on hand was equivalent to a 100-mesh screen and did not perform in this case. These filters are much more complicated to disassemble and clean. No further test was conducted with this unit since the simpler conventional strainer performed well.

A Hayward diverter valve was used to eliminate the need for four valves to switch strainers. This system was operated for 30 min. while the flow was diverted from one strainer to the other every seven minutes. Both baskets and the valve were inspected and looked clean. Standard duplex strainers are very heavy and bulky while their baskets are quite small. They are used for single-phase fluids where a large surface area is not required. The combination of two single strainers and a diverter valve works best for this slurry fuel application.

7. Valves

One-inch ball valves were used in the handling facility in most cases and they performed well. A similar performance was experienced at the commercial asphalt emulsion plant and the fuel handling system for the combustion tests. In these cases, the valve is either open or closed. Ball valves are not desirable for controlling the flow or applying backpressure to a system. Any sharp object exposed to the fuel can damage the fuel potentially. Evidence of this potential was observed in the colloid mill pilot plant where a ball valve was used to provide sufficient backpressure to prevent the loss of water at elevated temperatures.

A Moyno Pinch valve was supplied by Robbins and Myers for testing in the pilot plant. This is a handwheel-operated valve with a Buna N sleeve, rated for 200 psig and 200°F. It fits a 1-in. line but has an internal diameter of 1/2 in. The advantage of this valve is the lack of any sharp internals or a torturous path. This valve was used several times to apply backpressure to the lines. It provided easy control of the pressure once a value was set. Other valves require constant attention of the operator to make continuous corrections to hold a certain pressure.

8. Atomizer Nozzles

Slurry fuel atomization and nozzle design has been studied by many interested groups in recent years. Their interest is focused on coal/water fuels mainly. These fuels have some common characteristics but are also different in some aspects when compared to the SDA tar/water fuel. The atomizer design is similar to the ones used for heavy oil since slurry fuels are liquid. The burner design for coal/water fuel is similar to those used

for pulverized coal. This is not necessarily true for tar/water
fuel which may require a heavy oil burner design. The coal/water
fuel will become a solid fuel once the water is vaporized. The
tar/ water fuel will turn into heavy liquid fuel in the hot furnace. The nozzle orifice size for a slurry fuel should be larger
to prevent bridging of the solid particles which will lead into
nozzle plugging. On the other hand, if the orifice is too large,
the droplets will be large and the atomization quality will suffer. Droplet size below 300 microns has been reported to be the
preferred size. Other factors have also been considered in
slurry fuel nozzle design. For example, a high droplet momentum
reduces the residence time of the fuel in the furnace and affects
the combustion efficiency.

The SDA tar/water slurry fuel was originally being developed
to replace the crude oil used as fuel in steam-generating furnaces used by Chevron in steam flood operations to produce oil in
California. The focus was to try to use the existing fuel nozzle
and burner to minimize additional retrofit costs. Most steam
generators in that area use the 50-MMBtu/hr North American nozzle. A few use a 25-MMBtu/hr nozzle. The North American nozzle
was tested in the fuel handling pilot plant on a limited basis to
determine the potential for success. The more detailed test of
this nozzle, which included cold atomization and droplet size
measurements, as well as a full-load combustion test, is reported
separately (Reference 1). Since the North American nozzle
plugged after about three hours of operation under a full load,
and since the decision was made, subsequently, not to use the
slurry fuel in the steam flood operation in California, it became
necessary and desirable to test other fuel nozzles. Although
many manufacturers claim they now have "off-the-shelf" slurry
fuel nozzles, they are not willing to sell individual units.
They are concerned that if they are not involved in the development of a project, their product may be used in the wrong application and be reported as one which does not perform, resulting
in bad publicity.

An oversized North American nozzle and four John Zink nozzles were obtained for this study. Table VI includes the
description of these nozzles. The fuel handling pilot plant is
limited to cold atomization of very short duration without droplet characterization. The fuel flow can be directed through a
1-in. pipe to the back of the building where atomizing nozzles
can be attached and tested. The fuel is atomized into a
container and transferred into drums using an air-operated

double-diaphragm Model M-2 Wilden pump. A large quantity of the fuel is required and, without a combustion facility, collection and handling of the atomized material is not practical. The facility is only used to rule out a nozzle if it fails immediately or after a few minutes of testing. The nozzles that pass this initial test need to be tested at a facility such as the one used for the combustion test for final approval.

The results of some selected atomization runs conducted in the pilot plant are presented in Table VII. The 25-MMBtu/hr North American nozzle was tested first; both with air and steam atomization. This nozzle has a smaller orifice diameter but will also require a lower flow rate. In one case, it was tested for 60 min. with about 300 gal. of pilot plant fuel. The 50-MMBtu/hr North American nozzle was tested once with fuel which had not been screened prior to use. This did not seem to affect the nozzle. The John Zink standard R-type nozzle plugged after 10-12 min. in one test. All other John Zink nozzles performed well. In most cases, the strainer basket and the nozzle were relatively clean. The fuel was atomized into a receiver vessel and pumped into other drums for removal and disposal. The use of steam atomization with the higher fuel flow rates was not practical due to the limitations in the size of this system.

9. Viscosity

Three samples with solid contents of 60%, 62.1%, and 64% were tested. Each sample was tested at room temperature, 100°F, and 130°F. Results are presented in Figure 6. As expected, the viscosity values increase as the solids content increases. In all cases, the viscosity reduction with temperature follows the same trend. Although there is a reduction as a function of temperature, the decrease is not as steep as that of typical hydrocarbon mixture viscosities. The slope follows that of water, as shown in the figure.

10. Stability

Stability of a slurry fuel is very much a part of its handling characteristics. OXCE Fuel Company defines stability as the number of days with less than 1% soft-packed sediment under static conditions. There are many other definitions as well. Slurry fuels can also have dynamic stability problems. This is related to methods of transportation. The stability of the SDA tar/water fuel using the current additive package is excellent. The combustion test fuel was produced at the commercial asphalt

151

TABLE VI

SLURRY FUEL ATOMIZERS

Manufacturer	Size	Type	No. of Holes	Orifice Size, In.
North American	Standard	25 MMBtu/Hr	8	0.086
North American	Standard	50 MMBtu/Hr	8	0.125
North American	Oversize	50 MMBtu/Hr O.S.	8	0.188
John Zink	Standard	PM 6-17-29-29-7-15	6	0.173
John Zink	Oversize	PM 6-17-22-23-7-15	6	0.201
John Zink	Standard	R 8-10-70	8	0.194
John Zink	Oversize	R 10-A-80	10	0.234

TABLE VII

SELECTED ATOMIZATION RUNS IN THE PILOT PLANT

Nozzle	Flow Rate, Lb/Min.	Air or Steam/ Pressure, psig	Fuel Pressure, psig
NA25	21	Steam/50	
NA25	23	Air/30	35-40
NA25	20.5	Air/35	65
NA25	22	Air/35	70
NA25	70	Air/60	70
NA25	70	Air/65	75
NA50	70	Steam/35	40
NA50	70	Steam/40	55
NA50	76	Air/35-40	70
NA50 OS	76	Air/35	30
JZ PM OS	76	Air/45-50	55-60
JZ PM ST	76	Air/60	80
JZ R OS	76	Air/30	50
JZ R OS	76	Air/55	50
JZ R ST	76	Air/40	85
JZ R ST	76	Air/45	95

FIGURE 6
SDA TAR/WATER SLURRY VISCOSITY

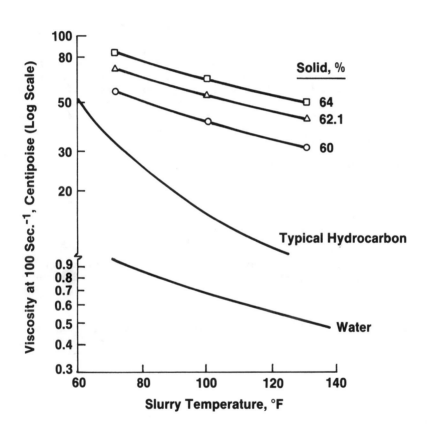

emulsion plant on November 13, 1986. It was shipped in trucks to Irvine, California. The trucks containing the unused fuel traveled back to Richmond. The product was loaded into 55-gal. drums, stored outdoors, and used gradually as needed for the handling tests in the pilot plant. The fuel was stable and useable after more than one year.

It is important to note that the stabilizer additive which enhances the storage stability increases the fuel viscosity. The additive level should be optimized to provide maximum stability at the lowest viscosity possible. Increasing the fuel viscosity will reduce the atomization quality. Larger droplets require more residence time which translates into a larger furnace or lower combustion efficiency.

V - CONCLUSIONS

A new Chevron slurry fuel has been developed successfully. This fuel contains up to 70% solids from the bottoms from a solvent deasphalting (SDA) process, a unique additive package, and water. The fuel is produced using an emulsification process similar to the process which makes asphalt emulsions. The additive package includes an emulsifier and a stabilizer and does not contain sodium. The resulting fuel has a median particle size as low as 2 microns with a viscosity as low as 100 cSt at room temperature. The fuel can be stable for more than one year and requires no heating before use.

The detailed study of the handling characteristics of the slurry fuel indicated that many existing and conventional fuel handling systems were adequate for delivering the fuel to the burner. A progressing cavity pump, a mass flowmeter, standard piping and valves, and strainers were tested successfully. Atomizer nozzles, however, should be similar to the ones used for coal/water slurry fuels and have larger orifice diameters.

The results of the large-scale atomization, combustion, and emissions tests are reported in a separate article (Reference 1).

REFERENCES

1. Nikanjam, M. and Kwan Y., "Atomization, Combustion, and Emissions Characterization of the Chevron Slurry Fuel," 5th Miami International Symposium on Multi-Phase Transport & Particulate Phenomena, December 12-14, 1988, Miami Beach, Florida.

2. Govier, G. W., and Aziz, K., The Flow of Complex Mixtures in Pipes, Krieger Publishing Company, Malabar, Florida, 1982.

3. Carleson, T. E.; Drown, D. C.; Hart, R. E.; "Comparison of Rheological Evaluation Techniques and Turbulent Flow Prediction of a Simulated Nuclear Waste Melter Slurry," 12th International Conference on Slurry Technology, March 31–April 3, 1987, New Orleans, Louisiana.

$$\underline{A} \ \underline{P} \ \underline{P} \ \underline{E} \ \underline{N} \ \underline{D} \ \underline{I} \ \underline{X} \quad I$$

RHEOLOGY

The shear stress of a fluid, τ, is a function of the rate of strain or shear rate, $\dot{S}$, and can be stated as: $\tau = \mu\dot{S}$, where μ is the viscosity of the fluid and is a constant for a Newtonian fluid.

Pressure drop, ΔP, in a pipe of diameter D and length L will yield a shear stress, $\tau = \frac{D\Delta P}{4L}$.   The shear rate is determined easily if the fluid is Newtonian and the flow is laminar. This is done as follows:

$$\tau = \mu\dot{S}$$

$$\tau = f \ \frac{\rho V^2}{2}$$

where f = 16/Re for laminar flow.

Since the Reynolds number, Re, is $\frac{\rho V D}{\mu}$, the shear rate for this case, $\dot{S}$, is simply 8V/D.

For the non-Newtonian slurry fuel, the relationship becomes more involved and is given by (Reference 2):

$$\dot{S} = [(1+3n')/(4n')][8V/D]$$

where $n' = \dfrac{d(\ln \frac{D\Delta P}{4 L})}{d(\ln \frac{8V}{D})}$

Laboratory viscosity versus shear rate measurements were conducted on the Haake Viscometer. The unit was matched with an IBM compatible computer for data acquisition and uses the Haake software. This is a concentric cylinder viscometer of the Searle type where the center cylinder, diameter D_1, rotates and the cup of diameter D_2 is stationary. The shear stress, τ, and the shear rate, $\dot{S}$, are determined by the torque, T, and the rotational speed, N. The torque is measured by the angular deflection, θ, of a torsional spring with a constant K, and is stated as $T = K\theta$. The shear stress is then given by:

$$\tau = \frac{2K\theta}{\pi D_1^2 h}$$

where h is the height of the fluid in the cup.

The shear rate is related to the speed and the ratio of the two diameters, $s = \dfrac{D_2}{D_1}$ and is given by: $\dot{S} = \dfrac{4\pi N}{1-s^2} F_{km}$.

F_{km}, the non-Newtonian factor, is expressed as:

$$F_{km} = 1 + \frac{s^2-1}{2s^2}\left(1 + \frac{2}{3}\ln s\right)\left(\frac{1}{n"}-1\right) + \frac{s^2-1}{6s^2}\ln s\left(\frac{1}{n"}-1\right)^2 + \frac{d(\frac{1}{n"}-1)}{D\,\log T}$$

where:

$$n" = \frac{d\,\ln T}{d\,\ln N}$$

The factor, F_{km}, is approximately one (Reference 3) if:

1. The gap between the cylinders is small, $1 < s < 1.1$.

2. The flow behavior index (the power law exponent), m, for the fluid is between 0.5 and 1.5. Typical values for most slurry fuels are well within this range.

:com/sms

156

Atomization, Combustion, and Emissions Characterization of the Chevron Slurry Fuel

M. NIKANJAM
Chevron Research Company
576 Standard Avenue
Richmond, California 94802, USA

Y. KWAN
Energy and Environmental Research Corporation
18 Mason
Irvine, California 92918, USA

Abstract

Chevron Research Company has developed a petroleum-based solid/water slurry fuel for potential application as a replacement for heavy resid fuel. A series of tests were performed at Energy and Environmental Research Corporation to evaluate the atomization, combustion, and emissions characteristics of this new fuel. Full scale cold atomization tests were conducted to determine the droplet size distribution, the spray angle, and the slurry fuel/atomizer compatibility. A 70 M Btu/hr bench scale furnace was used to study the NO_x emission levels and the effect of staged combustion to reduce these levels. Favorable results were observed compared to coal/water slurries and crude oil. Results are also presented for a full scale 50 MM Btu/hr combustion test. These include flame characterization, excess air requirements, carbon monoxide and nitrogen oxide emission levels, furnace exit temperature, and smoke numbers. These tests were carried out without major changes or optimization of an existing crude oil atomizer nozzle, burner, and furnace.

1. INTRODUCTION

Chevron Research Company has developed a slurry fuel which contains up to 70% solids from the bottoms of a solvent deasphalting (SDA) process, an additive package, and up to 30% water. This fuel, which can be stable for more than a year without appreciable particle settling, has passed a series of tests in a handling characterization pilot plant using mostly conventional equipment. The results of the slurry fuel handling characterization, as well as the details of its development, including the process and the additive package, have been discussed in a separate article [1].

An existing commercial plant, which produces asphalt emulsions routinely, was modified and used successfully to produce a trial batch of the slurry fuel at the rate of 80 gallons/minute. The fuel was loaded into tank trucks for transport to the facility where the atomization, combustion, and emissions characterization were to occur. These tests were performed by Energy and Environmental Research Corporation on contract, using the combustion facilities in Santa Ana, California. The SDA tar/water slurry fuel was originally being developed to replace the crude oil used as a fuel in the steam generating furnaces operated by Chevron for the steam flood operation to produce oil in the San Joaquin Valley. The goal was to try to use the existing fuel nozzle and burner to minimize additional retrofit costs. Therefore, a 50 MM Btu/hr North American nozzle and burner from the steam generation operation were used in this study.

2. SPRAY CHARACTERIZATION

The objective of these tests was to determine if the existing fuel nozzle would result in an acceptable atomization condition. Heavy fuel droplets larger than 300 microns reduce the combustion efficiency and should be avoided through the proper selection of the fuel nozzle.

Experimental

The atomization tests were carried out in EER's large-scale, cold flow spray analysis facility shown in Figure 1. The octagonal spray chamber is 4.8 ft across by 8 ft tall and has been used to characterize sprays at up to 10 gallons/minute. For straightforward atomization tests, screen air was co-flowed about the nozzle to prevent recirculation of fine droplets into the measurement volume. The system provides access for the Malvern 2600 HSD Particle Size Analyzer. For these tests, a 1000 mm focal length receiver lens was used. This lens covers droplet sizes from 19.4 to 1880 μ. To minimize obscurations, two measures were

FIGURE I
LARGE SCALE SPRAY CHARACTERIZATION FACILITY

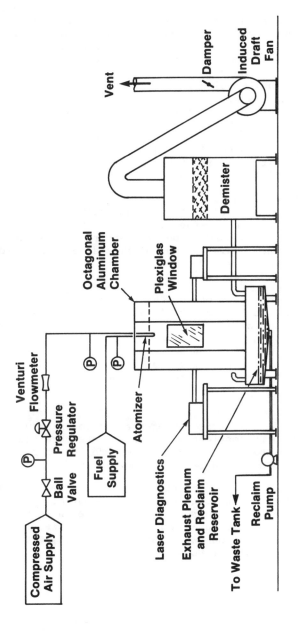

taken: (1) a movable light guide was constructed and installed to shield the laser beam from part of the spray and (2) the laser beam was lowered to an axial distance of 36 inches from the nozzle tip. The latter condition also allows slurry droplets to relax to the velocity of the free stream. With this setup, the typical observed obscuration value was below 20%. The Chevron SDA tar/water slurry fuel was tested with an unmodified North American oil burner nozzle. This nozzle is rated for 50 MM Btu/hr and is commonly used in steam generators for thermally enhanced oil recovery. The nozzle utilizes a prefilming process where the fuel exiting from a number of discrete holes impinges at right angles with a high velocity assist-fluid stream. The resulting fuel/assist fluid mix discharges into a secondary chamber undergoing further aerodynamic disintegration and finally exits from the nozzle through a single orifice drilled into the nozzle cap. The nozzle is designed for heavy oil, and a nominal atomizing air pressure of 70 psig is recommended by the manufacturer for full load operation.

Figure 2 provides a schematic of the fuel handling system. The major components of the system include: a progressing cavity pump, mass flowmeter, fuel reservoir, strainer, and discharge pump. Water was used to wet the entire fuel system prior to and after testing. This wetting minimizes dehydration of the slurry in the fuel system and also serves to purge slurry left in the system lines after testing.

The properties of the Chevron slurry are shown in Table I. The nominal solid loading of the slurry is about 64%, with a mean particle size of approximately 2 microns. The solid material in the slurry, the bottoms from the solvent deasphalting process, is low in ash and is believed to be more reactive in combustion than coal. This expectation is due primarily to the high volatile content, about 70% by weight, of the solids.

Results

Spray characteristics of the Chevron slurry fuel, using the standard North American oil nozzle, were obtained for two slurry flow rates, 8.4 and 4.2 gallons/minute, which correspond to full and half load firing rates, respectively.

All spray data presented hereafter were analyzed using a so-called Model Independent fitting program. This model yields a much more accurate representation of the tails of droplet size distribution, particularly for the large size fraction, than does extrapolation of the two-parameter data fit, that is, the conventional Rosin-Rammler Model [2].

FIGURE 2

FUEL HANDLING SYSTEM FOR ATOMIZATION TESTS

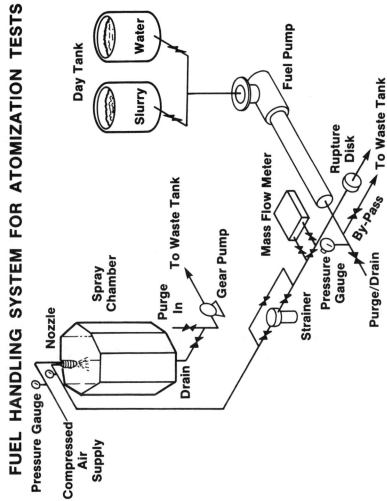

TABLE I: CHEVRON SLURRY FUEL PROPERTIES

Solids Loading, Wt%[1]	64
Median Particle Size, Microns	2
Specific Gravity, 60/60°F	1.04
Density, Lb/Bbl	384
Viscosity, cSt[2]	100
pH	10.1

[1]100% minus 100 mesh (150 μm).

[2]Haake viscometer, Shear Rate 100 Sec^{-1}, 77°F.

The tests were initiated by measuring the droplet size distribution at the centerline of the spray for a range of atomizing air pressures between 60 and 90 psig. Following this series of tests, spray characteristics away from the centerline of the spray were examined. For this second series of tests, the atomizing air pressure was set constant at 80 psig, which was found to produce relatively acceptable mean drop size. Figures 3 and 4 summarize the test results, expressed in terms of Sauter mean diameters (Dsm), mean mass diameter (D50), and weight fraction of droplets whose diameter is larger than 160 microns (Dp). The data show that at half load, mean drop size does not vary strongly for the range of atomizing air pressure tested. However, the weight fraction of droplets larger than 160 μm increased substantially as atomizing air pressure decreased. This latter observation was also noted for the full load condition. The test results also reveal that both Dsm and D50 increased toward the outer edge of the spray for the full load condition. For example, the Sauter mean diameters were computed to be 44 μm at the centerline and 85 μm at 6 inches away from the centerline of the spray. The corresponding mean mass diameters, D50, were 38 μm and 130 μm, respectively. The fraction of large size droplets (Dp> 160 μm), however, decreased near the outer edge of the spray for both loads.

In summary, drop size measurements from a section of spray located on the spray centerline 36 inches downstream from the nozzle tip indicate that full load performance at 80 psig atomizing air pressure produces an acceptable Sauter mean diameter of 44 μm. The weight fraction of droplets larger than 160 μm is found to be approximately 1.7%. The study also shows that the drop size distribution produced at the nominal design atomizing air pressure (70 psig) is marginal because of the amount of large droplets found in the spray (8.1% Dp> 160 μm).

3. COMBUSTION TESTS

The objectives of these combustion tests were:

1. To evaluate the capability of an existing crude oil burner to handle the slurry fuel and to establish minimum burner retrofit and operating changes required for a successful field demonstration.

2. To provide a relative comparison of the NO_x emissions from Chevron slurry fuel and Kern County crude oil in a well-controlled combustion environment.

FIGURE 3

EFFECT OF ATOMIZING AIR PRESSURE ON SLURRY DROPLET SIZE DISTRIBUTION

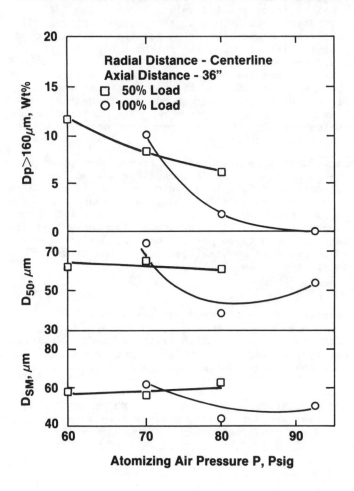

FIGURE 4

SLURRY DROPLET SIZE DISTRIBUTION AS A FUNCTION OF RADIAL POSITION IN SPRAY

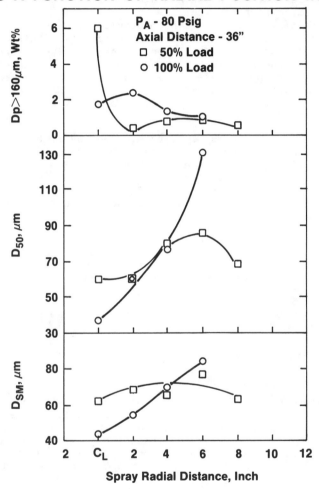

The rationale behind these tests was that the slurry has approximately twice the amount of fuel nitrogen contained in normal Kern County crude oil. Based on typical conversion of fuel nitrogen to NO_x when burning crude oil, NO_x emissions for the slurry fuel in conventional steam generators might exceed compliance in Kern County.

Experimental

A 70 M Btu/hr bench-scale furnace was used to study the NO_x emissions levels and the effectiveness of staged combustion in reducing these levels. This bench-scale furnace is a 9-ft refractory lined, down-fired tunnel with a main combustion chamber 6 inches in diameter. The furnace provides access for staged combustion experiments. A refractory choke isolates the fuel-rich first stage from the second (burnout) stage. The gas phase residence time in the fuel-rich zone can be varied from about 0.7 to 1.2 seconds over approximately 0.9 to 0.5 times the stoichiometric air/fuel ratio, respectively. This furnace has been used extensively in the evaluation of NO_x emissions from a wide range of solid and liquid fuels.

For the full scale tests, a 14-ft diameter by 17-ft long horizontal, cylindrical furnace was used. This furnace is designed for a firing rate between 30 MM to 80 MM Btu/hr whether fueled by gas, oil, or coal. The furnace is externally-cooled by water sprays. Part of the internal surface (approximately 60%) is lined with refractory to simulate the thermal environment of a field steam generator.

The burner, a North American oil burner which is commonly found in oil field steam generators, is mounted on one end of the research furnace and combustion products exhaust through the opposite end. The furnace exhaust passes through a long inclined stack to a scrubbing tower for particulate control.

The composition of the Chevron slurry and the Kern County crude oil is given in Table II.

Results

Pollutant Emissions - Bench Scale Tests

Figure 5 presents NO_x emissions produced in the bench-scale furnace by Chevron slurry and Kern County crude oil. NO_x emissions presented in this figure are corrected to 0% O_2 on a dry basis. Under normal excess air conditions, NO_x emissions from baseline crude oil increased from 731 ppm (0.82 lb/MM Btu) at 1% excess O_2 to 1038 ppm (1.17 lb/MM Btu) at 4.3% excess O_2. NO_x

TABLE II: ANALYSES OF COMBUSTIBLES

	Slurry[1]	Crude Oil
Elemental Analysis, Wt %		
C	86.73	86.88
H	9.66	11.21
N	1.16	0.76
S	2.28	1.04
Ash	0.08	0.02
O_2 (Diff.)	0.09	0.10
Heat of Combustion, Btu/Lb[2]	17,759	18,432
Metals, ppm		
Al	2	1
Ca	50	2
Fe	115	21
Na	30	3
Ni	149	66
Si	3	2
V	178	29
Viscosity, SSU		
at 140°F	NA	900
at 212°F	NA	106

[1]Solid

[2]Heating Value of Slurry: 11400 Btu/Lb

FIGURE 5

COMPARISON OF NO$_X$ EMISSIONS FROM KERN COUNTY CRUDE OIL AND SLURRY UNDER EXCESS OXYGEN AND STAGED COMBUSTION CONDITIONS (70 x 10^3 Btu/hr)

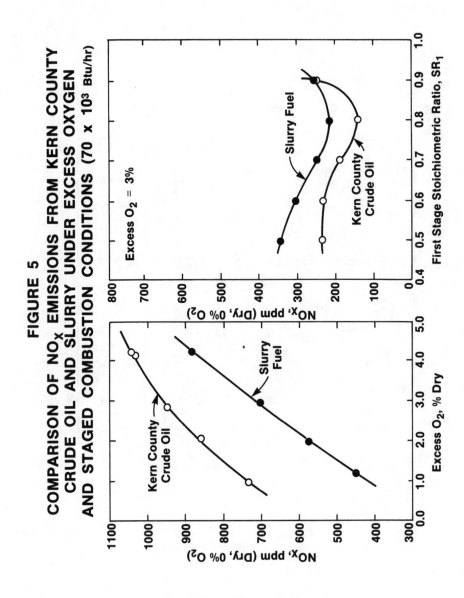

emissions from slurry also increased as excess air increased but were lower than the crude oil, ranging from 450 ppm (0.51 lb/MM Btu) at 1.2% excess O_2 to 877 ppm (1.05 lb/MM Btu) at 4.3% excess O_2.

Under staged combustion conditions, NO_x emissions decreased sharply as the fraction of the stoichiometric air/fuel ratio (SR1) was reduced, reached a minimum, and then increased as SR1 decreased further. The optimum SR1 for minimum NO_x was approximately 0.8 for both slurry and crude oil. Minimum NO_x was 139 ppm (0.16 lb/MM Btu) for crude oil while minimum NO_x for slurry was 214 ppm (0.24 lb/MM Btu). The staged combustion tests were conducted with a constant overall excess O_2 of 3%.

NO_x emissions from the slurry followed trends generally observed with conventional liquid and solid fuels; NO_x emissions decreased as excess O_2 decreased, and optimum first-stage stoichiometric ratio for staged operation was about 0.8. NO_x emissions from slurry were about 26% lower than those from crude oil under normal firing condition (unstaged) at 3% excess O_2. On the other hand, minimum NO_x was 54% higher for the slurry under staged combustion conditions. This result, which at first may seem surprising, can be explained by examining the impact of combustion conditions on NO_x formation.

Nitric oxide (NO) is formed primarily from two sources during the combustion of fossil fuels:

1.   Thermal NO is formed from oxidation of molecular nitrogen, N_2, and its rate of formation is temperature dependent.

2.   Fuel NO is a product of oxidation of nitrogen chemically bound to the fuel, and its formation is a function of local stoichiometry (air/fuel ratios).

Under normal excess air conditions, flame temperatures are suppressed with slurry fuel due to the heat absorbed in vaporizing the water. Reduced flame temperature has two potential impacts on NO_x emissions: (1) decomposition of the fuel and hence fuel nitrogen is further delayed and (2) thermal NO formation is reduced. Thus, although slurry fuel contains approximately twice the amount of fuel nitrogen contained in Kern County crude oil, NO_x emissions under normal excess air firing are lower because of decreased thermal NO formation and slower evolution of fuel nitrogen. The presence of carbonaceous material in the later stages of slurry combustion may also encourage heterogeneous reduction of NO.

The same factors that contribute to lower uncontrolled NO$_x$ emissions in "slow" burning fuels influence the effectiveness of combustion staging. The effectiveness of staged combustion in reducing NO$_x$ is diminished by: (1) decreasing fuel-rich zone (first stage) temperatures, which slows the rate of reactions which convert fuel nitrogen compounds to molecular nitrogen; and decreasing gas phase residence time of nitrogen species in the fuel-rich zone which decreases the extent to which equilibrium can be approached.

Since Chevron slurry contains approximately 36% water, combustion temperatures in the first stage are lower for slurry than for crude oil. The evolution of the combustible fraction of the slurry is considerably slower than that of the crude oil, and this decreases the time available for conversion of fuel nitrogen to N$_2$ in the fuel-rich zone. Further, gaseous fixed nitrogen species (NO, HCN, NH$_3$) and solid or liquid phase nitrogen compounds which escape the first stage will be partially converted to NO in the fuel-lean second stage. Thus, the lowest achievable NO$_x$ emissions under staged conditions can be expected to be higher for slurry fuel than for crude oil, based on the current understanding of the controlling mechanisms.

Burner Performance - Full Scale Tests

Initial attempts to fire the Chevron slurry fuel using a standard North American oil burner in a cold combustion chamber were unsuccessful. The flame just could not be stabilized with this burner configuration, even with the use of a continuous support fuel (pilot). For these brief attempts, flame standoff was typically between 2 and 3 ft long. The general appearance of the flame was long, narrow, and smoky. This appearance, together with the long flame standoff, suggested that the spray angle probably was not wide enough under full load combustion conditions. A narrow spray angle can reduce flame stability by limiting the effects of recirculating flow and by reducing fuel/air mixing.

In view of these observations, the standard North American oil burner was modified slightly. The modification consisted of replacing the original diffuser plate by a nonoptimized, fixed vane swirl generator. The goal of this change was to promote fuel/air mixing and, thus, increase flame stability.

The flame generated with the modified North American oil burner appeared to be much shorter. Flames were marginally stable with this nonoptimized burner configuration. However, additional improvement of the flame stability was seen when higher excess air was applied. This improvement is probably due to stronger tangential momentum imparted by the swirler under this operating condition. Another operating condition that showed good flame stability was during turndown. Lowering the firing rate produced a more compact and defined flame shape, as well as shortening flame standoff. This improvement could be attributed to lower fuel injection velocity. Table III summarizes the flame appearance of the tests.

The NO_x emission performance with the full scale combustor for the slurry fuel is compared to the performance of the Kern County crude oil in Figure 6. The data are corrected to 3% excess O_2 on a dry basis. As expected, NO_x emissions increased with higher excess O_2 for both fuels. For slurry fuel, NO_x emissions at the full load condition range between 200 and 270 ppm for excess O_2 of 3-5%, respectively. NO_x emissions for the Kern County crude oil were generally about 75 ppm higher for the same range of excess O_2, consistent with the trend seen in the bench scale tests.

Figure 7 shows the CO emissions for both the slurry fuel and the crude oil as a function of excess oxygen in the flue. In general, CO emissions for crude oil were less than 40 ppm and relatively insensitive to the range of excess oxygen tested. As for the slurry fuel, the trend of CO emissions was quite similar to that of the crude oil except when operated at the full load condition, where at 4% oxygen, a minimum CO emission was seen. It is believed that the slightly higher CO at 5% excess O_2 resulted from excessive heat extraction around the burner zone. The high CO level at a low O_2 condition is attributed to flame quenching due to the impingement of the flame tails on the furnace wall.

Throughout the experiments, Bacharach smoke number (BSN) tests were conducted for each test condition. These smoke numbers in general were below 3 for crude oil and an indication of good combustion. The BSN for slurry was somewhat higher, between 7 and 8, which indicates poor combustion and a dense smoke condition. Nevertheless, no smoke was observed at the stack during the tests. This discrepancy led to a careful inspection of the smoke spot under a microscope. The examination revealed that the constituent of the smoke spot was not soot but small beads of cenospheres. Care should be exercised when interpreting the results of the BSN for the slurry fuel. The selection of an optimum burner, atomizer, and furnace to provide the proper fuel residence time should relax this concern.

Furnace exit temperature was also monitored continuously during the experiments. In general, the flue gas temperature at the exit of the radiant section of the furnace for the slurry

TABLE III
SUMMARY OF THE FLAME OBSERVATIONS

Fuel	Load	Burner Configuration	Flame Pattern	Combustion Observations
Slurry	Half	Diffuser	*(flame diagram)*	Stable and compact flame; medium long. Fuel spray hit burner tile.
Slurry	Full	Diffuser	*(flame diagram)*	Very long and unstable flame. Typical standoff was more than 2.0 ft.
Slurry	Half	Swirl	*(flame diagram)*	Marginally stable, about 0.5 ft. standoff with pulsation; very bushy and occasionally smoky. Flame tails hit furnace side walls
Slurry	Full	Swirl	*(flame diagram)*	Luminescent and wide spread bushy flame. Marginally stable. Flame tails licked back wall occasionally. Typical standoff was about 1.0 ft.
Crude Oil	Half	Swirl	*(flame diagram)*	Very stable. Long, narrow flame; no standoff nor fuel spray impingement on burner tile
Crude Oil	Full	Swirl	*(flame diagram)*	Bushy, stable flame with 2.0 ft. standoff. Very luminescent flames hit side walls occasionally

FIGURE 6

COMPARISON OF EMISSIONS PERFORMANCE
BETWEEN SLURRY AND CRUDE OIL
IN FULL SCALE RESEARCH COMBUSTOR

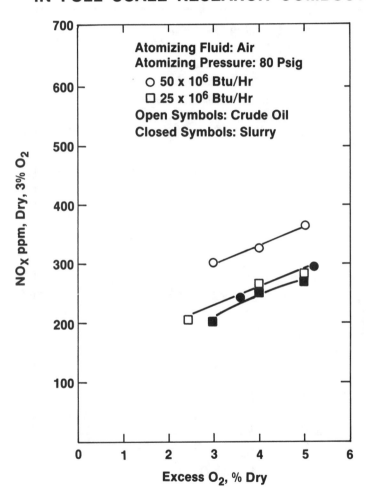

FIGURE 7
CO EMISSIONS

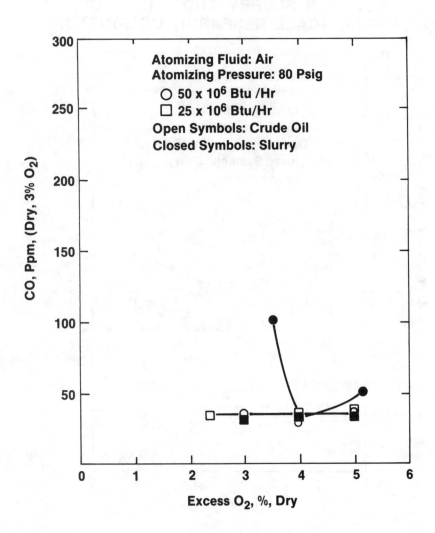

fuel and the crude oil were similar; about 1740°F and 1730°F, respectively, for the full load case at 4% excess O_2. The exit temperature of the flue gas for the half load condition for crude oil was somewhat higher than that of the slurry fuel, for example, 1588°F at 4% excess O_2 compared to 1517°F for the slurry fuel at similar combustion conditions. The lower flue gas temperature for the slurry fuel at the exit of the furnace is caused by the heat absorbed by the water. The higher flue temperature measured at the furnace exit for the slurry fuel at the full load condition is primarily due to the longer flame produced by this fuel.

Summary

    The combustion experiments conducted in the bench and full scale combustors indicate that the Chevron petroleum-based solid/water slurry with 60% solid loading could be burned satisfactorily in a field steam generator with comparable performance to crude oil. NO_x emissions from slurry are expected to be also comparable with existing uncontrolled field emissions levels for oil firing. In addition, the experiments also suggested that the standard North American oil burner currently used in the oil field would need to be modified to provide adequate flame stability for the slurry fuel. Potential burner modifications to correct this deficiency have been identified.

4. CONCLUSIONS

    The Chevron slurry fuel containing heavy bottoms from a solvent deasphalting process was produced successfully in an existing commercial plant. This fuel was shipped to Energy and Environmental Research Corporation without any problem.

    Atomization tests to characterize the fuel spray droplet size and distribution indicated the 50 MM Btu/hr North American burner nozzle used in Chevron's enhanced oil recovery steam generating furnaces in central California was adequate to produce droplets with an acceptable Sauter mean diameter of 44 μm. The weight fraction of droplets larger than 160 μm was around 1.7%. No droplets over 300 μm were generated for the full load atomization at an atomizing air pressure of 80 psig.

    NO_x emissions produced by burning the slurry fuel in a bench-scale test furnace under normal excess air conditions were 26% lower than the case of burning Kern County crude oil, with half the fuel nitrogen, under similar conditions. This was the result of the reduced flame temperatures and delayed fuel nitrogen evolution, relative to the crude oil. The application of staged combustion reduced NO_x emissions by 70% when burning the slurry fuel, compared to 85% for crude oil.

The full scale combustion test verified the bench scale test trends and resulted in lower NO_x emissions for the slurry fuel consistently. Carbon monoxide levels were less than 40 ppm for the slurry fuel at an optimum excess oxygen level of 4%. Some problems were experienced with flame stability and smoke numbers. The burner modification to include swirl to promote fuel/air mixing resulted in increased flame stability. Optimization of the burner and furnace will result in further improvements in flame stability and better combustion efficiency.

Acknowledgements

The authors wish to acknowledge the diligent efforts of many who have contributed to the success of this work. In particular, the authors would like to express appreciation to G. S. Flores of Chevron Research Company for preparing the slurry fuel and maintaining the fuel system operable. The authors also gratefully acknowledge the contributions of those at EER; R. Zimperman for assisting in the spray characterization tests; J. F. LaFond for conducting the bench-scale tests; and J. Johnsen, J. Pionessa, and E. Monsivais for assisting with the full scale combustion trials.

REFERENCES

1. Nikanjam, M., and Kohler, D. A., "Chevron Slurry Fuel Development and Handling Characterization," 5th Miami International Symposium on Multi-Phase Transport and Particulate Phenomena, Miami Beach, Florida, December 12-14, 1988.

2. Meyer, P. L., and Chigier, N., "The Atomization Process in Coal-Water Slurry Sprays," Eighth International Symposium on Coal Slurry Fuels Preparation and Utilization, U.S. DOE and Pittsburgh Energy Technology Center, Orlando, Florida, May 27-30, 1986.

Friction Coefficient of Moist Bulk Solids in a Centrifugal Field

F. REIF and W. STAHL
Institut fur Mechanische Verfahrenstecknik und Mechanik
Universitat Karlsruhe
Postfach 6980, 7500 Karlsruhe, FRG

Abstract

Provided the frictional behaviour of bulk solids is known, then their transport properties can be determined for various centrifuges, including scroll-type centrifuges, such as decanters.
The most important property of the bulk is its friction coefficient, of which only very little information is available. For these reasons an experimental centrifuge was developed: In a rapidly rotating drum a variable number of knives revolve at a higher rate. As a result of the differential velocity the product is conveyed circumferentially, hence forming a wedge of approximately triangular cross-section. The conveyance torque is measured by strain gauges, so that the circumferential force is easily derived. The ratio of centrifugal to circumferential force finally yields the friction coefficient.
The experimental work involves measurements of the effect of the centrifugal acceleration, the moisture content, the differential speed and the clearance between drum and knife on the friction coefficient of different products, PVC-powder and glass beads, with particle diameters from 20 to 300µm. Water and water-tenside mixtures are used as liquids, thus also incorporating the influence of the surface tension. The experimental settings can be chosen to represent the conditions in common industrial centrifuges, i.e. centrifugal acceleration up to 2600g and differential or shear velocities up to 2m/s.
If the friction coefficient is plotted as a function of the moisture content, all curves follow a common trend regardless of the experimental conditions: With increasing moisture content the friction coefficient initially increases, passes a maximum, on which it subsequently decreases. The coefficient of the totally saturated bulk is approximately equal to that of the dry solid. The maximum is reached at the equilibrium saturation.
Based on the experiments, a model was developed to predict the friction coefficient as a function of the above parameters. The theory assumes a superposition of the "outer" normal, i.e. centrifugal, force and of the sum of the "inner" forces due to liquid bridges within the bulk. The fluid bridge forces are calculated using Rumpf´s theory.
It could be shown that the predicted and measured coefficients of friction agree very well. Only the friction coefficient of the dry and in some cases of the wet solid must be determined experimentally; as far as we know this dry friction coefficient is a material constant, i.e only one experiment needs to be conducted for a wide variety of products, which enables an efficient design of centrifuges with respect to transport torque, dewatering capacity and power consumption.

177

1. INTRODUCTION

Most industrial processes involve solid-liquid separation, which is frequently conducted by centrifuges. Research is commonly found in the fields of dewatering and clarification, but hardly in the transport behaviour of moist solids, although the conveyance, and as a result, the residence time, is essential when solving these two other problems – at least concerning continuous centrifuges. Furthermore, the power consumption and the internal forces required for the transport can also be calculated, and for scroll type centrifuges, e.g. decanter, one may determine, whether the transport is guaranteed at all /1/ /2/.

Before the conveyance can be predicted, the friction coefficient of the moist solid, defined as the ratio of circumferential to centrifugal force, is needed. Therefore a device was developed which allowed this coefficient μ to be measured experimentally in a centrifugal field: In this device a variable number of knives revolve in a rapidly rotating drum at a higher rate than the bowl itself. As a result of the different velocities the product is conveyed circumferentially, thus forming an approximately triangular wedge. The conveyance torque is measured by strain gauges, from which the circumferential force can be easily determined. The design of this centrifuge enables the behaviour of the sheared bulk to be observed during the process of friction.

For a PVC powder and glass beads, the effect of the centrifugal acceleration, expressed in terms of a multiple of the earth acceleration, C, the saturation S (or the respective moisture content RF), the blade angle Γ_B, the clearance between drum and knives and the scraper velocity v_S were all investigated. The relevant data of the products are presented in Table I, the size of the experimental settings in Table II.

The aim of the conducted research was to determine the friction coefficient in a centrifugal field as a function of these parameters, whereby the results of other research areas, e.g. from soil or bulk mechanics, cannot be simply transferred since the following differences exist:
1. Soil mechanics usually only involve static problems.
2. In experiments with ring shear devices, neither saturation nor normal stress gradients exist.
3. Surface tension effects were not explicitly considered.
4. Effects of mixing in the shear zone are usually negligible.

From the experiments, one could observe, that the above factors may not be disregarded. In this presentation, the aspects 1,2 and 3 will be discussed in detail.

The developed theory takes the fluid bridge forces into account. These "internal" normal forces overlap the "external" force, i.e. the normal force given by the centrifugal field. Whereas the forces induced by surface tension increase the friction and its coefficient, buoyancy acts against them. In case of a thin primary layer, buoyancy may be neglected as done in this project.

The use of this theory offers the advantage that at least for an approximation only the dry bulk´s friction coefficient must be de-

TABLE I: RELEVANT PRODUCT DATA

Product	PVC	glass
Mean diameter d_p /μm	51	165
Mean Volume /μm^3	1.56×10^5	4.56×10^6
Particle form	rough, "potatoe"-shaped	smooth, spherical
Surface roughness /μm	1-20	0
Porosiy ε	0.5	0.38
Solid density ρ_s /kg/m^3	1380	2460
Liquid	water	
Liquid density ρ_l /kg/m^3	1000	
Surface Tension γ /N/m	72×10^{-3}	
Contact angle δ /$^\circ$	0	0

TABLE II: SIZE OF THE EXPERIMENTAL SETTING

C-Value	–	200-888
Rotational Speed n	/1/min	635-1600
Scraper Velocity v_s	/m/s	0.12-0.52
Saturation S	/%	0-100
Blade Angle Γ_B	/$^\circ$	78-90
Clearance c	/mm	1-3
Mass $m_{t,W}$	/kg	0.2-3.0
Number of Scrapers	–	1 and 6

(Included is only the range of these experiments, which are explicitly used in this paper.)

179

termined experimentally, which appears to be product but not size specific.

Fig.1 illustrates the applicability and quality of the presented theory and measurements: Based on them, the torque of an industrial decanter was calculated as a function of the mass throughput. In addition to the operational setting, like the pond level radius r_{pond}, and machine design parameters only the product, its particle size distribution and its dewatering capability, which is implied in the mass specific throughput $\dot{m}_S^+$, has to be known.

2. EXPERIMENTAL SET UP AND PROCEDURE

Experimental equipment

The portrayed experimental results are obtained with two similar machines, briefly described in the following: Figs.2 and 3 show the main components and the chief geometric dimensions of the new centrifuge C1 in front view and in side view respectively. A variable number of scrapers or knives - two, three or six - rotate at a higher rate than the rotor in which they are mounted. One end of the drum is sealed by a plastic window, supported by a spoked disc for stability, so that the processes within the drum including the primary product layer can be observed. The shaft, on which the scrapers are fixed, is borne within the hollow rotor axis twice and additional at the spoked end to suppress vibrations due to unbalances. The scrapers shown in Fig.4 posess a sharp edge to minimize the friction between the product layer and the knives. In the same figure the blade angle is included for clarity. Care must be taken that the clearances between the scrapers and the drum do not differ too much, otherwise the product wedge will form only at the scrapers with the smaller clearance, hence inducing unbalances and influencing the results in addition to causing problems with the machine´s material strength.

In Fig.5 the whole equipment is schematically shown. An electric motor drives the hollow shaft, which is connected to the drum and a cycloid disc gear. Its output is linked to the scraper axle located within the hollow shaft. To be able to change the gear ratio as desired the journal of the cycloid gear has to be driven by a variable speed belt drive. The output of the latter can be switched on permitting to get a differential speed suddenly, which is important in order to observe the interactions of the primary layer and the wedge with the aid of different coloured tracer particles.

The torque, under which the inner shaft is subjected, is measured by wire strain gauges clued to it. The signal is led out of the rotating system via a slip ring. This is amplified and then converted into torque by a PC.

The speed of the scrapers, the rotor and the journal are measured by the inductive transducers 1, 4 and 2+3 respectively. For safety reasons, the latter are used for: If their speeds differ, i.e., when the safety clutch declutches due to an overload, an alarm will sound.

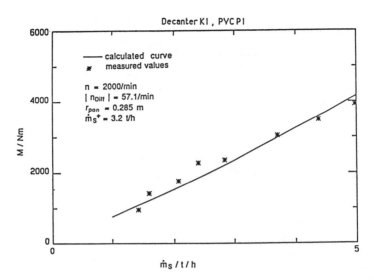

Fig.1: Torque of a decanter centrifuge versus mass throughput. The symbols mark the measured data, the line the calculated torque using the presented theory.

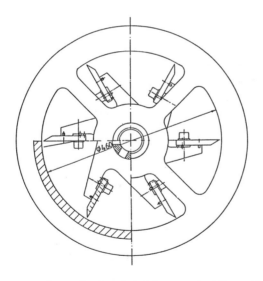

Fig.2: Front view of the friction centrifuge C1.

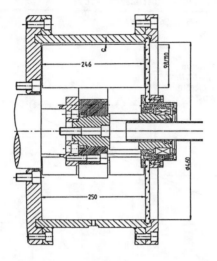

Fig.3: Side view of the friction centrifuge C1.

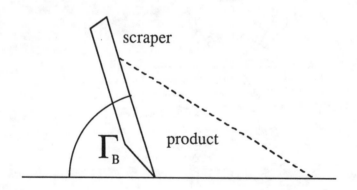

Fig.4: Scraper with the definition of the blade angle Γ_B.

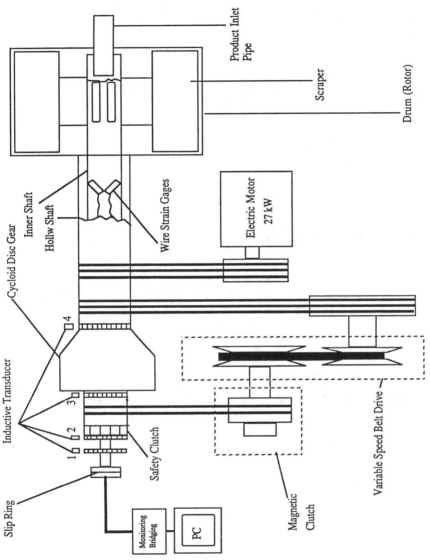

Fig.5: Schematic sketch of the complete experimental equipment.

The dimensions of the older centrifuge C2 are slightly different: the bowl diameter D is 630mm, its length L 315mm.

Experimental procedure

Product with a defined moisture content is fed through the inlet pipe into the rotating drum, where it dissipates, hence forming the primary layer. After feeding more product of the total mass Δm_t,

$$m_t = m_s + m_1 \qquad , \qquad (1)$$

that is the liquid and solid mass, then wedges will build up and a torque M can be measured. The friction coefficient μ is then simply defined as the ratio of circumferential to centrifugal force, i.e.

$$\mu = \frac{M/r_c}{\Delta m_t \; \Omega_s^2 \; r_c} \qquad (2)$$

3. EXPERIMENTS

The typical trend of the friction coefficient μ corresponding to eq.(2) as a function of the moisture content RF,

$$RF = \frac{m_1}{m_s + m_1} \qquad (3)$$

shall be discussed with the data obtained with the product PVC1 shown in Fig.6.
It should be mentioned, that instead of the moisture content RF, the saturation S

$$S = \frac{V_1}{V_v} \qquad (4)$$

is also frequently used as measure to describe the amount of liquid in a bulk. Between the saturation and the moisture content RF, the relation

$$S = \frac{RF \; \rho_s \; \dfrac{1-\varepsilon}{\varepsilon}}{\rho_1 \; (1-RF)} \qquad (5)$$

exists with the porosity ε

$$\varepsilon = \frac{V_v}{V} \qquad (6)$$

Independent of the operational setting the curves start at $\mu \approx 0.2$ for RF=0 and augment to a maximum at RF=17.5% or S=30% respectively. This increase may be regarded as a result of the fluid bridges, which strengthens the bulk solid in the primary layer due to the additional adhesive forces. At moisture contents above 17% the wedge dewaters, a liquid film will form in the layer; since within this film fluid bridges cannot exist anymore, the strength of the primary layer and thus μ decreases to a value of 0.1, when the bulk is totally saturated. Furthermore in the liquid film the particles experience a buoyancy force, which can be disregarded in the conducted experiments, since the height of the primary layer is essentially smaller than the height of the wedge or the normal force is mainly induced by the wedge respectively.

Because of the scattering of the measurements a regression curve of a 3rd order polynomial was fitted. For clarity, only these curves are drawn for the PVC product in subsequent diagrams.

For the purpose of comparison the same function is plotted for glass beads in Fig.7, whereby a similar trend can be observed: For dry product a friction coefficient of 0.2 could be evaluated. The maximum of this curve is already reached at a moisture content of approximately 2% or 9% saturation respectively. For the totally saturated bulk a value of about 0.17 was obtained for the friction coefficient. The slight increase at a moisture content of about 14% or a saturation of 70% may result of the capillary forces due to the capillary pressure within the primary layer of the bulk.

As it can be seen later on (compare Fig.12), for both products the saturation at the friction coefficients´ maxima are identical with the equilibrium saturation at the corresponding centrifugal field.

<u>Influence of the C-value</u>

As shown in Fig.8 the maximum of the friction coefficient decreases with augmenting centrifugal force, since the influence of the fluid bridge forces diminishes in comparison to the centrifugal force.

<u>Influence of the blade angle Γ_B</u>

One can observe in Fig.9 that the friction coefficient decreases for blade angles below 90°. This means that μ in eq.(2) is calculated with a normal force higher than actually acting between the interface primary layer-wedge, whereby this effect increases with declining angle. One can hence conclude that the scraper supports the solids forming the wedge.

<u>Influence of the clearance c</u>

The influence of the clearance c is shown in Fig.10. Slight differences appear in the friction coefficient for the three clearances, but no uniform trend is apparent. Whereas μ decreases on raising the clearance from 1 to 2mm, its value augments on increasing c further to 3mm.

<u>Influence of the differential speed</u>

In the next figure, Fig.11, μ is plotted versus the moisture content for different scraper velocities. No noticeable influence can

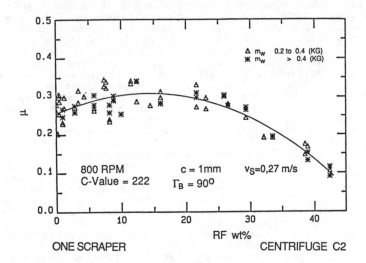

Fig.6: Friction coefficient μ vs. moisture content RF for PVC
/3/. Included is the regression curve of a third order
polynome /7/.

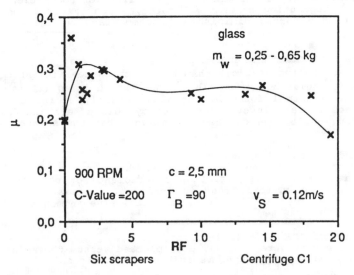

Fig.7: Friction coefficient μ vs. moisture content RF for glass
beads.

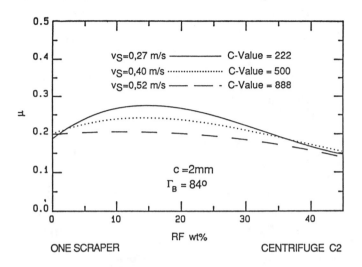

Fig.8: Friction coefficient μ vs. moisture content RF for PVC
for different centrifugal accelerations /3/ /7/.

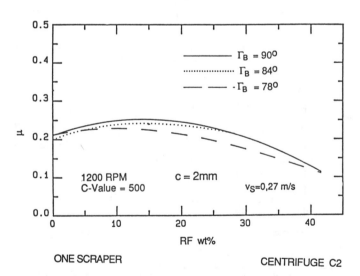

Fig.9: Friction coefficient μ vs. moisture content RF for PVC.
The blade angle Γ_B serves as parameter /3/ /7/.

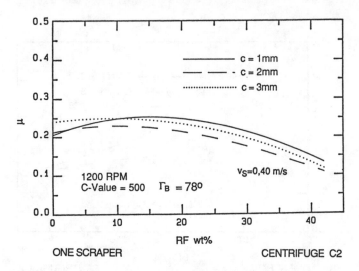

Fig.10: Friction coefficient μ vs. moisture content RF for PVC
for three different clearances c /3/ /7/.

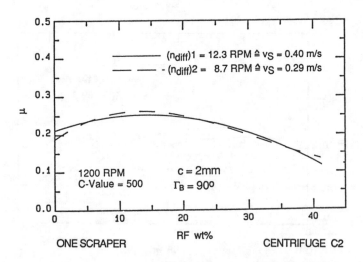

Fig.11: Friction coefficient μ vs. moisture content RF for PVC.
The scraper velocity serves as parameter /3/ /7/.

be observed. This behaviour coincides with experiments conducted with the shear ring cell /4//5//6/, where only with extremely viscous liquids an effect of the shear velocity on the friction coefficient could be found.

Influence of the surface tension

In this presentation it shall be only stated that for the glass beads a decreasing surface tension resulted in a smaller maximum, which appeared at a smaller moisture content. A detailed description is prepared for a later publication by the authors.

5. THEORY

The first step is to consider the dewatering and the support of the wedge by the scraper by determining or defining a new friction coefficient μ_{mod}, which relates then only to the physical effects of friction, i.e. eq.(2) will be modified to

$$\mu_{mod} = \frac{M - M_S}{R_S \, \bar{N}} \tag{7}$$

R_S : Radius at the scraper tip

M_S is the torque which originates at blade angles unequal to 90° and may be neglected in the conducted experiments /7/.
$\bar{N}$, the mean normal force, is defined by

$$\bar{N} = \bar{\sigma} \, l_w(R_S) \, w_W \tag{8}$$

$l_w(R_S)$: Length of the interface between primary layer and wedge

w_W : Width of wedge, identical with the length of the rotor L_R

whereby the mean normal stress is determined according to

$$\bar{\sigma} = \frac{1}{l_w(R_S)} \int_{R_S - h}^{R_S} \rho_b(S(r,C)) \, \Omega_S^2 \, r \, l_w(r) \, dr \tag{9}$$

with the density of the moist bulk

$$\rho_b = \rho_s \, (1-\varepsilon) + \rho_1 \, \varepsilon \, S(r,C) \tag{10}$$

With eq.(10) the dewatering of the wedge is already considered. The solution of eq.(9) yields

$$\bar{\sigma} = \frac{1}{l_w(R_S)} \frac{\rho_b \, \Omega_S^2}{\tan \varphi_{prod}} \left((R_S - h_k) \frac{h_w^2}{2} + \frac{h_w^3}{3} \right) \tag{11}$$

whereby the following assumptions were made:
1. The curvature of the rotor is considered negligible /7/.

2. The whole wedge may be regarded as a solid body.
3. The base of the wedge is coincident with the tip of the scraper.
All three assumptions lead to a wedge of triangular cross-section, which corresponds to the observations. Furthermore assumption two implies, that in case of an inclined scraper, part of the product is completely supported by the blade, i.e. no shear stresses are transferred within the wedge. Finally assumption three leads to the identity of the fed solid mass m_s and the mass of the wedge $m_{W,s}$ or that the wedge does not embody part of the primary layer respectively.
The height h_W and the length of the wedge $l_W(R_S)$ in eq.(11) can be obtained by simple geometric relationships together with its solid mass m_s, density ρ_s and porosity ε /7/.

Dependency of the dewatering on the centrifugal field

If the saturation of the fed product is higher than the equilibrium saturation S_∞, which depends on the applied centrifugal field, the wedge begins to dewater and its mass decreases by the amount of drained liquid, which accumulates at the bottom of the rotor. Thus the two products and their dewatering behavior, i.e. their equilibrium saturation, must be briefly described: The liquid in bulk solids as well as the saturation can be subdivided into capillary S_C, fluid bridge S_{Br}, surface S_a and inner saturation S_i /8/. In this row the fluid holding forces decrease, i.e. the capillary liquid can be removed with the least effort or in the lowest centrifugal field.
In Fig.12 the equilibrium saturation S_∞ is plotted as a function of the dimensionless number Bo_1

$$ Bo_1 = \frac{d_h \, h \, \rho_1 \, C \, g}{\gamma \, \cos\delta} \tag{12} $$

γ : surface tension
h : product height
d_h : hydraulic diameter

for both products. At values of Bo_1 approximately equal to four, the capillary liquid starts to dewater down to a plateau, where the saturation S_{plat} remains unchanged over a wide range of the Bo-number or the centrifugal acceleration respectively. At this plateau the fluid consists only of the three other liquid contributions. The shaded areas represent those Bo-numbers, where the experiments were conducted, i.e. only fluid bridges, surface and inner fluid are present in a particular ratio. For the forthcoming calculations it is assumed that in case of a saturation lower than the plateau, these fractions remain constant, i.e.

$$ \frac{S_{Br}}{S_{plat}} = const\,(product) \qquad for \ S \leq S_{plat} \tag{13} $$

For the PVC1 this fraction was evaluated as being 0.66 /7/, for the glass beads 1.0, since the latter possess a totally smooth surface and no internal capillaries.

Prediction of the modified friction coefficient

Following idea is developed: Consider two particles to be in contact as shown in Fig.13. If the upper particle is withdrawn, then the drag force F_T is the weight or normal force multiplied by the friction coefficient of the dry solid μ_0:

$$F_T = \mu_0 \, F_N \qquad (14)$$

If a liquid bridge is present, not only the weight must be considered as an "outer" force, but also the force due to this bridge F_{Br}, i.e

$$F_T = \mu_0 \, (F_N + F_{Br}) \qquad (15)$$

Then the general definition of the friction coefficient μ,

$$\mu = \frac{F_T}{F_N} \qquad (16)$$

and eq.(15) yield

$$\mu = \mu_0 \, (1 + \frac{F_{Br}}{F_N}) \qquad (17)$$

Instead of the single bridge force the sum of all fluid bridge forces N_{Br} and instead of the single, the total normal force $\bar{N}$ must be applied, if the whole bulk is to be considered. Furthermore, in the above equation, it is assumed that the friction of the material itself does not change with the moisture content. Hence, the above equation has to be modified in

$$\mu_{mod} = \alpha_1 \, \mu_0 \, (1 + \frac{N_{Br}}{N}) \qquad (18)$$

where the coefficient α_1 accounts for possible changes in the material friction. Since for dry products α_1 has to be necessarily unity, whereas for the totally wet bulk $(\alpha_1 \times \mu_0)$ must equal μ_1, so that

$$\alpha_1 = (1-S) + \frac{\mu_1}{\mu_0} \, S \qquad (19)$$

As in the known cases the ratio of μ_1 to μ_0 is close to one, α_1 might also be set to unity for simplification.

Calculation of N_{Br}

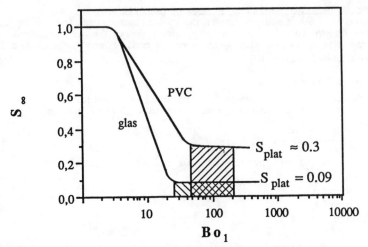

Fig.12: Equilibrium saturation S_∞ as a function of the Bond number Bo_1 for the used PVC and glass beads. The shaded areas indicate the range of the conducted experiments.

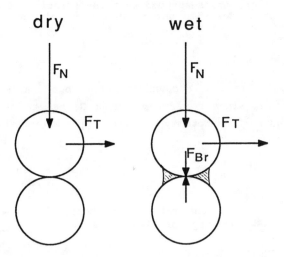

Fig.13: Forces acting on two particles during friction with and without a fluid bridge.

With the assumption that shearing only occurs beneath the wedge within the primary layer, the number of fluid bridges can be calculated by

$$n_{Br} = \frac{k}{2} \; n_P \; \frac{c-f}{c} \tag{20}$$

f : film height, height of the totally saturated area within the shear zone
k : coordination index
n_P : number of particles

The fraction (c-f)/f accounts for the fact that in the sheared volume fluid bridges cannot be existent within the developed film of dewatered liquid from the wedge. f is calculated by an elementary mass balance /7/.
It will be assumed that

$$k = \pi/\varepsilon \tag{21}$$

/9/ also holds valid for non-spherical and non-uniform particles. Rumpf stated for a related problem, for the evaluation of the tensile strength of moist bulks, that this assumption is at least applicable for first approximations /10/.

The number of particles n_P is simply the total sheared solid volume divided by the mean particle volume

$$n_P = \frac{V_{s,sh}}{\overline{V_P}} = \frac{(1-\varepsilon) \; l_W(R_S) \; w_W \; c}{\overline{V_P}} \tag{22}$$

The sum of the fluid bridge forces N_{Br} is then

$$N_{Br} = P_1 \; \frac{(1-\varepsilon)}{\varepsilon} \; \frac{c-f}{2 \; \overline{V_P}} \; l_W(R_S) \; w_W \; F_{Br}(S) \qquad c>f \tag{23}$$

$$N_{Br} = 0 \qquad\qquad\qquad\qquad\qquad\qquad\qquad c \leq f$$

The parameter P1 was determined by a regression, dependent on the clearance c. The product P1·c may be considered as the true shear depth s.
The mean fluid bridge force F_{Br} is averaged for the different particle sizes and the different distances assuming that only particles of the same size and roughness touch or that the bulk consists of regions, where only identical particles exist, respectively, i.e.

$$F_{Br}(S) = \frac{1}{i} \sum_{i=1}^{i} F_{Br,i}(d_{P,i}, S) \; \frac{n_P(d_{P,i}, d_{P,i}+\Delta d_P)}{\Sigma n_P} \tag{24}$$

with

193

$$F_{Br,i}(d_{P,i}, S) = \frac{1}{j} \sum_{j=1}^{i} F_{Br,i}(d_{P,i}, a_j, \beta_j) \qquad (25)$$

The ratio of the number of particles $n_P(\Delta d_P)$ with a diameter between d_P and $d_P + \Delta d_P$ to the total number can be easily calculated from any particle size distribution /11/.
As distance a the particle roughness of the particles was taken, measured with an electronic microscope /7/; this procedure implies that the particles are in direct contact with each other.
The force of a single fluid bridge in eq.(25) is calculated in the following manner: For the given saturation, the share of the fluid bridges are calculated with eq.(13). The liquid volume related to this saturation is then the sum of all fluid bridge-volumes, i.e. the geometry of a single bridge can be computed, with which the force $F_{Br}(d_P, a, \beta)$ may be calculated. A detailed description of the procedure was conducted in /7/. For the geometry as well as for the fluid bridge force, Rumpf's theory was used /12/. In his calculations, Rumpf did not solve the exact Laplace equation /13/, which can only be evaluated numerically; instead he approximated the projection of the liquid-gas interface as a section of a circle. It was shown that by this simplification, the error is less than 20% /14/ /15/. Here it shall be only stated that the fluid bridge force is directly proportional to the surface tension, i.e. a reduction of γ by a half decreases the fluid bridge force and hence the difference between μ and μ_0 by a factor of two: the maximum of the friction coefficient drops down.

6. RESULTS

In Figs.14 and 15 the modified friction coefficients are plotted as a function of the saturation for different experimental conditions together with the theoretically predicted curve, corresponding to eqs.(18),(19) and (23). The agreement between experimental and theoretical values is satisfactory.
The friction coefficient for the dry solids, μ_0, was determined as being
$\mu_0(\text{PVC1}) = 0.2$
$\mu_0(\text{glass beads}) = 0.2$
whereas for the wet solids
$\mu_1(\text{PVC1}) = 0.16$
$\mu_1(\text{glass beads}) = \mu_0(\text{glass beads}) = 0.2$
was determined. The identity of the dry and wet friction coefficient of glass is confirmed by Abel /15/ for a tri-axial device, and by Walser for a ring-shear device /16/. The difference between μ_0 and μ_1 may be regarded as a result of physical interactions be-

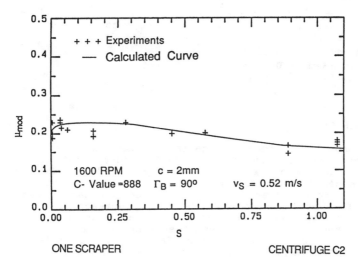

ONE SCRAPER   CENTRIFUGE C2

Fig.14: Friction coefficient μ vs. moisture content RF for PVC
for a particular experimental setting. The symbols mark
the measured data, the line the caclulated friction
coefficient using the presented theory.

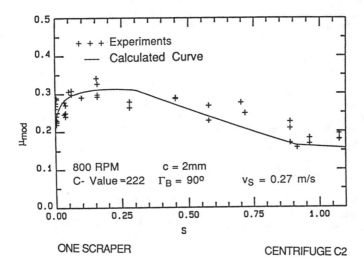

ONE SCRAPER   CENTRIFUGE C2

Fig.15: Friction coefficient μ vs. moisture content RF for PVC for
a particular experimental setting. The symbols mark the
measured data, the line the caclulated friction coefficient
using the presented theory.

tween the solids surface and the liquid e.g. dwelling. Both μ_1 and μ_0 were independent of the experimental conditions.

The parameter P1 was regressively adjusted for the different clearances 1, 2 and 3mm, but it was found that the product (P1·c) was almost constant:

P1(1mm)·1mm = 0.61mm
P1(2mm)·2mm = 0.54mm
P1(3mm)·3mm = 0.57mm

7. DISCUSSION

In the conducted project, no statement concerning the influence of the rotor's surface roughness and the velocity profile within the shear zone can be made. If the velocity profile reaches the drum, an influence of the surface must be observed. With similar products and similar clearances one could notice in decanter centrifuges that profile strips welded to the rotor increase the friction of the product, whereas after some operational time of the centrifuge the surface becomes polished and the friction diminishes; thus the shear zone reaches the drum's surface. The independence of the friction coefficient from the differential speed also favours this possibility.

In contrast, the constance of the product (P1·c) could be explained, if the shear zone is smaller than the clearance: Then this product could be interpreted as effective shear depth. Furthermore the ratio of (P1·c) to the mean particle diameter of the PVC1 yields approximately ten, which corresponds to values of the shear depth to particle diameter obtained with a ring shear device /8/.

A final decision is not yet possible. Therefore further experiments have to be conducted, in which mainly the clearances, the particle diameters and the rotor's surface roughness are varied.

New experiments seem to prove that the dry and wet friction coefficient are constant for the specific products, i.e independent of particle diameter and operational setting.

Since the friction coefficient of the glass beads shows a slight increase at about 14% moisture content or 70% saturation respectively, the effect of the capillary pressure should be introduced in eq.(18), so that μ_{mod} becomes

$$\mu_{mod} = \alpha_1 \, \mu_0 \, (1 + \frac{N_{Br}}{N} + \frac{N_C}{N}) \tag{26}$$

whereby N_C is an additional "internal" normal force due to the capillary pressure within the bulk. Here it shall be only mentioned that N_C becomes zero for S less than S_{Plat} and for S equal to unity.

8. SUMMARY

This publication initially introduced a centrifuge, which enables, the frictional behaviour of solids in a centrifugal field to
be determined. A variable number of scrapers rotate at a different
rate to the bowl. When product is fed into the bowl, wedges form;
the friction coefficient is then evaluated by means of the measured
wedges' mass, the speed and the torque.
In order to obtain only the influence of the physical effects on
the friction, a modified friction coefficient is presented, i.e.
the dewatering of the wedge and its corresponding mass reduction in
addition to the support of the bulk by the scraper is considered.
The influence of the blade angle, the clearance, the centrifugal
field, the moisture content and the shear velocity is shown. Independent of the experimental setting the friction coefficient plotted as the function of the moisture content, starts at a constant
value, passes a maximum and decreases again to a constant value for
a completely saturated product. This maximum decreases for an increasing centrifugal field.
The scraper velocity had no significant effect on the friction of
the investigated products.
A function of the modified friction coefficient was determined,
which describes the results satisfactorily. The general fundamental
idea is that the fluid bridge forces overlap the normal force within the bulk. The fluid bridge force is calculated using Rumpf's
theory. The trend of the friction coefficient could be predicted
for all operational conditions. Only the dry and wet friction coefficient had to be measured. Since no knowledge of the shear depth
was available, a parameter had to be introduced which depend solely
on the clearance. The product of this parameter and the clearance
was constant, allowing it to be considered as effective shear
depth.
Further research must be conducted to obtain informations about the
processes and the velocity profiles within the primary layer. Also
should the theory be proved for a variety of products.

9. NOMENCLATURE

a	distance between two particles	m
Bo_1	Bond 1 number	–
c	clearance	m
C	C-value: multiple of the earth's gravity	–
d_p	particle diameter	m
D	bowl diameter	m
f	film thickness	m
F	force	N
g	gravitational acceleration	m/s^2
h	height	m
k	koordination number	–
l	length	m
L	rotor length	m
m	mass	kg

M	torque	Nm
n	number	−
n	rotational speed	s^{-1}
n_{Diff}	differential speed	s^{-1}
N	normal force	N
N_{Br}	sum of the adhesive force between the particles caused by fluid bridges	N
P1	parameter by the regression with eq.(23)	−
r	radius	m
R	cylinder radius	m
RF	moisture content	−
s	effective shear depth	m
S	saturation	−
v	velocity	m/s
V	volume	m^3

Greek symbols:

γ	surface tension	N/m
δ	contact angle	−
Γ_B	blade angle	−
ε	porosity	−
μ	frictional coefficient	−
ρ	density	kg/m^3
φ_{prod}	frictional angle of repose	−
Ω	angular velocity	s^{-1}

Indices

Subscripts

b	relative to the bulk
Br	" liquid bridge
c	" center of gravity
C	" capillarity
D	" drum
l	" liquid
N	" normal force
plat	" plateau of the Bo-S-diargram
s	" solid
sh	" shear zone
S	" scraper

t	"	total amount
T	"	tangential force
V	"	voidage
W	"	wedge
0	"	dry product
∞	"	equilibrium

Superscripts

–		mean value

References

/1/ : F. Reif, W. Stahl: Transportation of Moist Solids in Decanter Centrifuges, Annual Meeting of the AIChE, New York 1987, accepted in Chem. Eng. Prog. 1988.

/2/ : F. Reif, W. Stahl: Feststofftransport in Dekantier- und Siebschneckenzentrifugen, Final Report to DFG Sta 199/4-1, German Research Council, Bonn-Bad Godesberg, 1987.

/3/ : S. Steiner: Reibwertuntersuchungen im Zentrifugalfeld, Diploma Thesis, Institut für Verfahrenstechnik, TU München, 1967.

/4/ : H. Buggisch, R. Stadler: Untersuchungen der Fließeigenschaften von Schüttgütern in Abhängigkeit vom Gutfeuchtegehalt unter hohen Geschwindigkeiten und unter hohen Normallasten. Final report to AIF-Project No. 5818, 1986.

/5/ : H. Buggisch, R. Stadler: Influence of the deformation rate on shear stress in bulk solids, Theoretical aspects and experimental results, EFCE Publication Series No. 49, Proceedings "Reliable Flow of Particulate Solids", Bergen, Norway, 1985.

/6/ : R. Stadler: Stationäres, schnelles Fließen von dicht gepackten trocknen und feuchten Schüttgütern, PhD Thesis, University of Karlsruhe, 1986.

/7/ : F. Meck: Aufstellung eines Modells zur Ermittlung des Reibkoeffizienten feuchter Schüttgüter im Zentrifugalfeld, Diploma Thesis, Inst. f. MVM, University of Karlsruhe, 1987.

/8/ : G. Mayer: Die Beschreibung des Entfeuchtungsverhaltens von körnigen Haufwerken im Fliehkraftfeld, PhD Thesis, University of Karlsruhe, 1986.

/9/ : W.O. Smith: Packing of homogeneous spheres, Phys. Rev., **34**, p.1271-1274, 1929.

/10/ : H. Rumpf: Zur Theorie der Zugfestigkeit von Agglomeraten bei Kraftübertragung an Kontaktpunkten, CIT, **42**, 8, p. 538-540, 1970.

/11/ : J. M. Coulson, J. F. Richardson: Chemical Engineering, Vol.2, 3rd edition, Pergamon Press, New York 1978.

/12/ : H. Rumpf, W. Pietsch: Haftkraft, Kapillardruck, Flüssig-
keitsvolumen und Grenzwinkel einer Flüssigkeitsbrücke zwi-
schen zwei Kugeln, CIT, **39**, 15, p. 885-892, 1967.
/13/ : H. Schubert: Kapillarität in porösen Feststoffsystemen,
Springer-Verlag, New York 1982.
/14/ : H. Schubert: Kapillardruck und Zugfestigkeit von feuchten
Haufwerken. CIT, **45**, 6, p.393-401, 1973.
/15/ : P. G. Abel: Zur Kapillarkohäsion feuchter Haufwerke, Publi-
cations of the Institute for Soils and Rock Mechanics, University of Karlsruhe, Paper 43, 1969.
/16/ : C. Walser: Untersuchungen zur Ermittlung der Fließeigenschaften von Schüttgut-Flüssigkeits-Gemischen in Abhängigkeit von der Flüssigkeitskonzentration, vom Druck und der
Gleitgeschwindigkeit, Diploma thesis, Institut für Mechanische Verfahrenstechnik und Mechanik, University of Karlsruhe, 1983.

Acknowledgements

The authors would like to thank both the Comittee For Industrial
Research in Germany (AIF) for financing this project, and the German Research Council (DFG) for the financial support of the travel
expenditures.

200

Effect of Strong Secondary Currents on Suspended Load Distribution

F. ARAFA and M. SHERIF
Department of Irrigation and Hydraulics Engineering
Cairo University
Cairo, Egypt

SUMMARY

An analytical model is presented to simulate the manner by which strong secondary currents at estuaries affect the suspended sediment distribution. Experimental work was carried out for verification of the results.

INTRODUCTION

Secondary currents have significant effect on estuary hydrodynamics. Along curved reaches and in bends, the interaction between the vertical gradient of streamwise velocity and the curvature of the primary flow generates a so-called secondary or spiralling flow. The velocities produced by this secondary flow increase the erosive attack on the banks and also undermine them.

Effective planning and protection of streambanks and structures along curved reaches and bends from erosion require determining the relationships between the primary flow, secondary flow and sediment motion.

The purpose of this study is to investigate the effect of strong secondary currents on suspended load distribution. A series of experiments is carried out for different values and modes of secondary currents and for different sediment particles diameters.

Analytical approach for the effect of strong secondary currents was developed assuming the amplitude of the side-wall perturbation to be small compared with the depth, deduced linearized equations for the perturbation velocities where the periodicity of the perturbated flow was determined. The experimental results were compared with the theoretical analysis.

THEORETICAL APPROACH

Curved reaches are always found at estuaries. Due to this curvature a strong secondary currents are developed (Engelund, et al., 1973). As a first step to investigate the effect of these secoundary currents on suspended sediment, consider a steady incompressible flow in a rectangular cross section where the side walls are assumed to meander according to the sinusoidal law

$$b = b_o e^{ikx} , \quad i = -1 \tag{1}$$

where the meandering wavelength $L = 2 /k$ is much larger than the half width B.

The amplitude b_o of the side wall perturbation is assumed smaller than the water depth D. The bed elevation d and the surface fluctuations are also considered smaller than the depth D. See definition sketch, Fig.(1).

The governing equations of motion read

$$\underline{V} \cdot \nabla \underline{V} = \frac{1}{\rho} \nabla p + \nu \frac{\partial^2 V}{\partial z^2} \tag{2}$$

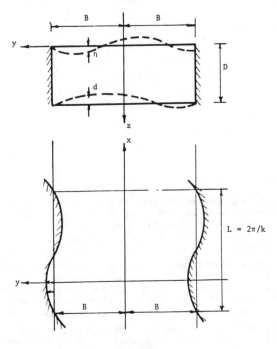

Figure 1 Definition sketch

and

$$\nabla \cdot \underline{V} = 0 \tag{3}$$

where $\underline{V} = (u,v,w)$ is the local velocity vector, P, ρ and ν are pressure, density and eddy viscosity, respectively. The flow is now considered to consist of a primary flow and a secondary flow due to the perturbation of the side walls, neglecting the side wall friction leads to

$$u = U_o(z) + u', \quad v = v', w = w', P = P_o + P' \tag{4}$$

Inserting (4) into (2) and (3) and neglecting second and higher order terms, to obtain

$$\frac{\partial P'}{\partial x} + \frac{dP_o}{dx} = -U_o \frac{\partial u'}{\partial x} - w' \frac{dU_o}{dz} + \nu \frac{\partial^2}{\partial z^2}(u' + U_o),$$

$$\frac{\partial P'}{\partial y} = -U_o \frac{\partial v'}{\partial x} + \nu \frac{\partial^2 v'}{\partial z^2} \quad ,$$

$$\frac{\partial P'}{\partial z} = -U_o \frac{\partial w'}{\partial x} \quad ,$$

$$\frac{\partial u'}{\partial x} + \frac{\partial v'}{\partial y} + \frac{\partial w'}{\partial z} = 0 \qquad (5)$$

For the primary flow, Engelund (1973) established that

$$U_o = U_s \cos(\alpha\bar{z}) \qquad (6)$$

where $U_s = U_b + \gamma$ = surface velccity,

$$U_b = \sqrt{gmD} \ [1.9 + 2.5 \ \ell n \frac{D}{q}] = \text{bottom velocity,}$$

$$\alpha^2 = \frac{14\sqrt{gmD}}{U_s} \quad ,$$

$$\gamma = mgD^2/2\nu \qquad (7)$$

and m, g and q are the mean slope, the acceleration of gravity and the equivalent sand roughness, respectively. For the secondary flow

$$\frac{\partial P'}{\partial x} = -U_o \frac{\partial u'}{\partial x} - w' \frac{dU_o}{dz} + \nu \frac{\partial^2 u'}{\partial z^2} \quad ,$$

$$\frac{\partial P'}{\partial y} = -U_o \frac{\partial v'}{\partial x} + \nu \frac{\partial^2 v'}{\partial z^2} \quad ,$$

$$\frac{\partial P'}{\partial z} = -U_o \frac{\partial w'}{\partial z} \quad ,$$

$$\frac{\partial u'}{\partial x} + \frac{\partial v'}{\partial y} + \frac{\partial w'}{\partial z} = 0 \qquad (8)$$

The basic assumption is that the secondary flow velocity components and
the pressure vary sinusoidally in the longitudinal direction exactly with the
same manner as the side wall meandering i.e.

$$P' = \rho U_b P(\bar{y},\bar{z}) e^{ikx} \quad ,$$

$$u' = U_b U(\bar{y},\bar{z}) e^{ikx} \quad ,$$

$$v' = i U_b V(\bar{y},\bar{z}) e^{ikx} \quad ,$$

$$w' = i U_b W(\bar{y},\bar{z}) e^{ikx} \quad ,$$

$$d = D\bar{d}(\bar{y}) e^{ikx} \quad ,$$

$$\eta = D\bar{\eta}(\bar{y}) e^{ikx} \quad , \qquad (9)$$

where $\bar{y} = y/D$, $\bar{z} = z/D$, P,U,V,W,$\bar{d}$,$\bar{\eta}$ are unknown complex non-dimensional functions.

Inserting these expressions in (8), we get :

$$\varepsilon \frac{\partial^4 W}{\partial \bar{z}^4} + \frac{\partial^2 W}{\partial \bar{y}^2} + \frac{\partial^2 W}{\partial \bar{z}^2} = [(KD)^2 + \frac{U_{ozz}}{U_o}]W \ ,$$

$$\varepsilon \frac{\partial^2 V}{\partial \bar{z}^2} + V = -\frac{1}{U_o} \int_0^{\bar{z}} U_o \frac{\partial W}{\partial \bar{y}} \, d\bar{z} + \beta \frac{dP}{\bar{y}} (\bar{y}, 0) \tag{10}$$

where $\varepsilon = i\nu(KD^2 U_o)^{-1}$, $\beta = U_b(KDU_o)^{-1}$ and $U_{ozz} = d^2 U_o / dz^2$

The solution of these equations can be written as

$$V(\bar{y}, \bar{z}) = \sum_{n=0}^{\infty} F_n(\bar{y})[C_{1n} e^{\bar{S}z} + C_{2n} e^{-\bar{S}z}] - \frac{KDU_o}{U_b} \cos(\frac{nD}{B} \bar{y}) - \frac{i}{S^2} \sum_{n=0}^{\infty} A_n \tag{11}$$

$$W(\bar{y}, \bar{z}) = 2K\alpha^{-2} \frac{B}{\eta} \bar{z} \sin(\frac{nD}{B} \bar{y}) + \sum_{n=0}^{\infty} H_n(\bar{y})[C_{3n} e^{r_n \bar{z}} + C_{4n} e^{-r_n \bar{z}} + C_{5n} e^{h_n \bar{z}} +$$

$$+ C_{6n} e^{-h_n \bar{z}}] - \bar{z} \sum_{n-0}^{\infty} J_n \tag{12}$$

where $S = (M\theta)^{\frac{1}{2}}$, $M = iKD^2 U_b \nu^{-1}$, $\theta = \alpha^{-2}/\alpha^2$,

$$\alpha 2 = -\frac{d^2 U_o}{dz^2} U_o^{-1}, \quad \alpha^{-2} = -\frac{d^2 U_o}{dz^2} U_b^{-1},$$

$$F_n(\bar{y}) = \cos(\bar{y}e_n D), \quad H_n(\bar{y}) = \sin(\bar{y}e_n D),$$

$$e_n = \eta(n+\tfrac{1}{2})B^{-1}, \quad t_n = [(k^2 + e_n^2)D^2 - \alpha^2]^{-1},$$

$$A_n = e_n KDM[\theta^2 + \frac{\alpha^2 \theta 2}{M} + \frac{\alpha^{-2}}{M} + (1+\theta)^2(e_n D)^2 - (\frac{nD}{B})t_n] + \frac{e_n D^2 g R_n(0)}{i\nu K U_s},$$

$$r_n = [\frac{S^2}{2}(1-L_n)]^{\frac{1}{2}}, \quad h_n = [\frac{S^2}{2}(1-L_n)]^{\frac{1}{2}}, \quad L_n = [1-4S^2 t_n]^{\frac{1}{2}}, \quad \text{and}$$

$$J_n = \frac{16(-1)^n KB}{\Pi^2(1-2n)(3+2n)} \alpha^{-2}$$

$$U(\bar{y}, \bar{z}) = \frac{-1}{KD} (\frac{\partial v}{\partial \bar{y}} + \frac{\partial w}{\partial \bar{z}}) \tag{13}$$

Under steady conditions, the sediment concentration equation is given by Chin and McSparran (1966)

$$\frac{\partial}{\partial y}(Cv) + \frac{\partial}{\partial z}(w+w_s)C = \frac{\partial}{\partial y}(\varepsilon_y \frac{\partial C}{\partial y}) + \frac{\partial}{\partial z}(\varepsilon_z \frac{\partial C}{\partial z}) \tag{14}$$

where ε_y, ε_z are the y and z components of the diffusion coefficient for sediment, and w_s is the fall velocity of the representative particle under gravity.

$$\varepsilon_y = K^2(\frac{\partial U_o}{\partial y})^3 / (\frac{\partial^2 U_o}{\partial y^2})^2 ,$$

$$\varepsilon_z = K^2(\frac{\partial U_o}{z})^3 / (\frac{\partial^2 U_o}{z^2})^2 \qquad (15)$$

Using equation (6)

$$\varepsilon_y = 0 ,$$

$$\varepsilon_z = [\frac{2\gamma z^3}{D^2}] e^{ikx} \qquad (16)$$

where K is the Von-Karman constant.

The concentration C will be distributed as

$$C(x,y,z) = \bar{C}(\bar{y},\bar{z})e^{ikx} \qquad (17)$$

Equation (14) then becomes

$$\frac{\partial}{\partial y}[\bar{C}V(\bar{y},\bar{z})] + \frac{\partial}{\partial z}[\bar{C}(w_s + W)(\bar{y},\bar{z})] = \frac{\partial}{\partial z}[\bar{C}.2\gamma z^3(\frac{1}{D})^2] \qquad (18)$$

where $V(\bar{y},\bar{z})$ and $W(\bar{y},\bar{z})$ are given by equations (11) and (12).

Equation (18) is a non-homogeneous partial differential equation including complex functions, so the exact solution for this equation is very difficult. The solution may be obtained using finite element method and we will present it in next paper. Therefore, this paper presents as experimental work conducted to simplify the effect of the strong secondary currents on suspended sediment distribution.

EXPERIMENTAL WORK

a. Experimental Procedure

To study this effect, few experiments wew conducted in the Hydraulics Laboratory at Cairo University. The general Layout of the experimental flume is shown in Fig. (2).

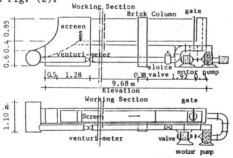

Figure 2 Experimental flume

Four types of side walls were used to produce different modes and values of secondary currents. These four types were straight, corrugated sheets, inclined perforated plates and inclined solid plates. Three different uniform sediment materials were used in the experimental work; medium, fine and very fine of 0.6, 0.2 and 0.1 mm particle mean diameter respectively.

For the four types of side walls the velocities were measured at the working section 3.80 meters apart from the flume inlet using a pitotsphere, where a constant water depth of 0.28 meter was maintained. Seventy points were chosen to measure the velocities for each type.

The experimental work was divided into four groups. Each group consists of three runs conducted with constant side wall type but different bed materials. The concentration of suspended sediments were measured using sediment sampler. The concentrations were measured at twenty four points for each case.

b. Experimental Results

A summary of the experimental results is given in Table 1. The measured and calculated concentrations are shown in Figs. (3), (4) and (5). The concentrations are calculated using equations.

$$\frac{C}{C_a} = [\frac{(a)(d-z)}{(z)(d-a)}]^Z \qquad \text{FOR} \quad z/d < 0.5 \tag{19}$$

$$\frac{C}{C_a} = [\frac{(a)}{(d-a)}]^Z e^{-4Z(\frac{z}{d} - 0.5)} \qquad \text{for } z/d \geq 0.5 \tag{20}$$

where C, C_a are the concentration at any level z and at reference level a above the bed, respectively, and Z the suspension parameter

$$Z = \frac{w_s}{\beta K u_*} \tag{21}$$

where w_s is the fall velocity for suspended particles, β is the coefficient related to diffusion of sediment particles, K is the Von Karman constant and u_* is the total bed shear velocity.

c. Discussion of Experimental Results

There is a difference between the measured and calculated concentrations using equations (19) and (20). This is due to the existence of secondary currents. This difference can be approximated in terms of the intensity of secondary velocities by the following factors:

$$62.3 \, (\frac{V_{sec}}{V_m})^{3.5} \quad \text{for medium sand } (D_s=0.6mm) \tag{22a}$$

$$2.2 \, (\frac{V_{sec}}{V_m})^{1.4} \quad \text{for fine sand} \quad (D_s=0.2mm) \tag{22b}$$

and

$$196.9 \, (\frac{V_{sec}}{V_m})^{5.2} \quad \text{for very fine sand } (D_s=0.1mm) \tag{22c}$$

where V_m and V_{sec} are the mean and secondary velocities.

206

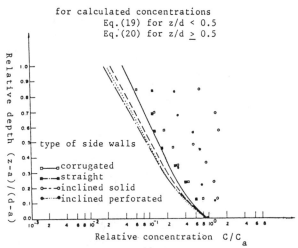

for calculated concentrations
Eq.(19) for z/d < 0.5
Eq.(20) for z/d > 0.5

type of side walls

□——□corrugated
■——■straight
○—--○inclined solid
●——●inclined perforated

Relative concentration C/C_a

Figure 3 Relative concentrations for medium sand

In Fig. (6), for example the lines of equal secondary velocity and the sediment isopleths (lines of constant concentration) for a flume with inclined perforated plates and fine sand as bed material are represented. The sediment isopleths are lowered in regions of high values of downward vertical components of secondary velocities. The transverse components of secondary velocities tend to increase the sediment concentration in their direction.

CONCLUSIONS

The secondary currents affect the pattern of the isopleths of suspended sediment in a remarkable way, Fig. (6).

TABLE I

SUMMARY OF EXPERIMENTAL RESULTS

Run No.	Sediment Particles diameter mm.	Discharge Q lit/sec.	Water surface slope	$\dfrac{\bar{V}_{sec}}{V_m}$ %	Average concentration $\bar{C}$ gm/lit.	Remarks
1	0.10		0.015		13.360	
2	0.20	100.0	0.015	17.10	12.528	Group 1
3	0.60		0.0143		2.975	Straight flume
4	0.10		0.040		20.954	Group 2
5	0.20	100.0	0.040	22.50	18.090	Flume with corrugat-
6	0.60		0.0545		6.436	ed sheets
7	0.10		0.030		14.041	Group 3
8	0.20	100.0	0.025	29.30	6.986	Flume with inclined
9	0.60		0.025		1.145	solid plates
10	0.10		0.030		13.241	Group 4
11	0.20	100.0	0.025	20.50	12.066	Flume with inclined
12	0.60		0.025		1.966	perforated plates

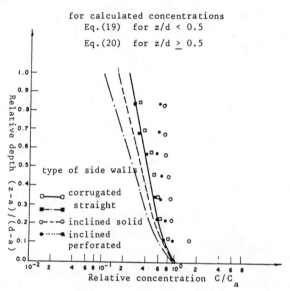

for calculated concentrations

Eq.(19)   for z/d < 0.5

Eq.(20)   for z/d $\geq$ 0.5

Figure 4 Relative concentrations for fine sand

The relative concentrations increase with the increase of secondary currents intensity, Figs. (3), (4), and (5)

The relative concentrations increase with the increase of sediment particles diameter for constant secondary currents intensity, Figs. (3), (4), and (5).

Secondary velocities may be calculated using equations (11), (12), and (13), while the sediment concentration may be calculated using equation (18).

The mean difference in relative concentrations can be calculated using equation (22).

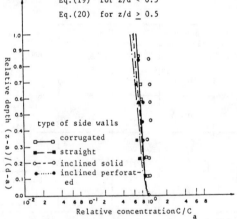

Eq.(19)   for z/d < 0.5

Eq.(20)   for z/d $\geq$ 0.5

Figure 5 Relative concentrations for very fine sand

REFERENCES

ARAFA, F. (1983). Influence of Strong Secondary Currents on Suspended Load Distribution in Open Channels. Department of Irrigation and Hydraulic Engineering, Cairo University.

CHIU, C.L. and McSPARRAN, J.E. September (1966). Effect of Secondary Flow on Sediment Transport. Journal of the Hydraulics Division, Vol. 92, No. HY5.

CHIU, C.L., and HSIUNG, D.E. July (1981). Secondary Flow, Shear Stress and Sediment Transport. Journal of the Hydraulics Division, Vol. 107, No. HY7.

COLEMAN, N.L. (1980). Velocity Profiles with Suspended Sediment. Journal of Hydraulic Research, Vol. 19, No. 3.

EINSTEIN, H.A. and LI, H. December (1958). Secondary Flow in Straight Open Channels. Transactions, Americam Geophysical Union, Vol. 39.

ENGELUND, F., and SKOVGAARD, O. (1973). On the Origin of Meandering and Braiding in Alluvial Streams. Journal of Fluid Mechanics, Vol. 57, Part 2.

GESSNER, F.B., and JONES, J.B. (1965). On Some Aspects of Fully-Developed Turbulent Flow in Rectangular Channels. Journal of Fluid Mechanics, Vol. 23, Part 4.

COTTLIEB, L. August (1974). Flow in Alternate Bends. Progressive Report 33, Tech. University, Denmark.

IMAMOTO, H., ASANO T. and ISHIGAK, T. (1977). Experimental Investigation of a Free Surface Shear Flow With Suspended Sand Grains. Proceedings of the 17th Congress of the International Association for Hydraulic Research, Vol. 1, PP. 105-112.

LAU, Y.L. (1983). Suspended Sediment Effect on Flow Resistance. Journal of Hydraulic Engineering, Vol. 109, No. 5,

LIGGETT, J.A., CHIU, C.L., and MIAO, L.S. (1965). Secondary Currents in a Corner. Journal of the Hydraulics Division, Vol. 91, No. HY6.

VAN RIJN, L.C. November (1984). Sediment Transport, Part II: Suspended Load Transport. Journal of the Hydraulic Engineering, Vol 110, No. 11.

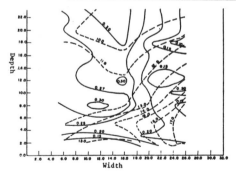

— Lines of Constant Secondary Velocity. $\left(\dfrac{V_{sec}}{V_m}\right)$

--- Isopleths of Sediment (gm/lit)

Figure 6 Lines of equal secondary velocity and sediment isopleths.

SUSPENSIONS

The Application of the General Energy Theory to the Studies of Chemical Engineering: Some Fundamental Aspects of the Theory of Sedimentation

K. PARTHASARATHY
Department of Electrical Engineering
Annamalai University
Annamalai Nagar 608 002, India

ABSTRACT :

In this paper, first the details of the electrokinetic phenomena are discussed from the GET point of view. The study also
includes the adsorption at the solid-liquid interface. In the light
of these, the studies are extended to the discussion on the sedimentation stability of the particles. A fuller discussion on the
coagulation of the colloidal system would follow these. The analyses made in the light of the principles of the GET will show that
a clear understanding of the various physical mechanisms involved
can be greatly brought out.

1. INTRODUCTION :

In the companion papers on colloids and aerosols, many of the
general characteristics of fine particles in solution and gas have
been brought out. Details on the electrical double layer, the adsorption on the solid-gas and solid-liquid interfaces, the optical
and other related physical aspects of the particles are some of
them. In this paper, details on the electrokinetic phenomena, sedimentation and stability of sols, coagulation etc., of these fine
particles are discussed. These will adequately cover the various
aspects of the colloidal sciences.

2. ELECTROKINETIC PHENOMENA :

The four familiar electrokinetic phenomena are well-known. These
have been analysed to a certain extent already by the GET (21, 22).
In the present context, we find that the electroosmotic effect that
would pull the solvent(mainly water)molecules towards the cathode
is largely due to the p- character of the solvent(water)molecules.
On the other hand, the drift of the solute molecules which generally are p+ is due to the repulsive character of the excessive p+
that is contained in the cathode, constituting electrophoretic
effect will become obvious. If a solvent is forced through a suit-

213

able membrane, the seggregation of the charges that could occur
would give the streaming potential. If in a column of liquid,
(charged)particles are allowed to fall towards the bottom, the
sedimentation potential(Dorn effect) is realised.

Although the above phenomena have been extensively quoted in
the literature, a more realistic physical interpretation appear to
be lacking . The application of the thermodynamic laws do not bring
out the physical picture in entirety. However, these can be understood as matters of simple routine , through the principles of the
GET. In considering the behaviour of the molecules in phenomena
such as above, the nature of the charge generation must be clearly
understood. In this regard, the details of the structure of the
atom , the proton and the electron together with that of the doublet
etc., will become very essential. Sufficient information on these
have already been presented elsewhere(9-45). In as much as the
electrical , chemical and thermal effects are considered, it can be
shown that the basic doublet (proton+neutron)of the nucleus together
with the electron would constitute a total of 10.22p+ units(1 unit
= 10^{-12}erg) of energy which is mainly shared by the spectral and
electrical counterparts of the atom. Thus for example, if an atom
had a spectral counterpart of 8p+ units, only the remaining 2.22p+
would be available for electrical energy generation.Using this basic
principle, we could compute the electrode potentials of various
types of electrodes that are used in electrochemistry(33,34).

When an attempt is madeto dislodge an electron from its location in the atom, it must gather sufficient quantities of the energy particles as defined above . These particles will be p+ in
character and would load the normal buffer of the electron.As much
as $1.1x10^{19}$ particles would be required by the electron to generate
a potential of one volt. Such an excited electron, when jumping
from one atom to the next(passage of electric current)would pump
out all its p+ clouds and would become normal. One of the electrons of this second atom will in turn pick up all but a small
quantity of this cloud and transfer it self to the next atom. The
small difference so rejected manifests itself in the generation of
the magnetic field and the irreversible joule heat.

By computing the nature and magnitude of the charges that can
be picked up by the atoms constituting the +ve and -ve clectrodes
of a battery, it can be seen that the cathode will be always at a
higher p+ than the anode, the difference in the value leading
directly to the generation of the cell voltage. In as much as the
electron starts its journey from the cathode, it can be seen that
it does so because of the mutual repulsion it experiences. On the
same score, an anion experiences a repulsive force at the cathode
because of its p+ buffer and the cation experiences an attraction
because of its p- buffer.

The foregoing general concepts will now enable us to study the
various electrokinetic phenomena in detail. In the case of the electroosmosis , the cathode by virtue of its higher p+ character
will tend to pull the water molecules which are highly p- (8.5).

In this regard, the internal p- of the water molecule is more
contributory than the outside p+ buffer . This arises because,
due to the contact angle of attachment of the H-atoms to the O-
atoms being 109°, an unsymmetric structure for the water molecule
will be made exposing much of the internal structure of the O-atom
at the opposite end that is powerfully p- in character(with a
corresponding p+ buffer that is weak). At the same time, the hydrogen (made of two hemispheres of buffer clouds containing the proton and the electron unlike the floral structure of all other atoms)
attaches itself to the Oxygen atom through its electron-end which
is the basic requirement for a chemical bond(electronic circles of
the two atoms must always engage for the bond). This throws out open
the p- buffer of the proton. Thus appreciable zone of the buffer
of the water molecule will have this oriented p- at one end and
p+ at the opposite end (cause for the dipole generation). The molecule thus gets pulled into the cathode compartment more by the
exposed Oxygen end(p-). This pull of the internal p- is more pronounced than the repulsion that could be caused by the p+ buffer
end of the oxygen atom. On this score, the attraction of the water
molecule by an anode is not possible. This is deduced from the general fact that in an electrolyte into which the anode (p_1+) and
cathode(p_2+) are existing we find that $p_2+ > p_1+$ so that any anion
(p- internal charge and p+ buffer) left between the electrodes
will experience a greater repulsion via its p+ buffer from p_2+
than from p_1+ and hence will move towards the anode. On the same
token, the cation will have a p- buffer and experiences more pull
from the p_2+ than from p_1+ and so moves towards the cathode.Hence
when water is dissociated into $H^+$and OH^-their buffers are net p-
and p+ and so the ions drift towards the cathode and anode respec-
tively.But undissociated water, as it were, by virtue of its exposed
p- as explained above , will always experience a pull more towards
the cathode only rather than towards the anode. It should also be
remembered that, in general, undissociated molecules can, by virtue
of their resultant charge, will slowly drift towards the proper
electrode while the same molecules when dissociated will generate
ions that could move very fast and reach the appropriate electrodes.
This will be clear, for example, from the behaviour of colloidal
particles undergoing electrophoresis . The reason for the slow
diffusion rate of the undissociated molecules will be, that, in
spite of their high p+ or p- (spectral and electrical) absorption , these clouds will be basically internally bound and hence
masked by the electronic circle in that they do not become highly
effective. But, in dissociation, electronic addition or subtraction
is involved. This permits the free moving of the electron(from one
atom to another and enable it to dump effectively all its P+ which
will either enhance or counterbalance the atom's existing buffer
so as to generate powerful reaction with the charges of the electrodes. In this respect, in electroosmosis, the undissociated water
molecules will be found to diffuse only at a slow rate (on this
score, the solute molecules will also move slowly as explained
below). Thus in the elctroosmosis, the water molecules can freely
drift across the separating membrane into the cathode compartment.
The solute molecules (undissociated), on the other hand, will be
generally p+ in character. Hence they will be repelled by the

cathode and pushed into the anode compartment. But this action is largely hindered by the semipermeable membrane . This repulsive action of the membrane is due to its inherent p+ owing to the general proteinic character of it(mostly these are biological membranes and in cases of synthetic membranes also it can be shown to have a net p+). The drifting solute molecules will then have to pass through the membrane pores against this repulsive action.On this score, the solute molecules already present in the anode compartment cannot drift in the opposite direction(to the cathode end) owing to the repulsive forces of both the membrane molecules and the cathode. Likewise some of the water molecules will tend to get pushed back into the anode compartment consequent on the hydrostatic head developed in the cathode compartment. But equilibrium is always established only with sufficient hydrostatic head at the cathode end. Finally, the electrophoresis effect will also thus get completely visualised from the foregoing details.

As a passing mention, it can be stated that as per the conventional literature, the observation of electrically neutral water molecules and solute molecules drifting toward the electrodes will remain rather puzzling. The electroosmosis and electrophoresis are often explained through recourse to the theory of electrical double layer. But although discussed at length, the generation of the surface charges in the double layer model remain rather elusive. The GET on the other hand, deduces the electrophoretic and electroosmotic phenomena from the creational point of view and hence will be found to be highly superior. The basic aspects of the double layer theory as visualised by the GET is presented in the companion paper on colloids to which refernce should be made.

The streaming and sedimentation potentials can be similarly visualised. The sedimentation potential being generated(Dorn effect) becomes immediately apparent when we consider the slow sedimentation of the solute molecules under the action of gravity. In as much as these molecules are individually charged and usually p+ , their accumulation at the bottom of the vessel will generate on an electrode present at that point a -ve polarity.Obviously,an electrode located sufficiently above will become +ve and the establishment of the sedimentation potential must be evident.

In this connection, the accumulation of the p+ charges at the bottom electrode should not be construed to give the electrode (embedded at that place)the property of obtaining a perrennial source of p+ . Since the atoms constituting the particles will tend to hold their spectral energy as captive, their contribution becomes depleted soon and the potential drops to zero. Rejuvanating the atoms is normally not possible because of the solvent layer around being water(p-).

Again, as envisaged in the conventional literature, if the principle of the double layer could be applied to the charge around a given particle, there remains the basic difficulty of explaining the transfer of this charge on to the electrodes.Normally,the very physical nature of the charges lack perception, excepting that we

216

understand their existence through their(field)effect like potential
development, force generation and the like. In view of the inability
to fix up these matters from the point of view of physical visualisation, the mechanism of charge transfer from the particle to the
electrode will also remain elusive(one may point out to the development of the potential which could arise only out of the charge transport, but what about the modus operandi would have to be settled).
Also the sheath of the double layer as it forms around the falling
particle is independent in itself to call for electrical neutrality
(particle-wise)and on this basis, as to how the electrodes are sup-
plemented with charges of the same polarity always by these various
falling particles should be clearly known. It should also happen
that the particles must dispense away with their charges simultaneously to both electrodes(otherwise they will be briefly charged
singularly) from whatever spatial position they hold with respect
to the electrodes.

The streaming potential will be found to be developed by the
transport of charges between the given pair of electrodes under
external simulation. It should be significant to note that this
occurs with electrically neutral molecules. The forcing of the P+
charges of the solute molecules(in the experiment on electrophoretic
phenomena) in the reverse direction against the repulsive action
of the semipermeable (p+) membrane will be found to just accumulate
the charges which will lead to the development of the -ve electrode.
On the same token, the chargewise depleted electrode becomes +ve
and thus the streaming potential appears with the correct polarity
to give the same direction of current as it would have been obtained
in the normal electrophoretic experiment.

The streaming potential is also interpreted through the double
layer theory. The measured potential is often presumed to correctly
specify the zeta potential .There is still considerable differences
on this matter and some would prefer the measurement to indicate
the electrophoretic mobility rather than the zeta potential which
is not possible to be measured(46). According to Lamer, the zeta
potential corresponds to a fixed (stationary) double layer while the
styeaming potential involves motion. The electrophoretic mobility
is more concerned with the charges and not potentials(46).

3. ADSORPTION AT THE SOLID. SOLUTION INTERFACE :

In as much as this topic is highly important from the point
of view of formation,stability and disintegration of lyosols,we
shall briefly document some of the basic aspects for consideration
by the GET. The study can become more complicated since in addition
to the molecules of the adsorber and the adsorptive , we have also
those of the medium that may compete in adsorption.

Substances consisting of polar molecules on being wetted by
polar liquids produce great thermal effects . Similar effect is
observed with non-polar substances and non-polar liquids.It is
also assumed that the better a medium dissolves an adsorptive,the

worse is the adsorption in the medium. On this basis, the Duclaux-
Traube rule will be reversed. It is found that while non-polar
adsorbants adsorb greatly non-polar adsorptives, the polar adsorbers
do well with polar adsorptives. Porosity is a deciding factor in
adsorption. When the adsorbers have narrow pores, adsorption decre-
ses with the increased size of the adsorptive. The rule of polarity
formulated by Rehbinder may be used to study the adsorption of the
medium and adsorptive. Accordingly, a substance C can be adsorbed
into the interface of A and B if C could equalise the difference in
the polarities of A and B. In terms of the dielectric constants,
this will imply $\epsilon_A > \epsilon_C > \epsilon_B$ or $\epsilon_A < \epsilon_C < \epsilon_B$. According to this rule, diphilic
surfactant molecules must orient such that their polar part orients
towards the polar phase while the non-polar part faces the non-polar
phase of the interface. This implies that a polar hydrophillic surface adsorbs the surfactant well from a non-polar liquid and vice-
versa.

The adsorption of the solute occurs very slowly. As temperature increases, adsorption from a solution decreases. In ionic adsorptions,
ions capable of being polarised are adsorbed on the surfaces that
consist of polar molecules or ions(polar adsorption). A surface
bearing a definite charge adsorbs oppositely charged ions. Electrolytic ions of opposite sign are not adsorbed directly here but remain near the adsorbed ion by electrostatic attraction forming the
electric double layer. The radius of the ions strongly affect their
adsorbability. For the given valency , ions of larger radii have
better adsorption. The cause for this is attributed to hydration
which hinders adsorption and it is more for smaller ions than for
bigger ones. The Hofmeister series for monovalent ions can be put
in the following way(according to adsorption capacity) as $Li^+ < Na^+ <$
$K^+ < Rb^+ < Cs^+$. Bivalent cations are arranged as : $Mg^{2+} < Ca^{2+} < Cr^{2+} < Ba^{2+}$.
Monovalent anions would give : $Cl^- < Br^- < NO_3^- < I^- < NCS^-$. Also as the
valency increases, the ion strongly adsorbs to the oppositely charged microregions of the surface.

We now apply the principles of the GET to visualise the data
presented above. When polar substances are wetted by polar solvents,
in view of the identical nature of their buffers , it is obvious
that the buffers will coalesce and this in turn will bring the molecules (and their atoms) closer together . This coalescence will naturally involve the merging of the A-circles of the molecules so as
to generate much smaller ones(spherically symmetrical). The reduction in the total volume of the A-circle particles will mean that
the free energy is reduced (21,22) which is the characteristic of
wetting. Thus the observation is valid and a similar situation
exists with the non-polar solvents and substances. In this connection, the basic aspects of the Duclaux-Traube rule and its reversal
have already been discussed in the companion paper on Aerosols to
which reference can be made. For the same reasons given above, the
adsorption of the non-polar adsorptives on non polar adsorbants and
similarly the polar substances on the polar adsorbers must be made
very easy as found(buffer coalescence). When pores are involved,the
interspace of the pore will be filled with the buffer particles of
the surface atoms. Hence as an adsorptive molecule tends to enter

into the pore-space, buffer coalescence and consequent density enhancement must occur .On this account, the adsorption must decrease as the size of the adsorptive increases(increasing buffer sizes), which is clearly a reversing factor for the Duclaux-Traube rule .

The polarity rule of Rehbinder would also be in harmony with the GET. For equilibrium, the charges of the buffers must always be compatible at a given interface. Thus when a polar solvent is brought into contact with a polar adsorbant, the buffers coalesce leading to the atoms of one tending to go into the other or, in other words, the polar phase grows (if permitted by the system otherwise).When a non-polar adsorber is involved with a polar adsorbate it is easy to see that the buffers do not interact freely(although of opposite polarity, the buffers are equilisers of charge conditions of the parent atoms and hence more governed by the nature of the atoms and affected only marginally by the surrounding influence). It is then logical to see that when a diphilic surfactant molecule is brought into the interface between two substances ,the orientation of the ends of the surfactant molecule must always occur to satisfy the buffer relations stated above. In this respect, the polar to polar and non-polar to non-polar alignment must always be made . Considered as a whole, the entire volume in space including the interface zone must be smoothly graded densitywise and hence the character of the surfactant must necessarily fit in with those of differing dielectric constants (a measure of the energy particle distribution or density gradation)of the adsorber and adsorptive.

Adsorption at the solid-liquid interface must necessarily be slow,when compared with the solid-gas systems. This is because, the buffer equilization of a given liquid molecule has already been settled with its neighbours in the bulk of the liquid.Hence when it is being adsorbed by the solid, the rearrangement in buffer distribution at the surface-liquid zone would be such as to generate local buffer variations. On this account, the gradual equilization of buffers occur with consequent time-increase. The effect of temperature in reducing the adsorption is because the heat photons ($p+$) required by both the adsorbant and adsorptive atoms will lead to the mutual repulsion which enhances with temperature.

In as much as the ionisation tends to increase the strength of the buffer(for both electron loss or gain) , it is obvious that it will inprove the adsorption of ions of opposite polarity.Polar molecules will have to be oriented to the surface with either one of the dipole end and will have again the same property. The surfaces generating charged zones and the adsorption of various types of ions(potential determining and counterions) have already been discussed at length in the companion paper on colloids to which reference should be made.

The deduction on the Hofmeister series can now be made. Consider the monovalent cations , for the series Li^+, Na^+, K^+, Rb^+ and Cs^+. For these, we have the atomic radii(in $A°$) and the corresponding spectral absorption(in brackets) as : 1.27(4.2p-); 1.57(3.3p-) ;

2.03(2.6p−) ; 2.16(2.5p−); 2.35(2.4p−) . Thus the corresponding
buffer requirement reduces and hence the equivalent buffer sheath
thickness also must decrease with the increasing radius. Clearly
the atom's ionic absorption ability becomes more as the buffer's
influence reduces(for increasing radii). On the same basis, for
the bivalent cationic series Mg^{2+}, Ca^{2+}, Sr^{2+} and Ba^{2+}, the values
of radii and energies are given as: 1.36(7.2p+); 1.74(5p+);1.91
(4.6p+); and 1.98(4.4p+) respectively.Thus this series also behaves
as above. With the anionic series Cl^-, Br^-, NO_3^-, I^-, and $NCS^-$the
values are : 0.99(4.2p+); 1.14(4.3p+); 1.3(4.9p+); 1.33(9.5p+)
and 1.81(10.2p+) respectively. While it may appear that both the
energy and and radii are increasing , it can be seen that the radii
increase at a faster rate than the energies giving the same conclusions as for the cationic series. For the given atom , as the valency increases, the internal structure of the atom is more exposed
and hence the atom's influence is enhanced.

4. STUDIES ON THE SEDIMENTATION STABILITY AND COAGULATION OF FINE
PARTICLES :

Perrin studied the equilibrium distribution of colloidal particles and found that as the vertical height increased arithmetically
the number of particles decreased geometrically. (under the influence of gravity). However, in highly concentrated dispersions, the
particle density tends to remain the same with height. The normal
sedimentation rate of fine particles in liquid is very slow compared
with the aerosols. This will be illustrated from the following data
available(49). For quartz suspensions(in water) of radii 10^{-3},10^{-4},
10^{-5}, 10^{-6} and 10^{-7} cm, the time required for sedimentation of the
particles through a distance of one cm, will be 31 seconds, 51.7
minutes, 86.2 hours, 359 days and 100 years respectively. This may
be compared with the data for aerosols with the initial numerical
concentration per cc, of particles varying as 10^{12},10^{10},10^8 and
10^6 with the time for reducing the concentration by two orders as
fraction of a second, 15−30 seconds, 30 minutes and several days
respectively.(these two different sols are to be compared relatively
only). However, the hypsometric law(geometric law of Perrin) is
obeyed for lyosols under conditions of equilibrium distribution.

Smoluchewski studied the rapid coagulation of the colloids(49).
The particles coalesced into larger ones when they came close within
the critical distance (cf: companion paper on colloids). He showed
that at high concentration of the electrolytes the time−rate of
coagulation was linear. Based on these concepts, the potential energy of the system of particles are deduced. These were used to explain many of the phenomena associated with coagulation. Examples
of these will be the thixotropy (conversion of gels into sols on
standing and subsequent transition of sols into gels on long standing), the Brownian motion of small particles adhering to a rigid
wall(Buzagh effect), the array of small particles moving around
large ones being well preserved in aqueous dispersions of rubber
(Langeland effect) etc., .

220

Modern theory of stability and coagulation envisages the role of ions. Stability may also be caused by thin shells of water on particles (solvation). As per Rehbinder, such shells do not stick to one another and they cannot form in lyophobic colloids. Solvation appears to disturb non-polar dispersion medium. The concept of the double layer was also used to explain the stability, but however, these attempts have run into difficulties. Some presumed that the potential determining ions were solvated across the double layer.

The solvated shells around the molecules will be presumed to have no net force on them due to the simultaneous attraction of the molecules and those of the medium. According to Rehbinder, the structural mechanical stability offered by the highly saturated shells of adsorbed surfactant molecules forming two dimensional films and gels would be mainly contributory in stabilising the sol. He would presume the paucity of time for the gelated shells to interact during collisions caused by the Brownian motion. But when there is a cross-linked lyophillic shell, the particles may stick together. This is observed in floatation when the surfactants are adsorbed by the polar groups on the surface of the solid hydrophillic particles. The h.c. chain facing the aqueous medium are bound with one another by peculiar local sticking together of the hydrophobic shells. Accordingly, the structural mechanical barrier is the strongest stabilizing factor and its need is apparent in producing highly concentrated stable sols.

Simha etal., showed that sufficiently long and flexible molecules of surfactants and polymers can adsorb by their individual units on a solid . The ends of these units will show high Brownian motion generating the stability of the system. The causes for this are variously attributed as due to mutual repulsion, dimunishing of the conformation due to crowding of protruding units and osmotic forces.

5. THE APPLICATION OF THE GET TO THE STUDIES ON SEDIMENTATION STABILITY :

The Perrin law stated above will be seen to follow the same type of behaviour of the colloidal solutions which we see in connection with light scattering(Bougeur-Lambert-Beer law) excepting that in this case, it will appear apparently (from conventional angles) that the exciting light source of the sun will seem to pass through the atmospherically-graded air column in a reverse manner (from a layer of lower density to one of higher density).But if the real nature of the earth's gravity is taken into consideration we will see that the two cases) are just identical to one another. We notice that the very effect of the gravity arises due to the manifestation of the buffer distribution of the earth from its surface(9,21,22). Referring to figure-1a, we find that when a column of energy particles above an atom like X is concerned, it is the difference in the vertical thrust generated from below and the force of the column standing above X of the energy particles will account for the net gravitational force. Obviously this will be a maximum

on the atom like A situated on the surface of the earth . In as much
as this force is due to the density variation of the energy particles, it also stands as a measure of the density of equivalent light
energy. Thus for two atoms X and Y separated as in the light scattering experiment by a distance corresponding to a layer thickness
t , clearly the radiations received by X from Y will vary as per
the geometric distribution which we had already established. However,
in highly concentrated solutions , the atoms like X and Y are brought so close together that their envelopes of influence coalesce
thereby tending to modify the above geometric law. Reference to
figure-1b would show that $\delta_2 = \delta_1/2$ (as deduced earlier) for isolated
buffers but when the buffers coalesce generating common zones like
Z , the density will be δ_2 $\delta_y/2 + \delta_1/2$ $= \delta_1$, indicating that close
to a given layer where X and Y are distributed a layer containing
Z-like atoms will also have the same density distribution . This
is clearly the observed result.

The nature of this buffer coalescence leading to the equaliza-
tion of density of suspended particles as noted above can further
be understood by comparing the sedimentation characteristics of the
lyosols with the aerosols . For an aerosol of typical particle radius of about 100 A° and a numerical concentration of 10^{10} p/cc,the
time taken to sediment such that for a concentration reduction by
two orders is quoted as about 15-30 seconds. This concentration is
fairly very dilute so that each particle behaves independently and
has a sheath volume given by $4 \Pi R^2 d$ where $d = 1/\sqrt{R}$ (companion paper
on aerosols) . With the radius of R= 100 A°, this volume is 1.08×10^{-12}
cc. As the concentration is reduced,by two orders, the number of
particles undergoing settlement by receiving energy is 10^{-8} so that
the net energy particles required will be $(1.08 \times 10^{-12}) \times (10^8) \times (1.24 \times$
$10^{42})$ $= 1.34 \times 10^{30}$. It can be seen that the particles tend to receive
the radiation value of density to commence interaction with the
neighbours as they come closer to each other(intermolecular distance
= 2.5 R, for which buffer sheath thickness is d) . The radiation
imparted by the light beam of average λ = 5500 A° on the particle
will be $(100 \times 10^{-8}/3.3 \times 10^{-11})$ $\times (5.5 \times 10^{14}) \times (0.88 \times 10^5)$ $= 4.47 \times 10^{28}$.
Hence the time for loading the given 10^8 atoms for interaction and
subsequent sedimentation will be $1.34 \times 10^{30}/4.47 \times 10^{28}$ = 30 seconds.
On the other hand, considering a lyosol particle of the same size
we notice that due to the nature of the liquid medium of dispersion
all the particles in a given plane will tend to coalesce and hence
during sedimentation will descend equally. In this sense, they will
tend to show a coalescence in the two dimensions(plane).The particles are so close that they will obey the intermolecular interaction
law(R= D/2.5) . Thus for a particle having D=4A° , the radius R=
4/2.5 = 1.6A° . This will accumulate energy particles all into the
atoms upto the electronic radii and allowing 20 % reduction for this
effect, the net radius of the particle (equivalent aggregate of
atoms)is about 1.3 A° . It should be seen that no sheath effect
exists in this case due to the coalescence-effect but the volume-
effect is exhibited at a higher density of 10^{45} p/cc. The number
of particles that we can have per cm^2 is obviously $(1/2 \times 10^{-6})^2$ =
2.5×10^{11}, so that the net particle absorption at $\epsilon =10^{45}$ p/cc will
be $(4 \Pi/3) \times (1.30 \times 10^{-8})$ $\times (2.5 \times 10^{11}) \times 10^{45}$ $= 2.27 \times 10^{33}$. The beam

absorption per particle will be $(4x10^8/ 3.3x10^{-11})^2$ $x5.5x10^{14}$ $x(0.88x$ $10^5)= 7.26x10^{25}$. Hence the time for the particles to move through one cm, (so that the given number of particles will generate the 1 cc-volume that is required for the density value of 10^{45} p/cc) is $(2.27x10^{33})/(7.26x10^{25} x8.64x10^4) = 359$ days. We also see that in as much as the radiation received by the 1.6 A°-particle (core molecule, cf: companion paper on colloids)remains the same in all cases of concentration , the number of particles per cm^2 of the surface alone will contribute to the variation of the sedimentation times. Thus the particle sizes of radii of 10^{-7}, 10^{-5}, 10^{-4} and 10^{-3}cm will have particle counts differing from the previous value of $2.5x10^{11}/cm^2$ by factors of 100, 1/100 , $1/100^2$ and $1/100^4$ respectively. These will then correctly generate the time factors as about 98.4 years, 86.16 hours, 51.7 minutes and 31 seconds respectively. The size of 1.3 A° for the nucleating core is justified from the fact that when the core rises to 10^{45} p/cc from the surrounding value of $2x10^{40}$ p/cc, the radius of control will be $1.3x\sqrt{10^{45}/2x10^{40}}=47.8A°$ and hence for the buffer influence factor of 2 , the radius of the overall size of the particle that could be generated by the core atom is 95.6 A° ($\doteq$ 100 A°) .

The result of the Smoluchewski's investigation that the time rate of the coagulation is linear will be apparent from the above. The phenomenon of thixotropy is also logical in that as long as stirring is present, the buffer coalescence is hindered and hence there is no gelation permitted for particles located in a plane (for directional effect as noted above). On the same score, small particles can be stably be buffer-linked(sticking in conventional sense) with an already well-oriented array(of wall) particles (or molecules). As the particles move along, they have a rolling buffer contact so that they simulate the Brownian motion also. The stability of a group of such particles moving or rolling(buffers are closely rolling on one another like steel balls would do ,but without complete coalescing due to the like character of them all)around bigger ones will also be obvious.

Whatever the success of any modern theory of stability based on the intermolecular forces would be, it still stems on the ever-unknown characteristics of the solvated shells. In order that the stability existed, it must be presumed that coagulation by aggregation must be absent and so the individual particles must be bestowed with the neutral character of not interacting with one another . The only way which the conventional theories could do this is to presume both atteactive forces and repulsive forces as stated in the last section. However, the neutral effect of the solvated particles will be automatic , since all the buffers will have the same polarity. Particle aggregation can occur only when there is a stress improving coalescence , which, as for example, can be brought in by increasing the concentration . In this respect, the ideas of Rehbinder in discrediting the paucity of time needed for coalescence will not be tenable, but his visualising of the behaviour of the saturated shells of surfactants tending to form two dimensional coalescence-sheaths will be found to be the same as given by the GET. The peculiar sticking observed with the surfactants adsorbing to polar

groups of a surface will again be due to the buffer interaction.
When the polar ends of the surfactants are so-engaged to the surface , the hydrocarbon(non-polar)ends projecting into the aqueous
medium must each be heavily repelled by the buffer of the medium
and in the process the adjacent buffers of the closely held h.c.
ends(of the surfactants) must slowly coalesce manifesting in the
peculiar sticking property. Hence the presumed structural mechanical
stabilizing factor as yet vaugely defined will be seen to account
for the effect of the buffer characteristics of the particles as
portrayed by the GET.

Since the polar ends of surfactants will generally be clumped
regions , they would greatly exhibit spherical buffers around them
and on this account can easily coalesce with a sufficiently strong
and symmetrical buffer associated with a bigger solid particle. Thus
well-anchored, these surfactant molecules with their free h.c. ends
(being same in buffer polarity) can undergo considerable vibration
around their mean position. Coalescence at the farthest(tip)end of
these celia-like projections cannot easily occur and hence a parent
particle covered with these protruding flexible molecules will be
highly stabilized.

6. COAGULATION OF COLLOIDAL SYSTEMS :

Coagulation of colloids can occur under varied conditions.
Factors such as aging, change in concentration or temperature ,
mechanical effects, action of light etc., are mainly contributory.
In addition, the significant effects of the electrolytes have also
been found. Stabilizing electrolytes must have sufficient concentration. Hardy found that the ions of the electrolytes bearing the
charge of the same sign as that of the counterion of a micelle (or
a charge opposite to that of the colloidal particle) only could
contribute to coagulation. In this respect, a minimum concentration
(coagulation threshold)must be attained by the electrolyte in the
sol. The beginning of coagulation is determined usually by the
onset of the colour change , appearence of turbidity, commencement
of precipitation etc., .

Schulze and Hardy found that the coagulation power of an ion
increased with its valency(Schulze-Hardy rule), with the coagulation
rate depending on the valency to as high as the 6-th power(49) .
For example, for mono, di and trivalent cations this dependency
varies as 1 : 20 : 350 . (a wide variation on these values is reported by others due to experimental difficulties) It is also found
that the coagulating power of the ion of the same valency grows
with the ionic radius. The coagulating action of organic ions is
remarkable in that the +vely charged monovalent ions of the alka-
loids and dyes act much more strongly than inorganic electrolyte
ions of the same valency.

The coagulation usually begins not at zero value of the zeta
potential (isoelectric point) as Hardy believed, but usually at a
certain critical value of the zeta potential. For many electrolytes

this value is around 30 mV. For some systems, however, a much higher variation in this value is also reported . When the critical zeta potential reduces, the sol may not coagulate and its stability may even increase. It is often found that the coagulating ions also always precipitate along with the dispersed phase.

Many theories on coagulation exist. The chemical theory of coagulation, the adsorption theory, the electrostatic theory and the physical theory(DLVO theory) are some of these. The last one is the currently favoured. It is based on the balance between the molecular forces of attraction and ionic forces of electrostatic repulsion.

Distinction must be made between two extreme cases of coagulation, viz., the neutralisation coagulation and the concentration coagulation. Neutralisation coagulation is observed for sols with weakly charged particles which have low double layer potential jump. Concentration coagulation occurs for sols having highly charged particles as the concentration of an indifferent electrolyte in a system increases. Adsorption phenomena in the coagulation of the colloids is very important. Radiometric data show that adsorption by ions bearing a charge of the same sign as that of the colloidal particles is very small except when the ions have strong adsorbability . The effect of the adsorption of the surfactant on a colloidal particle is also important. For example, for the hydrosol of arsenic sulfide, the highly hydrophobic surface of the colloidal particles become hydrophillic due to adsorption of non-ionogenic surfactants.

The degree of stability of sols was found to depend on their age and concentration. The AgI sol had the greatest stability as the particles attained a monolayer covering of polyethylene glycol. There are data to show that ions of same sign as that of the colloidal particles can also be adsorbed if they had high adsorption potential. Considerable diverse opinion appears to exist on this aspect in current literature(5,46,49).

A number of interesting phenomena are observed in coagulation. These are the phenomena of irregular series, the antagonism and synergism of ions in coagulation , the accustoming of the sols to the action of the electrolytes and colloidal protection. The phenomenon of the irregular series occurs, when various ever increasing quantities of polyvalent ions of opposite charges to the colloidal particles are introduced into the sol. The sol first remains stable but coagulates within a definite concentration range , after which it becomes stable again only to be finally coagulated under the excessive addition of the electrolytes. This is mainly attributed to the charge neutralisation brought in by the polyvalent ions. It is now believed that the charge of -ve particles is reversed not by the polyvalent ions but by their products of hydrolysis containing +ve ions of oxides or their hydrates. As proof for this, it can be shown that the particle charge is not reversed in the acid medium but only in weakly alkaline solutions. While in a weak alkaline solution , a lower concentration of the polyvalent electrolyte

could do so , in high pH the recharge becomes impossible. The reversal of charge can also occur when a divalent electrolyte is added
provided the solubility of the appropriate hydroxides is low. The
phenomenon can also be observed when the potential determining ions
are added to the sol.

Antagonism and synergism occur when sols are coagulated with
a mixture of (for example, the stabilizing and coagulating) electrolytes. Such mixtures may act independently and may call for a
definite quantity for each electrolyte (addition action), counteract with one another thereby calling for more addition individually
than required for the addition action (antagonism) or aid each
other resulting in a reduced quantity than would be needed for the
addition action(synergism). Several theories are proposed to
explain these.

Sometimes colloidal systems will require larger amounts of
coagulating electrolytes when added gradually than when it is all
added at once. The system then becomes positively accustomed to
thw electrolytes . Sometimes the -ve accustoming occurs when the
quantity slowly added happens to be less than that needed for the
rapid addition. Krestinskaya(49) proposed chemical changes as the
cause for these. For example, when HCl is added to Fe(OH)3 , positive accustoming occurs but when alkalis are added to Fe(OH)3 then
negative accustoming results. Positive accustoming occurs very
rarely in an electrolyte in very low concentration.

The stability of the colloidal systems can be enhanced by forming an adsorption layer on the particles using high molecular weight
substances . Zsigmondy proposed a gold number to characterise such
systems(the amount of high molecular weight substance needed to
prevent the bluing of the normally red gold sol on the addition of
sodium chloride will be the gold number). Others specify the use
of iron oxide, rubin, silver etc., numbers. It is significant to
note that electron microscopy has revealed the presence of such
shells.

Organosols having high dielectric constant exhibit electro -
phoresis. This implies that an electric double layer must be associated with such organosols. These obey the regularities exhibited
by the hydrosols and also the Schulze-Hardy rule and phenomena like
the additivity and antagonism etc., . However, owing to the lower
dielectric constant (when compared with the hydrosols) these sols
generally have a highly diffused and thick double layer so that
they are less stable than hydrosols. Hetero coagulation(or mutual
coagulation) was originally attributed as due to the action of the
opposite charges on particles in both sols. However, later investigations showed that such coagulation can occur even when the particles had the same charges.

Coagulation often occurs under physical factors like mechanical action, heating and freezing , dilution or concentration ,action
of visible and UV light , ultra sound etc., . Long-time storage

226

often destroys the stability of a colloidal system(spontaneous action) . Coagulation under the action of an electric field is also observed. Polar molecules strongly orient in the electric field and form dentrite-like structures . The use of low intensity alternating voltages are suggested in this regard when the aggregates orient along the lines of force.

It is far more difficult to explain coagulation when colloidal systems are concentrated. But heating or even boiling is found to have only a small effect on the hydrosol stability.

7. THE APPLICATION OF THE GET TO THE STUDIES OF THE PHENOMENA OF COAGULATION :

The various factors contributing to the coagulation of sols will all be such as to improve the coalescence of the buffers of the particles involved. In as much as the internal absorption by the molecules(or their atoms) of the particles can be enhanced by an increase in the concentration and temperature will be to enlarge the size of the buffers and hence promote coalescence.However, in a homogeneous sol, the various particles will have identical buffer sizes and so they would tend to naturally repel and marginally prevent the coalescence unless the stress brought on them is high. On this account the rate of coagulation must be marginally dependent upon the temperature and concentration(for the pure sol and unaided by coagulating agent). On long standing, the buffers coalesce due to a slow merger of the buffers and thus would show aging effects. Vigourous mechanical stirring should again result in improvement of buffer coalescence(as the particles are made to collide with each other with sufficient force). Finally the action of light in improving the energy content of the particles and thereby promoting coagulation had already been discussed at length earlier.

When electrolytes are employed in coagulation , their tendency to improve buffer coalescence of sol particles will obviously be more when they carried charges opposite to that of the parent particles . In this case, each electrolyte buffer would easily act as a bridge between adjacent sol particles to accelerate aggregation and hence coagulation. When such an action is contemplated, in as much as each individual electrolyte molecule has its own zone of influence(36.8 R), it becomes obvious that for a given sol , the numerical concentration of the electrolyte must bear a definite minimum (threshold) value. It will also be clear that the onset of coagulation which is initiated by an increased absorption of the energy particles (usually p+) must result in the change of the colour of the sol. Turbidity is again a measure of the sol in losing its clarity (by allowing more light scattering) and hence on the absorption of energy particles. Precipitation is obviously the result of coalescence of particles during coagulation and so must indicate the starting of the process.

The Schulze-Hardy rule follows from the fact that when an atom like A is considered alone, its combining power(valency) with the other atoms will depend primarily upon its nuclear buffer mass(extra over the $2Z$ -value ; Z- the atomic number) and the nature of its charge($p+$ or $p-$) . As the charge present in the nucleus is increased, the atom can attract more electrons (one electron for every combining atom) radially displaced at $\theta = 360^\circ/n$ where $n-$ is the valency. Also the nuclear attraction will be enhanced so that the combining atoms will come closer towards the parent atom A. Thus the contact area at the electronic surface would increase(as the valency tends to increase)so that the contact zone develops a buffer volume which is proportional to $r^2 xt$ where $r-$is the radius of the contact circle and $t-$is the thickness that will be a constant(electronic radius). Therefore the energy particles that can be loaded into this volume will be proportional to r^2 and in as much as the energy density is proportional to r^3 (cubic gradation law) we find that the coagulating stimulus (due to the accumulation of the energy particles)tends to increase as $(r^2)^3 \alpha r^6$ .Also, higher the strength of the stimulus , the lower would be its quantity required to effect the coagulation and hence the coagulation threshold values for the mono, di, trivalent atoms must be inversely proportional to r^6 as justified by the Deryagin-Landau equation(49) . It will also be seen that for a given valency(requiring the same number of contact areas as explained above) , each area in itself will increase as the radius of the atom is increased and hence finally the energy of activation (of coagulation) must also be increased.

By virtue of the $p+$ charge character of the basic CH_2 radical of an organic substance , these molecules will be highly $p+$ in character. . Thus their buffer coalescences leading to the chain growth is always much easier. On the other hand, for inorganic compounds, such a facility isnot provided. On this score, the organic ions introduced into the sol will provide easy nucleation centers for rapid coagulation as observed otherwise.

The commencement of the coagulation in a given sol is due to the tearing away of the individual particles along their buffer contact planes . We observe that as each particle is associated with a basic charge in it, with the coalesced buffer extending all around the environment of an electric double layer is duplicated. Hence as the particle tends to move, it has to establish its slipping plane at about 2R where R-is the radius of the atoms of the particle (cf: companion paper on colloids). In as much as the zone of interaction of two neighbouring particles will correspond to 2.5 R these two effects being present simultaneously,will tend to rupture the coalesced buffer at a mean of 2.25R . Then the zeta potential corresponding to this will be around $400/(2.25)^2 = 35$ mV as reported. In view of the fact that the potential of the double layer in itself can vary from about 400-2000 mV(depending upon the accumulation of charges like the electron and the free energy clouds etc., ; cf ; companion paper on colloids), the reported wide variation in the above value will be justified. Clearly, for very low zeta potentials at the shear plane , it implies that the interacting energy clouds are weaker (so that they produce only a lower double layer potential) and therefore they will not be successful in generating

228

the shear. Hence there cannot be any coagulation and in fact, the buffer of such a system will have to be only highly stabilized(as there is no tendency for the shear at all) . In as much as the coagulating ion has the same type of buffer environment, it can also undergo density enhancement effect leading to aggregation and consequent coagulation so that such ions will always be precipitated along with the particles of the sol.

In connection with the concept of the coagulation process, we had already found (cf: companion paper on Aerosols) that the conventional theories cannot directly identify the role of the buffer clouds of particles. Hence to evoke balance between sedimentation stability and coagulating tendency , mutually opposing forces for the given particle in its surrounding must have to be presumed as found in the DLVO-theory. Since the modeling envisages almost the same behaviour of the buffer in equivalent terms, its applicability to a few cases will become obvious.

By virtue of a weakly charged nature , ions will tend to possess only a correspondingly weaker buffers . Such ions present in a sol will have buffer interaction simulated in the usual way but at a very slow rate . Aggregation of individual particles can therefore definitely occur but at much slower pace. Coagulation will therefore follow which is otherwise identified as the neutral coagulation effect. It must be seen that this is presumed as due to reduced interaction (repulsive) forces arising out of the weakly charged nature of the particle. But the exact mechanism will be only as stated above. On this score, we notice that strongly charged ions will have stiff buffers and tend to remain stable but are not permitted to do so by crowding them otherwise with the introduction of higher concentration of the indifferent electrolytes (which now tend to stress each buffer to a greater measure so that they are actually brought in the coagulation)as found in the concentration coagulation.

The adsorption of an ion (with specified buffer field)will mainly depend on the polarity of the surface atom as we had already seen (cf: companion paper on Aerosols).Hence when the polarities of the adsorbed and adsorbing atoms involved are of opposite sign, better interaction can always occur between their buffers. On this basis, the counterions of opposite polarity tending to adsorb better on the given surface atoms will become clear. For the counterions of the same polarity as that of the surface ions, it will then apparently appear that adsorption of the former on the latter will be highly reduced. However, the spectral absorption of the two atoms will not always be the same and therefore the nature of such absorbed clouds can be either of the same polarity (in spectral region as UV or IR) but with differing energy values (since the absorption wave lengths differ) or of opposite character. This implies that ultimately the effective buffer of the adsorbing atom will always be relatively different with respect to the adsorptive atom(for example, if two atoms like oxygen and hydrogen had absorption as 2.5p- and 3.0p- respectively, then, the hydrogen is still relatively more -ve with respect to the oxygen and hence the two will tend to attract to form water). This is how the concept of the strong adsorbability of a given ion of the same polarity as that of the adsorber comes into the observation.On this basis, by suitably altering the polarity and strength of the buffer at the adsorbing surface it becomes logical that we could promote adhesion by atoms of widely differing polarity and strength.Hence when the hydrophobic nature of arsenic sulfide is considered, its buffer can be moderated by the introduction of the non-iongenic surfactants (of opposite character of buffer due to their spectral absorption)so that the resultant surface of the arsenic sulfide is made hydrophillic in character . Hence the improvement in stability of the AgI particles (when provided with a monolayer of high p+ of the polyethylene glycol tending to repel each other and hence discourage coagulation) will also be evident.

The controversies cited in the literature on the combining nature of the ions (5,46,49) , arise primarily because, by conventional angles , one has no way of perceiving the nature of the combining characteristics of the ions. Hence if one were to go by the overall charge of the anion or cation, it should appear as logical for the anion to always join with a cation and not with an ion of same sign. Hence data indicating the latter case(of combining of ions of same sign) when made available , it may appear as highly contradictory. However, in the light of the GET, it will now become clear that this situation is perfectly understandable from the above analysis.

The various phenomena such as the irregular series , antogonism, synergism and accustoming etc., will all be due to the outcome of the buffer interactions. Considering the first, let us presume that a given sol particles had a basic p+ characteristics and its absorption corresponding to p+ . Clearly their mutual buffer repulsion would tend to discourage coagulation and hence the sol will remain stable when we added electrolytes of opposite (p-) character till some p - is reached, we find that the above repulsive forces are gradually reduced ($|p_1+| \doteq |p-|$) , when due to the interaction forces(D= 2.5R)the sol particles will coagulate. When the p- is still increased, beyond a certain addition, the sol particles become all p- so as to exhibit mutual repulsion and remain as clear colloid. Finally, when the p- is so high that coagulation (precipitation) sets in more by a heavier concentration of the electrolyte (and consequent compaction of individual particles due to crowding). In this connection, it must be seen that the addition of any amount of acid (p-) to a sol containing -ve particles cannot bring in reversal of polarity .But when an alkali (p+) is added, the reversal(of the buffer charge of the particle) is easy. If however, the pH of the base is high(very high p+),the electrolyte molecules will experience repulsive forces in among themselves to such an extent that they cannot come closer together (and all around) near the main colloidal particle so as to interact with it and there by modify the charges. It can be seen in this respect, that the potential determining ions will also behave identically as above.

When two electrolytes involved had almost neutral buffers(equal

spectral absorption in the UV and IR regions) their addition to the
sol particles will bring effects that will be independent of one
another. In this way, the additive property is established. Buffers
of the same character , when present simultaneously , would tend
to counteract with one another simulating antagonism, while buffers
of opposite character will aid each other simulating synergism. The
nature of the +ve and -ve accustoming also would arise because of
buffer characteristics only. Consider the addition of a fixed amount
of electrolyte to the sol. In as much as the addition can be presumed to be made of a total of N drops , then each drop can be
assumed to be uniformly spread out from its neighbour so as to influence the sol particles surrounding it in an identical manner. In
this way, each electrolyte molecule , left to itself freely, would
gradually give off its energy starting from its initial
spectral absorption value of I_0 so that with time, its effective
intensity is only I and where $I < I_0$. On this score, if the addition
of the molecule of a given electrolyte occurs at a slow rate then
each may have only an interaction corresponding to I only and at
this rate a certain minimum quantity of the coagulant will be needed.
This act is independent of the character of the charge on the sol
and electrolyte molecules . However when specifically the two particles having the same type of charge are involved , sudden addition
of the electrolyte into the sol would involve a greater attraction
(due to the spectral charges available being greater than I but
less than I_0) and hence the quantity of electrolyte required for coagulation will be lesser than for the previous case. Clearly this
is +ve accustoming. But if the two particles involved had the same
types of charge, sudden addition would enhance repulsion and hence
bringing about coalescence can be done only at the cost of crowding
which in turn is brought about by excessive addition of the electrolyte . Then this is the so-called -ve accustoming.

The colloidal particle sheath generated by the thin shell of
the high molecular weight polymers follows directly from the fact
that generally such substances are all highly p+ and so would
promote mutual repulsion between sol particles coated with them.

Organosols which are highly p+ in character will generally
have a lower dielectric strength. This is due to the fact that being
located in a terrestrial environment which is highly p- , the p+
substances would require greater p+ excitation (as no part of p+
can come from the background) to pump a given amount of p+ energy
particles into them so as to produce the requisite field equivalent
to electronic expulsion or voltage generation(electrostatic effect).
On the other hand, the p- substances by virtue of their having
already absorbed a high amount of p- from the terrestrial background , would require only a lower amount of p- excitation to
generate the same current in the materials will manifest itself as
the dielectric constant and clearly for organic substances (p+)
the dielectric constant must be lower than for a substance like
water(p-). Hence the organosols being differentiated by buffer
considerations only from those of the hydrosols should still exhibit
the properties of the latter but to a lesser extent as reported. On

this basis, the behaviour of the mutual coagulation exhibited by two electrolytes will also become clear.

Coagulation under the influence of physical factors like stirring etc., will all involve in attempts to coalesce the buffers by forcing their interaction. The action of light(white and UV) has already been shown to account fully for the coagulation. The buffer coalescence on long-term storing will also be due to the gradual absorption of the light. The action of the electric field will naturally be such as to highly orient the p+ and p- buffers of particles (generating dendrite like structures) and thereby bring about the coagulation. Concentration must necessarily bring in more effective buffer interaction(by reducing intermolecular distances). But when hydrosols(p-) are heated or even boiled, the contribution of the heat photons(p+) will be only to neutralise so as to marginally reduce the overall buffer charges but expand their envelopes thereby reducing the tendency on the hydrosol stability only marginally.

8. CONCLUSIONS :

From the foregoing analyses, it will be found that the principles of the GET can be greatly applied to study any of the complex behaviour of colloidal solutions. In as much as the GET always permits quantitative computations to be made in easy terms , it will also be clear that such detailed investigations can now be made according to the requirements.

9. REFERENCES :

1 . Bockris,J.O'M., and Srinivasan,S., Fuel cells: Their Electro-chemistry., McGraw Hill Pub.Inc., New York, 1969.

2 . Clark, John,B., Mathematical and physical Tables, Oliver and Boyd Co, London.

3 . Coulson,J.M, and Richardson, J.F., Chemical Engineering, volume III, Pergoman Press, New York, 1971.

4 . Frederick Seitz., Modern Theory of Solids, Mc Graw Hill Co, New York, 1940.

5 . Glaston,S and Lewis,D., Elements of physical chemistry, 2 nd Ed, Mac Millan Co , London, 1970.

6 . Keith J. Leidler., Physical chemistry with Biological Applications, The Benjamin Cumminng Pub. co, california,1980.

7 . Louis,F. Fieser and Mary Fieser., Organic chemistry, Reinhold Pub.co, New York, 1956.

8 . Max Born, Atomic physics, Blackie and sons Ltd, London, 1963

9 . Parthasarathy,K., The General Energy Theory of Matter, part I, journal of steam and fuel users association of India, volume 31, no 1, pp 46-50, 1981.

10. Parthasarathy,K., The General Energy Theory of Matter, part II, journal of steam and fuel users association of India, volume 31, no 2, pp 43-50, 1981.

11. Parthasarathy,K., The General Energy Theory of Matter,part III, Application to energy crisis,National seminar, journal of steam and fuel users association of India, volume 31, no 3, pp 15-34 , 1981.

12. Parthasarathy,K., The General Energy Theory of Matter, part IV, journal of steam and fuel users association of India, volume 31, no 4, pp 12-47 , 1981.

13. Parthasarathy,K., The significance of the General Energy Theory in relation to the development of modern sciences,journal of the Institution of Engineers (India), Neyveli chp, annual number, pp 35-38, 1982-83 .

14. Parthasarathy,K., The application of the General Energy Theory to the studies of the problems of DNA synthesis, journal of the Annamalai University Agriculture Research Association of India, 1985.

15. Parthasarathy,K., The application of the General Energy Theory to the studies of the problems of photosynthesis,journal of the Annamalai University Agriculture Research Association of India, 1985.

16. Parthasarathy,K., The identification of the petroleum zones of the earth using the General Energy Theory of Matter, 7 th International conference on Alternative energy sources, University of Miami, Florida, USA, Dec, 1985 , Ed T. Nejat Veziroglu., Alternative Energy sources VII , hydrocarbons/ energy transfer, volume 5, pp 27-50, Hemisphere Publishing co, New York, 1987.

17. Parthasarathy,K., The application of the General Energy Theory to the evaluation of the coal conversion processes in relation to the natural oil resources, International conference on Development of Alternative Energy sources and the lessons learned since the oil embargo, Great Plains Forum on energy and minerals management and policy analyses, University of North Dakota, USA, May, 1986.

18. Parthasarathy,K., The application of the General Energy Theory to some theoritical aspects of the problems of Geothermal energy production, International conference on Development of Alternative Energy sources and the lessons learned since the oil embargo, Great Plains Forum on energy and mineral management and policy analyses, University of

North Dakota, USA, May , 1986.

19. Parthasarathy,K., The application of the General Energy Theory
    to the studies of chemistry: estimation of chemical bond
    lengths, 23 rd Annual convention of Chemists(Indian
    Chemical society) India, Dec, 1986.

20. Parthasarathy,K., The application of the General Energy Theory
    to the studies of Electrochemistry, First National semi-
    nar of the Electrochemical Teachers of India, The Central
    Electrochemical Research Institute, Karaikudi, India,
    Dec, 1986.

21. Parthasarathy,K., The application of the General Energy Theory
    to the studies on the physical and chemical realiites
    of some Thermal properties and the specific heats of
    atoms, 4 th International symposium on Multiphase Trans-
    port and particulate phenomena, University of Miami,
    Florida, USA, Dec, 1986, Ed- T. Nejat Veziroglu, parti-
    culate phenomena and multiphase transport, volume 5, pp
    473-501, Hemisphere Publishing co, New York, 1988.

22. Parthasarathy,K., The application of the General Energy Theory
    to the deductions of the fundamental proofs of Thermody-
    namic laws and the Onsager relations , 4 th International
    symposium on Multiphase Transport and particulate pheno-
    mena, University of Miami, Florida, USA, Dec, 1986, Ed-
    T.Nejat Veziroglu, particulate phenomena and multiphase
    Transport, volume 5, pp 503-533, Hemisphere Publishing
    co, New York, 1988.

23. Parthasarathy,K., The application of the General Energy Theory
    to some problems of future energy sources : DST National
    cum futurology workshop on Technology towards 21 st
    century-forecasting and assessment, University of
    Jodhpur, India, April, 1987.

24. Parthasarathy,K., The application of the General Energy Theory
    to the studies of the problems of the energy sources
    and energy conversion, National symposium on Electro-
    chemicals, Energy and Industry, University of Rajasthan,
    India, April, 1987.

25. Parthasarathy,K., The application of the General Energy Theory
    to the studies of the problems of the energy sources
    and energy conversion,some fundamental aspects of the
    creation of the structure of the chlorophyll and its
    photosynthetic activity, National symposium on Electro-
    chemicals, Energy and Industry, University of Rajasthan,
    India, April, 1987.

26. Parthasarathy,K., The application of the General Energy Theory
    to the studies on Alternative Energy sources:A new
    approach to the utilisation of the solar heat of the

234

deserts for domestic purposes, Fifth All India conference on Desert Technology, Jndian Society of Desert Technology, India, Oct,1987.

27 .Parthasarathy,K., The application of the General Energy Theory to the studies of solar energy utilization: some fundamental aspects of the various photovoltaic cells, International symposium-workshop on silicon technology, development and its role in the sun belt countries, National Institute of silicon Technology, Pakistan, june,1987.

28. Parthasarathy,K., The application of the General Energy Theory to the studies of Electrochemistry: some fundamental aspects of the electroorganic chemistry, National symposium on Electrochemicals, The central Electrochemical Research Jnstitute, Karaikudi, India, july, 1987.

29. Parthasarathy,K., The application of the General Energy Theory to the studies of Electrochemistry: some fundamental aspects of the behaviour of catalysts in chemical reactions, National symposium on Electrochemicals, The Central Electrochemical Research Institute, Karaikudi, India, july, 1987.

30. Parthasarathy,K., Ashok kumar,P., and Nirmala,P., The application of the General Energy Theory to the studies of Biochemistry: some fundamental aspects of the creation and dynamics of human cells, 8 th Annual conference on Environment and Health, The Indian Association of Bio-Medical scientists, India, Oct,1987.

31. Parthasarathy,K., The application of the General Energy Theory to the studies of solar energy utilisation: The genetical aspects of the solar radiations and its influence on the energy transactions of terrestrial chemical elements, 8 th International conference on Alternative energy sources , University of Miami, Florida, USA, Dec, 1987.

32. Parthasarathy,K., The application of the General Energy Theory to the studies of Electrochemistry: some fundamental aspects of the structure of the double layer,electrosorption and electrocatalysts, The CECRI Research conference on Electrocatalysis, The Central Electrochemical Research Institute, Karaikudi, India, Jan, 1988.

33. Parthasarathy,K., Tha aaplication of the General Energy Theory to the studies of Electrochemistry: some fundamental aspects of physical and chemical characteristics of the electrolytes used in the fuel cells, International symposium on Batteries and Fuel cells, The Central Electrochemical Research Institute, Karaikudi, India, Feb, 1988.

34. Parthasarathy,K., The application of the General Energy Theory

to the studies of Electrochemistry: some fundamental
aspects of the energy transfer modes in fuel cells,
International symposium on Batteries and Fuel cells,
The Central Electrochemical Research Institute, Karaikudi, India, Feb, 1988.

35. Parthasarathy,K., The application of the General Energy Theory
to the studies of Biology and Biomedicine: some aspects
of the magnetic field interaction in Human cells,
National seminar on the effects of the Electromagnetic
fields on the Biological systems, The Madras Institute
of Magnetobiology, India, Feb, 1988.

36. Parthasarathy,K., The application of the General Energy Theory
to the studies of the problems of energy sources and
energy conversion: The genetical aspects and physical
limitations of the terrestrially available major energy
sources, The second International congress on energy,
Tiberias, Israel, june,1988.

37. Parthasarathy,K., The application of the General Energy Theory
to the studies of solar energy utilisation: some fundamental aspects of the mechanism of the radiation damage
in solar cells and associtaed phenomena, Cairo International symposium on renewable energy sources, cairo,
Egypt, june,1988.

38. Parthasarathy,K., The application of the General Energy Theory
to the studies of Electrochemistry: An insight into the
phenomenon of superconductivity in metals, National
symposium on Electrochemical material sciences, The
Central Electrochemical Research Institute, Karaikudi,
India, Nov,1988.

39. Parthasarathy,K., The application of the General Energy Theory
to the studies of Electrochemistry: An insight into the
phenomenon of electrolytic conduction with reference to
the electrodeposition of metals, International symposium
on Inductrial metal finishing, The Central Electrochemical Research Institute, Karaikudi, India, Feb,1989.

40. Parthasarathy,K., The application of the General Energy Theory
to the studies of Electrochemistry: some fundamental
studies on the elctrodeposition of alloys, The Central
Electrochemical Research Institute, Karaikudi, India,
Feb, 1989.

41. Parthasarathy,K., The application of the General Energy Theory
to the studies of petrolcum and coal sciences : some
evaluation of the current techniques of coal to oil
conversion, Manila International symposium on the development and management of energy resources, Manila,
Philippines, jan, 1989.

42. Parthasarathy,K., The application of the General Energy Theory
    to the studies of Electrochemistry: some fundamental
    aspects of the formation and behaviour of typical mem-
    branes , 6 th National conference on advances of sci-
    ence and technology of membranes, The membrane society
    of India, Feb, 1989.

43. Parthasarathy,K., The application of the General Energy Theory
    to the studies of chemical engineering: some fundamental
    aspects of the behaviour and characteristics of the
    colloids, 5 th International symposium on multiphase
    transport and particulate phenomena, University of
    Miami, Florida, USA, Dec, 1988.

44. Parthasarathy,K., The application of the General Energy Theory
    to the studies of chemical engineering: some fundamen-
    tal aspects of the properties and dynamics of aerosols,
    5 th International symposium on multiphase transport
    and particulate phenomena, University of Miami, Florida,
    USA, Dec, 1988.

45. Parthasarathy,K., The application of the General Energy Theory
    to the studies of chemical engineering: some fundamental
    aspects of the theory of sedimentation, 5 th Internat-
    ional symposium on multiphase transport and particulate
    phenomena, University of Miami, Florida, USA, Dec, 1988.

46. Samuel D.Foust and Joseph V. Hunter,,principles and applica-
    tions of water chemistry, John Wiley, New York, 1967.

47. Sienko and Plane., Chemistry: principles and properties,
    McGraw Hill co, New York, 1966.

48. Theodore Cutting., manual of spectroscopy, chemical publishing
    co, New York, 1949.

49. Voyutsky,S., colloid chemistry, Tr.by Nicholas Bobrov, Mir
    publishers, Moscow, 1978.

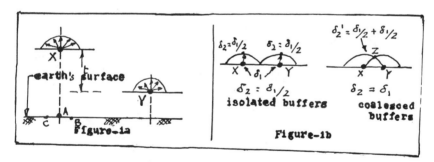

The Use of Three Parameter Rheological Model to Describe Annular Turbulent Drilling Suspensions Flow

C. C. SANTANA and M. A. CREMASCO
Chemical Engineering Department
State University of Campinas
SP, Brazil

Abstract

This paper analyzes the friction coefficient in the laminar, transition and turbulent annular flow of drilling fluids which exhibit a non-newtonian behavior with an yield stress. The Robertson and Stiff model is used in this work to obtain a generalized Reynolds Number for the empirical correlation of experimental data covering values of Fanning friction factor from 0.007 to 0.05 and generalized Reynolds number from 350 to 15500.

A comparison is made between the predicted and experimental pressure drop showing the good accuracy of the calculation procedure.

1. INTRODUCTION

The description of shear stress versus shear rate data for the laminar flow of fluids that exhibit an yield stress as well as non linear behavior can be made with success through the use of the three-parameter model proposed by Robertson and Stiff[1] of the form

$$\tau = A \ (\dot{\gamma} + C)^B \tag{1}$$

The evaluation of this model for a variety of fluids have been made by Robertson and Stiff and also by Beirute and Flumerfelt[2] and Santana et al.[3]. All these evaluations showed consistently superior results of this model in comparison to the Bingham and Power Law rheological data fitting for several drilling fluids and cement slurries, pointing in this way to a new trend that improves the accuracy of pressure drop predictions for laminar, transitional and turbulent flow. This model is now being spread out in the oil well drilling literature as shown by Whittaker[4].

Otherwise, several experimental data for laminar, transitional and turbulent annular flow have been published by Langlinais et al.[5], with the evidence of some poor results with the use of the parameters of classical models for Bingham and Power Law fluids inserted in correlations to predict the pressure drop, especially in the turbulent flow.

239

The present work treats the Langlinais et al. experimental data through an adimensional approach aiming the improvement of pressure drop predictions.

2. PRESSURE DROP ON LAMINAR FLOW

The correct development of relations describing velocity fields and wall shear rates for the Robertson and Stiff model both for tubes and annuli was carried out in Reference 2, resulted in the following set of equations for flow rate / pressure drop relationship:

PIPE FLOW (figure 1):

$$Q = \pi \left\{ \left(\frac{1}{2A} \left(\frac{\Delta P}{L} \right) \right)^{\frac{1}{B}} \left(\frac{B}{3B+1} \right) \left(R^{\frac{3B+1}{B}} - \lambda^{\frac{3B+1}{B}} \right) - \frac{1}{3} C \left(R^3 - \lambda^3 \right) \right\} \quad (2)$$

$$\lambda = \frac{(A\,C^B)}{\frac{\Delta P}{L}} \quad (3)$$

In equations (2) and (3) the parameter λ is concerned with the plug flow part of the velocity profile shown on Fig. 1.

NARROW ANNULI FLOW (figure 2 - parallel plates approximation)

$$Q = 2b \left\{ \left(\frac{1}{A} \frac{\Delta P}{L} \right)^{\frac{1}{B}} \left(\frac{B}{2B+1} \right) \left(R^{\frac{2B+1}{B}} - \lambda^{\frac{2B+1}{B}} \right) - \frac{1}{2} C \left(R^2 - \lambda^2 \right) \right\} \quad (4)$$

If Eq.(3) is substituted in Eq.(4) we obtain an implicit equation for $\frac{\Delta P}{L}$ in terms of Q. Only for those cases where $\lambda = 0$ (fluid with no yield stress or the approximation of very high pressure drop), Eqs. (2) and (4) can take the forms proposed originally by Robertson and Stiff as follows:

For pipe flow :

$$Q = \pi \left\{ \left(\frac{1}{2A} \left(\frac{\Delta P}{L} \right) \right)^{\frac{1}{B}} \left(\frac{B}{3B+1} \right) R^{\frac{3B+1}{B}} - \frac{1}{3} C R^3 \right\} \quad (5)$$

240

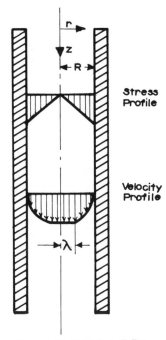

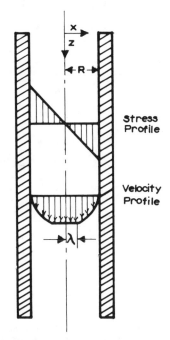

FIG.1- FLOW IN A PIPE

FIG. 2- FLOW BETWEEN PARALLEL PLATES

and for annuli, with the slot approximation:

$$Q = 2b \left\{ \left[\frac{1}{A} \frac{\Delta P}{L} \right]^{\frac{1}{B}} \left[\frac{B}{2B+1} \right] R^2 - \frac{1}{2} C R^2 \right\} \tag{6}$$

The main advantage for the use or Eqs (5) and (6) as an approximation for small values of λ is due to the explicits forms that can be obtained for $\frac{\Delta P}{L}$ in terms of Q, although this is not a difficult problem with the nowadays computation devices.

3. ADIMENSIONAL FORM AND GENERALIZED REYNOLDS NUMBER

Since the intent is extend the studies for the annuli beyond laminar flow, it is useful to modify Equation (6) in order to obtain adimensional forms. In this way, introducing the relations for the wall shear stress τ_R and the average velocity $\overline{V}$ given in terms of an equivalent diameter D_{eq} results:

$$\tau_R = \left(\frac{\Delta P}{L} \right) \frac{D_{eq}}{4} \tag{7}$$

$$\overline{V} = \frac{Q}{D_{eq} \, b} \tag{8}$$

we have for the annuli, from Eq. (4):

$$\tau_R = A \left\{ \frac{\frac{2B+1}{B} \left\{ \frac{4\,V}{D_{eq}} + \frac{1}{2} C \left[1 - \left(\frac{4\,\lambda}{D_{eq}} \right) \right]^2 \right\}}{1 - \left[\frac{4\,\lambda}{D_{eq}} \right]^{\frac{2B+1}{B}}} \right\}^B \tag{9}$$

Introducing the Fanning factor definition and imposing the classical laminar flow form as lollows:

$$f = \frac{\tau_R}{\frac{1}{2} \rho \, \overline{V}^2} \tag{10}$$

$$f = \frac{24}{Re_{RS}} \tag{11}$$

242

we can obtain the Generalized Reynolds number

Re_{RS} for the Robertson and Stiff Model :

$$Re_{RS} = \cfrac{12 \, \rho \, \overline{V}^2}{A \left\{ \cfrac{\dfrac{2B+1}{3B} \left[\dfrac{12 \, \overline{V}}{Deq} + \dfrac{3}{2} \, C \left(1 - \left(\dfrac{4 \, \lambda}{Deq} \right)^2 \right) \right]}{1 - \left(\dfrac{4 \, \lambda}{Deq} \right)^{\frac{2B+1}{3B}}} \right\}^B} \qquad (12)$$

A simpler form for the Re_{RS} for annular flow is,   obtained
with the approximation $\lambda = 0$

$$Re_{RS}^* = \cfrac{12 \, \rho \, \overline{V}^2}{A \left\{ \dfrac{2B+1}{3B} \left[\dfrac{12 \, \overline{V}}{Deq} + \dfrac{3}{2} \, C \right] \right\}^B} \qquad (13)$$

4.   ANALYSIS OF EXPERIMENTAL DATA AND CONCLUSIONS

Using the four Fann Rheometer readings reported by Langlinais
et al. it is possible to treat the data with the general   rotary
rheometer equations presented by Van Wazer [6]   to   obtain   the
rheological parameters for three different clay muds named Mud 1,
Mud 3, and Mud 4, fitting the Bingham, Power Law and   Robertson
and Stiff models. The results are summarized in Table I, together
with the mud densities. The absolute mean deviations   S   shown
also in Table I shows the best fitting of Robertson   and   Stiff
model especially for the more viscous fluid. This fact   agrees
with the previous results of references 2 and 3 for other   types
of fluids.

For the study of the annular frictional pressure   gradients
including the laminar transitional and turbulent flow   it   is
important to plot the experimental data for the Fanning friction
factor in terms of the Generalized Reynolds Number Re_{RS}.   Aiming
to identify the best performance of the several   equivalent
diameter computing techniques suggested by the literature,   the
values of Re_{RS} given by Eq. (12) were calculated embodying   the
following three definitions:

HIDRAULIC RADIUS CRITERIA :

$$(Deq)_1 = D_o - D_i \qquad (14)$$

243

TABLE I

PROPERTIES OF CLAY - WATER MUDS

MUD PROPERTY	MUD 1	MUD 3	MUD 4
Density, ρ (g/cm³)	1.030	1.060	1.060
Plastic viscosity, μ_p (cp.)	5.12	7.89	13.83
Yield Stress, τ_y (dyne/cm²)	3.26	6.12	13.38
Flow Behavior Index, n (adim.)	0.825	0.827	0.765
Consistency Index, k (eq. cp.)	18.19	28.6	77.38
Parameter A (eq. cp.)	0.0961	0.2996	0.2541
Parameter B (adim.)	0.9144	0.8210	0.9193
Parameter C (s^{-1})	40.3064	2.8414	77.6921
Absolute mean deviation S for Bingham model	1.89%	3.22%	1.11%
Absolute mean deviation S for Power Law model	2.01%	0.63%	2.22%
Absolute mean deviation S for Robertson and Stiff model	1.58%	0.64%	0.36%

$$S = \left(\frac{\Sigma x^2 - \left(\frac{\Sigma x}{n} \right)^2}{n - 1} \right)^{1/2}$$

where $x = \dfrac{\tau \text{ CALC.}}{\tau \text{ EXP.}}$

n = number of exp. data

$$(D_{eq})_2 = 0.816 \ (D_O - D_i) \tag{15}$$

CRITENDON CRITERIA :

$$(D_{eq})_3 = \frac{1}{2} \left(D_O^4 - D_i^4 - \frac{\left(D_O^2 - D_i^2 \right)}{\ln \left(\frac{D_O}{D_i} \right)} \right)^{\frac{1}{4}} + \frac{1}{2} \left[D_O^2 - D_i^2 \right]^{\frac{1}{2}} \tag{16}$$

The typical adimensional plot of flow data obtained by Langlinais et al.is depicted in Figs. 3,4 and 5, being each graph for one of the three equivalent diameter definitions. From these Figures the following comments may be done:

a) For $Re_{RS} \leq 2800$ (laminar flow), the pressure drop may be obtained from the set of equations (11), (12) and (3). A comparison between the computed and experimental results leads to the conclusion that the laminar flow slot equivalent diameter (Eq. 15) gives the best fitting for the experimental data tested.

b) A transition between laminar and turbulent flow is evident in the range of Re_{RS} from 2800 to 5000. This region is characterized by a scatter of data arround the prolongation of the turbulent proposed fitting.

C) For the turbulent regime (Re > 5000) we can propose the following Blasius type equation:

$$f = \frac{0.079}{Re_{RS}^{0.26}} \tag{17}$$

Equation (18) fits the experimental data for annular flow of Muds 1 and 3 with a maximum absolute mean deviation of 15% using the equivalent diameter given by Eq. (15).

For Mud 4 is evident that a true turbulent regime was not reached with the available experimental data. A correlation obtained with the transitional data that can be carefully extended for the turbulent regime:

$$f = \frac{0.079}{Re_{RS}^{0.30}} \tag{18}$$

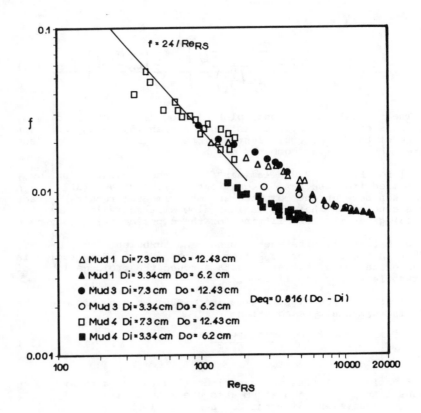

Figure 3: Adimensional plot and correlation for the friction factor with the equivalent diameter given by the flow slot approximation

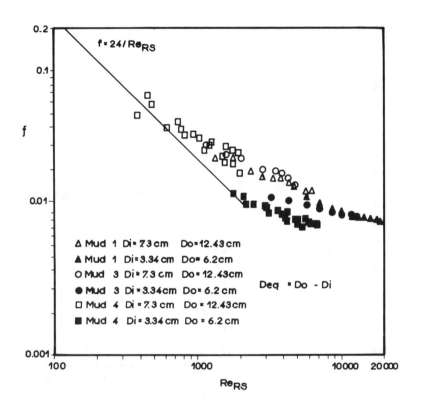

Figure 4: Adimensional plot for the friction factor with
the equivalent diameter given by the hidraulic
radius criteria

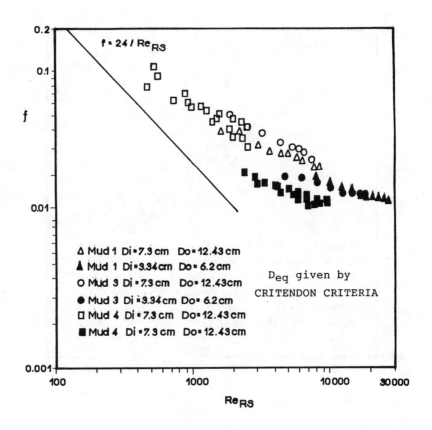

Figure 5: Adimensional plot for the friction factor with
the equivalent diameter given by the Critendon
criteria

248

A direct comparison between calculated and experimental pressure drop is depicted in Figures 6,7,8 and 9.

A good estimative of the transitional velocity V_T is obtained making Re_{RS} from Eq (14) equal to 2800, resulting the transcendental Equation

$$\left(\frac{12 \, \rho}{2800 \, A}\right)^{1/B} V_T^{2/B} - \left(\frac{2B+1}{3B}\right) \frac{12}{D_{eq}} V_T - \frac{3}{2} C = 0 \qquad (19)$$

The Eq. (19) can be solved easily by numerical methods like the Newton-Raphson procedure.

5. CONCLUSIONS

From these results it is quite evident that the experimental data treatment outlined in this work enables a good accuracy of the pressure drop for annular flow of drilling fluids, reducing the flow data to an adimensional plot that can be generalized like Dodge and Metzner [8] proceeded with the power-law fluids.

6. NOMENCLATURE

A   - Rheological fluid parameter, Eq. (1)
b   - Width of paralell plates
B   - Rheological fluid parameter, Eq. (1)
C   - Rheological fluid parameter, Eq. (1)
D_{eq} - Equivalent Diameter, Eqs. (14), (15) or (16)
D_i - Inner diameter of annulus
D_o - Outer diameter of annulus
f   - Fanning friction factor, Eq. (10)
L   - Tube lenght
P   - Pressure
Q   - Volumetric flow rate
R   - inner radius of pipe or one-half of gap between paralell plates
Re_{RS} - Generalized Reynolds Number, Eq. (12)
$\overline{V}$ - Area mean value of velocity
V_T - Transition velocity

Greek Symbols

$\dot{\gamma}$ - Velocity gradient
λ - Parameter of velocity profile, Eq. (3)
ρ - Slurry density
τ - Shear stress

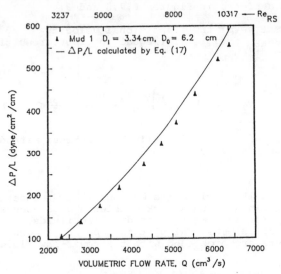

Figure 6: Comparison between calculated and experimental
pressure drop for Mud 1
(D_i = 3,34 cm , D_o = 6,2 cm)

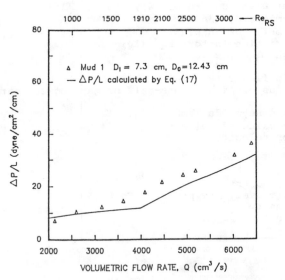

Figure 7: Comparison between calculated and experimental
pressure drop for Mud 1
(D_i = 7,3 cm , D_o = 12,43 cm)

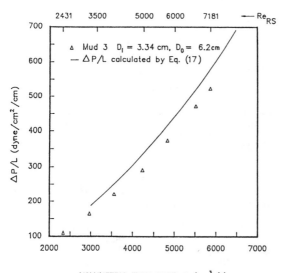

Figure 8: Comparison between calculated and experimental
pressure drop for Mud 3

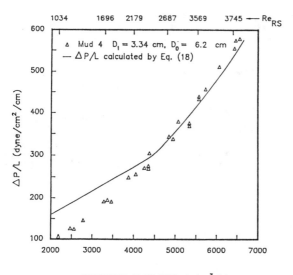

Figure 9: Comparison between calculated and experimental
pressure drop for Mud 4.

251

Acknowledgements

The authors wish to acknowledge Dr. Julius P. Langlinais from Louisiana State University for the valuable discussions and use of his experimental data. We are also grateful to CAPES and Fundação de Apoio à Pesquisa do Estado de São Paulo (FAPESP) for the research funds.

REFERENCES

1. Robertson, R.E. and Stiff, H.A., An Improved Mathematical Model for Relating Shear Stress to Shear Rate in Drilling Fluids and Cement Slurries, Society of Petroleum Engineers Journal pp. 31-36. February, 1976.

2. Beirute, R.M. and Flumerfelt, R.W., An evaluation of the Robertson-Stiff Model Describing Rheological Properties of Drilling Fluids and Cement Slurries, Society of Petroleum Engineers Journal pp. 97-100, April, 1977.

3. Santana, C.C., Ataíde, C.H. and D'Ávila, J.S., Critical Velocity and Turbulent Flow of Suspensions Using a Three--Parameter Theological Model, In: Particulate and Multiphase Processes, Vol.3, Eds. T.Ariman and T.Nejat Veziroglu, Hemisphere Pub. Co., pp. 225-232, 2987.

4. Whittaker, A., Ed., Theory and Application of Drilling Fluid Hidrulics, The EXLOG Series of Petroleum Geology and Engineering Handboods, D.Reidel Pub.Co., Boston, 1985.

5. Langlinais,J.P., Bourgoyne Jr., A.T. e Holden,W.R., Frictional Pressure Losses for the Flow of Drilling Mud and Mud/Gas Mixtures, SPE Paper 11993, October, 1983.

6. Van Waser, J.R., Kim, K.Y., Lyons, J.W. and Colwell, R.E., Viscosity and Flow Mensurement: A Laboratory Handbook of Rheology, Insterscience Publishers, N.Y., 1963.

7. Hanks, R.W., Low Reynolds Number Turbulent Pipeline Flow of Pseudohomogeneous Slurries, paper C2, Hidrotransport 5, pp. C2-23 - C2-34, 1978.

8. Dodge, D.and Metzner, A., Turbulent Flow of Non-Newtonian Fluids, "AICHE Journal 5, No. 2, pp. 189-204, 1959.

Studies Concerning Effective Utilization of Marshes of Rann-of-Kutch, Sind, Pakistan

P. HABIBULLAH and W. ALAM
E/M Engineering School
CATI, Pakistan

ABSTRACT.

Samples of the mud from marshes of Rann-of-Kutch, situated at 100 miles south-East of Hyderabad, Pakistan were taken and analysed at soil Research Laboratory of Agriculture Research Institute Tandojam, pakistan. According to the soil report the mud has clayey texture and moderate alkaline in nature. Sample of mud flaps from district of JATI contains total N = 0.0412%, P_2O_5 = 6 to 10.81 ppm K_2O=252 ppm, organic matters = 0.825%, T.S.S=2.96% and soils pH is 7.8 to 8.10. Composition of samples from Rahimki Bazzar varies largely and is given in the text.

Mineralogical analysis of the mud has been conducted at Zeal Pak Cement Factory, Hyderabad to know that this kind of marshes may be used as the replacement for the slurry used for manufacturing of cement. Principal mineralogical constitutents of these marshes are SIO_2=55 to 61.12%, Al_2O_3 = 10.13 to 18.57%, Fe_2O_3=1.87 to 5.25% Lime=8.46 to 8.76% Mgo=1.15 to 2.39% Sulphuric Anhydride 0.55 to. 1.92% chlorides = 1.70 - 1.94%.

During magnetic treatment of mud, Lorenze force acts on Mg^{++}, Ca^{++} Na^+, K^+, H^+, $sio3^{2-}$, $Si_2O^{2-}5$, OH^--radicals nd these ions may reach in the state of resonance. Thus mechanical properties e.g. compression strength and resisitance to abrasion, are considerably increased. These activated clays have extensive application in foundries, as a binder in the moulding sands.

253

1.   Introduction [5]

The Great Rann covers an area of about 7,000 sq mil (18,000 sq km) and lies almost entirely within Gujarat state, India, along the border with south of Pakistan. The little Rann of Kutch extends east and south from the Gulf of Kutch and occupies about 2,000 sq mi in Gujarat state.

Originally an extension of the Arabian Sea the Rann of Kutch has been closed off by centuries of silting. During the time of Alexander the Great it was a navigable lake, but it is now an extensive mud flat, inundated during monsoon seasons.

In 1965 a dispute arose about the boundary line between India and Pakistan toward the western end of the Great Rann. Fighting broke out in April between the regular forces of both countries, and ended only when Great Britain intervened to secure a cease-fire. On the report of the secretary-- general of the United Nations, who visited the scene, to the Secretary Council, the dispute was referred to an international tribunal. In 1968 it awarded about 10 percent of the border area of Pakistan and about 90 percent to India; the partition was effected in 1969.

Today the Rann of Kutch is a dry barren country which turns into swamps after rains, when water brought by Luni from the east, Puran from the north, and some stream of Kutch from the south fill it up, to the depth of a few feet. The water does not dry till November.

2.   History

Present investigations counducted by boring at Wirawah, Nagarpark (Rann of Kutch) have greatly helped in tracing the history of Rann of Kutch (See Map). The salts found in ground water were of the sea origin rather than calcareous nature. From this, it has been concluded that a branch of the Arabian Sea had been extended in the interior of Rajisthan, possibly along the Luni river, upto Panchabordra in the recent times, and probably upto Samber lake in the pre-historic period. An in land sea described in Mahabharata was probably this arm of Sea [11]

The silting of present Rann started in the recent times by both Hakra and Indus flowing into it. At the time of Alexander's conquest, according to Greeek writers, the Rann was a shallow sea. When mahmud Ghaznavi was returning from his march to Somanath and was pursuing a Hindu chief to the islands north-east of Kutch in 1006 A.D. He was told that sea waves would wash away his boats. This clearly shows that the Rann of Kutch was not dry in the early eleventh century. The town of Pari Nagr, a sea port in nagarparkar Taluka, was destroyed in 1226 A.D. This was the year when Hakra dried up near Umarkot. This also indicates that the Rann of Kutch was an arm of the sea and was fed by hakra. Feroz Shah Tughlak crossed the Rann in 1361 A.D. when it was dry. [11]

Balmir was a sea port on Luni River near Nagarparkar in historic times, which shows that in those days Luni was also navigable. Rann of Kutch, at one time a sea creek, was silted up slowly, possible due to silt brought by Luni from Rajisthan, and Hakra and Indus from the north.

3. Coastal Wetlands - Flux of Carbon, Nitrogen and Phosphorus

A variety of physical factors including the geomorphology of the marsh drainage, the areas of marsh and adjacent coastal waters and the magnitude of the water flux appear to be important determinants of whether specific wetlands show significant export or import of dissolved or particulate substances. Two studies appear to have particular relevance at this time in the progression of our understanding of the dynamics of coastal marshes. Both studies examine the nature and apparent origin of detritus exported from coastal salt marshes. harris (1980) noted that the suspended particulartes collected from ebb tide waters in a Florida coastal marsh are composed not of vascular plant fragments but of amorphous aggregates, derived primarily from organic films produced by benthic microflora. The second study, by haines (1977), examined the carbon isotope composition of seston in Georgia estuaries and concluded that this material si derived not from the vascular plants of the salt marsh but from algal production in the estuaries. This implies a minor contribution of marsh particulates to the organic carbon of coastal waters and suggested that important outflux may be in dissolved form.

Materials are brough to and removed from wetlands largely as a function of water movement, and the pattern of water movement is the primary determinant of the direction and thrust of nutrient flux in these complex systems. Nutrient discharge from the land tends to be correlated with water discharge (Brehmer, 1958; Kevern, 1961; vannote, 1961), and in most wetlands the major portion of the annual nutrient load anters in the spring.

The general pattern of reduced water discharge during the summer results in increased detention of water in wetlands during this period. However, water seldom runs through wetlands in a uniform manner, but is channeled such that the mean detntion time of the water varies greatly throughout the wetland. In those areas where the detention time is increased, the water temperature rises and pH increases because of changes in the carbonate--bicarbonate equilibrium caused by both the warming of the water and increased photosynthetic extraction of carbon dioxide from the alkalinity.

Sediments, and particularly inorganic sediments, play significant roles in the ability of wtetlands to retain phosphorus and heavy metals. The equilibrium adsorption capacity of sediments for such materials varies as an interactive function of both sediment type and chmical characteristics of the water. An example of these interactions is seen in the equilibrium phosphorus adsorption capacity of three different clays shown by Edzwald (1977) to vry as a function of clay type and water, Edzwald concluded that the type and amount of the free metal content of the clay played a major role in determining its phosphorus adsorption potential.

255

Nutrients cycle in wetlands as elsewhere as a function of biotic activity limited by interacting physical, chemical and biological factors. On the broad scale, macrophyte in wetlands varies as afunction of light, temperature and nutrient availability while the remainder of the community is limited by the production and introduction of organic carbon. Within these broad limits, however, biotic interaction can be alter the entire ecological structure of shallow water systems. Nutrient cycles in wetlands are controlled largely by chemical thermodynamics and mediated by biotic activity relative to the inputs of water and material. Seasonal variations in inputs of both water and materials in north temperate wetlands coupled with seasonal variation in light and temperature ensure marked seasonal variation in the nutrient dynamics of wetlands. Samples taken during the summer indicate that the wetland is a sink for many nutrients.

Samples taken during the summer indicate that the wetland is a sink for many nutrients while samples collected during the early spring would suggest that the wetland is a nutrient source. Some times even in a single season there will be a large variability in nutrient flux in most wetlands. For example, during the summer there is tremendous spatial variation in nutrient dynamics withing wetlands governed largely by interactions between plant photosynthesis and hydrology. The reasons for such variation can be illustrated with the effects on the nitrogen cycle caused by interactions between plant activity and the hydrology. (Fig. 1)

Ammonification is the process whereby nitrogenous compounds in plant and animal tissues are decomposed to produce ammonia which is changed by nitrification into nitrite, and then into nitrate, each stage being accomplished by specific microorganisms. The formation of ammonia is accomplished by heterotrophic bacteria but the two other stages are brought about by autotrophic bacteria. Ammonia is oxidised by Nitrobacter, nitrosomonas and Nitrococcus and the nitrite is oxidised by Nitrobacter. These processes require acrobic conditions; if the soil is saterlogged for any length of time the nitrogenous compounds are reduced by detitrification to nitrogen which is lost to the atmosphere.

Nitrogn ixation is the process during which soil bacteria take up nitrogen from the soil atmosphere to form their body protein. The organisms include Azotobacter, clostridium psteurianium and Beijeerinckia which upon death enter the nigrogen cycle and are decomposed to form nitrate for plant uptake.

There are also a number of bacteria which enter the roots of certain plants, particularly members of the Leguminoseae. There they multiply, form nodules and fix atmospheric nitrogen, which then passes into the conducting system of the plant as an essential element.

Bowden (1984) recently estimated the annual net ammonium production rate of a freshwater tidal marsh in Massachusetts to be about 1.7 mol $N/m^2/Yr$ (33.8 g $N/m^2/hr$) in the top 10 cm of sediment. King (1979), reporting on a series of two meter deep ponds in Michigan dominated by submerged macrophytes and charged with a good quality secondary effluent, noted a 97 percent nitrogen removal in a detention time of 120 days.

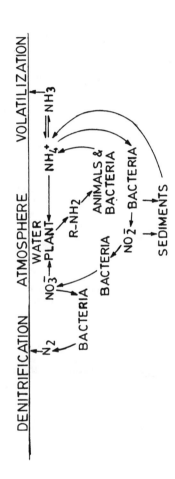

FIG. 1 NITROGEN CYCLE IN AQUATIC SYSTEMS

4.   Experimental:

4.1   Soil Analysis

Samples of marshes from Jati coastal areas (Musafirkhana) were taken from different areas and depths. They were mixed in equal ratios.

Samples were also taken from sandy, cleyey and gravelled strata of dried mud flaps near coastal areas of Rahimki Bazar.

All these samples were packed in plastic envelopes tagged and delivered to Soil Research Lab. for analysis.

4.2   Minerological Analysis

Contents of minerals e.g. SiO_2, Al_2O_3, Fe_3O_4, CaO, MgO, were also determined in the samples of marshes and dried mud flaps of Jati and Rahimki Bazar.

4.3   X-Ray analysis

All the above mentioned samples were analysed by X-ray diffraction.

4.4   Survey by Satellite

SUPARCO was requested to supply serial photographs of marshy area near Rahim bazzar and Jati.

5.   Results

5.1   Soil Analysis

Results of soil analysis are given in table 2 & 3. According to the Soil report the mud from Jati has clayey texture and moderately alkaline in nature, while contents of N,P_2O_5 and K_2O in four samples from Rahimkibazar Varies largely and is given in table 2. Texture is reported to be silty clay, Fina gravel, very fine sandy loam and clay loam.

Content of sand, silt and clay in lower Sind plain at Badin adjacent to the Rahimkibazar is given in table 4 [14]. Percentages of these constituents in soil of Tharparkar desert are given in table 5.

Table 1. Some of typical marshes of the world [5]

.NO.	COUNTRY/LOCATION	SEA/RIVER/ DELTA	EFFECTIVE UTILIZATION
1.	Camargue	Rhone delta	Bird sanctuaries
2.	Spain	Guadalquivin	--do--
3.	Romania	At mouth of Danube	--do--
4.	Egypt	Nile delta	
5.	Iraq	Tigris - Euphartes	
6.	Vietnam	Mekony delta	
7.	Brazil	Amazon delta	
8.	U.S.A.	Mississippi river delta	
9.	Poland/Russia Boundry	Pripet marshes	Natural boundry between USSR & Poland
10.	Sudan (Southern part)	Papers marshes if upper Nile	
11.	Bostmana (Kalahori desert)	Okavango marshes	
12.	Florida (Evergludes)		
13.	U.S.A. (Utah)	Great salt lake	

Table 2   Soil analysis of sample from JATI.

1.   T.S.S. %   =   2.96

2.   Soil pH   =   8.10 (moderate alkaline)

3.   Texture.   =   Clayey

4.   Organic Matters%   =   0.833

5.   Available Nitrogen % =   0.0466

6.   Available Phosphorus =   10.81 PPm.

7.   Available Potassium =   252 PPm.

Table 3   Soil Analysis of the samples taken from
               different zones of Rann-of-Kutch, Sind*

S.NO.	ZONE OF RANN-OF-KUTCH, sind.	SALINITY %	pH	ORGANIC MATTER %	NITROGEN %	PHOSPHO-ROUS AVA-LABLE,PPm.	POTA-SIUM, PPm.	TEXTURE
1.	Rahimki-bazar	3.20	7.81	0.825	0.0412	6.0	2.52	Silty clay
2.	---do---	0.304	7.78	0.238	0.0119	2.2	120	Finally gravel
3.	---do---	1.04	7.93	0.428	0.0214	2.4	100	Very fine sandy loam.
4.	---do---	2.16	7.37	01.333	0.0666	4.4	380	Clay loam.

* Conducted in Soil Research Lab., Agriculture Research Institute,
   Tandojam, Pakistan.

260

Table 4

Table 4

Analyses of Soil from Lower Sind **

LANDFORM	LOCATION	Analysis (before removal of calcium carbonate & organic matter & before dispersal of mineral fraction)			Analyses (after removal of calcium carbonate & organic matter & after dispersal of mineral fraction)			Approx. precentage of calcium carbonate	Percentage Clay Agg. in Untreated Sample.
		Sand%	Silt %	Clay %	Sand %	Silt %	Clay%		
Cover Flood Plain	Lower Sind Plain at Badin.	13.2	73.2	14.4	14.6	59.8	25.6	14.5	43.8

Note: Sand particles have diameters between 2 mm and 0.05 mm.
       Silt particles have diameters between 0.05mm, and 0.002mm
       Clay particles have diameters less than 0.002mm.

*   Source: Colombo Plan Report, p. 47 [14]

** Place of sampling.

Table 5

Analyses of soil: Tharparkar Desert

	Surface Percentage	Subsoil Percentage
Clay and finest silt ...	5.23	10.71
Fine Silt ...	0.74	1.44
Medium Silt ...	0.40	0.26
Coarse Silt ...	1.93	2.24
Fine Sand ...	30.41	25.12
Coarse Sand ...	61.29	50.23

5.2. Mineralogical analysis

Mineralogical analysis of two representive samples of Rahimkibazar and Jati are given in tble 6.

5.3 X-ray analysis

X-ray diffraction shows that all samples from Rahimkibazar and Jati contain mainly Quartz, Kaolinite, and Wustite. All these samples are identical except Silicious sample of Rahimkibazar (A-1) contains quartz in larger quantity as compared to other samples. These samples also contain halloysite, and Calcite while Lepidocrocite FeO (OH) is also suspected.

5.4 Study by Satellite

Marshy area near Rahimkibazar at boundry of Pakistan, scanned by the satellite is illustrated in the photographs. This information is given to the reader for understnding the geographical situation of marshy areas of Pakistan.

Conclusion

1. Research is showing that wetlands have a very high economic value. Salt marshes are believed to serve as a nursery grounds for over half the species of commercially important fishes in the south-estern united sttes. The rate of which wetlands are being destroyed is illustrated in the following example; of the nearly 30,000 acres of salt marsh once present in connecticut, less than half remains today. As a result, a $2 million in equivalent dollars. [12]

2. A very importnt characteristic of wetlands is their ability to remove inorganic nitrogen compounds and metals from groundwater polluted by land sources. Most o the removal is probably achieved through adsorption on clay-sized particles. Some of the nitrogen compounds trapped in the sediment are decomposed by dentrifying bacteria, releasing trhe nitrogen to the atmosphere as nitrogen gas. Much of the remaining compounds is used for plant production in this environment, which is one of the most productive in the world. With the death of the plants, the organic nitrogen compounds re either incorporated into the sediment and coverted to peat or are broken up and become food for bcteria, fungi, or detritus-feeding shell and fin fish. [12]

3. The soils of the Indus Delta are made up of sands, silts and clays. Insoluble silicates in sand and silt form more than 66 per cent of the constituents, the remaining 34 percent are alumina, ferric oxide, salt, managanese and other minerals. Different places have different surface texture. Generally the areas of higher relief, like the natural levees or overbank deposits, are charcterized by medium txtured soils, while areas of comparatively lower relief have moderately heavy-textured soils. The finer materials are carried by the river into the tidal areas, where they are either deposited as mudflat material, or carried to the sea. [14]

Table 6   Mineralogical analysis of Mud collected from
          different zones of Rann-of-Katch.*

| | | RAHIMKI BAZAR | | JATI | |
		Silicious Sample A-1	Clayey Sample A-2	Silicious Sample B-1	Clayey Sample B-2
Loss on Ignition	L.O.I.	8.10	7.52	7.00	5.38
Silica	SiO_2	80.57	55.97	68.74	61.12
Alumina	Al_2O_3	5.10	18.57	10.13	14.80
Ferric Oxide	Fe_3O_4	0.50	5.25	1.87	3.62
Lime	Cao	5.74	8.46	8.36	8.76
Magnesia	Mgo	1.30	2.39	1.81	1.15
Sulphuric Anhydride	SO_3^-	Traces	0.89	1.92	0.55
Chloride	Cl^-	1.68	1.79	1.70	1.94

*   Conducted in Chemical Lab. of Zeal Pak Cement Factory, Hyderabad, Pakistan.

4. Composition of clay and slurry used by Zeal pak Cement Facotory, Hyderabd PAKISTAN, for manufacturing of cement was compared with mud of the marshes from Rann-of-Kutch.

Zeal Pak

Rann of Kutch
(Rahim-ki-Bzar)

Clay

Slurry

SiO_2 = 55-60 % SiO_2 = 13-14 % SiO_2 = 55-80 %

Fe_2O_3 = 6-10 % Fe_2O_3 = 2-2.5 % Fe_3O_4 = 0.5 to 5.25 %

Al_2O_3 = 15-20 Al_2O_3 =3.5 - 4.5% Al_2O_3 = 5.10 - 18.57%

CaO = 7.4 %   CaO = 42 - 43 %   Cao = 5.79 - 8.76%

Mgo = 1.42   Mgo = 1.15 - 2.34 %
(Sulphuric anhydride = 0.41 %   Sulphuric anhydride=0.55-1.92%

By comparing the composition of clays of Zeal-Pak and Rann-o-Kutch zones, it is clear tht compositions of both of the clays are nearby and mud from the marshes of Rann-of-Kutch can be utilized as-such or by small modifications for cement manufcturing.

Future Research

This type of the clay is also useful for manufacturing activated bentonite, whis is entensively utilized in metal casting, as a bonding material for making sand moulds. When clays of above mentioned composition are treated mangetically the Lorenze force acts on Mg^{++}, Na^{+}, K^{+} H^{+}, SiO_2^{2-}, $Si_2O_5^{2-}$ and OH^{-} radicals[4] Further research in this domanion will open a channel for manufacturing of activated bentonite from the mud of marshes.

Acknowledgement

Authors are grateful to scientists of SPARCENT for managing satellite photographs of marshy areas of Rann-of-Kutch and to Mr. Dabir Ahmad, P.SO. PCSIR for X-ray diffraction analysis of the samples.

Authors are also thankful to the Dr. Ijaz Shahikh, Principal Investigator, Soil Laboratory, Agricultture Research Institute, Tandojam for soil analysis and Engr. Ijaz Ahmed, Dy. General Manager (Chemical) Zeal Pak Cement Factory, Hyderabad for Mineralogical analysis of Samples. Samples from Jati Coastl area and Rahimki bazar were collected by permission of coast guards and Indus rangers. Author are grateful for their help.

Bibliography

1. Brody B. - The nature and properties of soils Eurasia Publishing House (Pvt.) Ltd. New Delthi-1, repritn 1964 p.

2. Davar L.D. - Soil Physics, John Wiley and Sons, Inc. Newyork, Th.Ed., p. 225-257, 317-333, 363-379

3. Tisdale S.L. & Nelson W.L. - Soil fertility and fertilizers, Macmillan Publishing Co. Inc. Newyork, Th.Ed., p.

4. Buzila Simon - Proietarea si Executarea formelor, Editura Did. & Ped., Bucharesti Romania, 1976 p. 99-102.

5   *The Encyclopaedia Britannica vol 7, 15th Ed., p. 48, 876

6. Thar Minerals can earn valuable foreign exchange. Daily the Star vol-xxvi, Regd. No. M-21 dated 9th June, 1988.

7. Zaki A. - Directory of Mineral depostis vol-15, 3, Geological Survey of Pakistan, 1969 p. 65-77.

8. Farshori M.Z. - The Geology of Sind Deptt. of Gelogy, University of Sind, Jamshoro, Pakistan, 1972, p.59-65.

9. Strahler A.N. - Physical Geography, ed., 4th, p.216, 305, 310, 311.

10. Hilton K - Macmillan Geography, Coastal Environment, (Chapter-7) pub. Macmillan education.

11. Panhwar M.H. - Ground water in Hyderabd and Khairpur Divisions, 1969 Pub. W.P. Goct. Press, Khairpur, Pakistan, p.59-67

12. Thurman H.V. - Essentials of Occeasnography, p. 203-216

13. Fitzpatric E.A. - an introduction to Soil Science, 2nd Ed., Longman Scientific and Technical, p. 21-79

14. Rahman M. - A geography of Sind province, Pakistan. The Karachi geographers Association p.27 - 31.

15. Carp E. - Directory of wet lands o international importnce in the western palearctic. Prepared by UNEP Niobi, Kenya and WWF Gland, Switzerland.

16. Classification of wet lands and deep water habitats of United States (1979) US deptt of interior.

17. Sager P.E., Richman S, Harris H.J, Fewless G. - Preliminary observations on the seiche-induced flux of carbon, Nitrogen & phosphorous in a great lake coastal marsh. Coastal Wetlands, Lewis Publishers Inc.

18. Kig D.L. - Nutrient cycling by Wet lands and possible effects on the water levels. Coastal wetlands, Lewis publishers Inc. p. 69-85.

19. Mc Cullough G B - wet land threats and losses in lake st. clair. coastal wet land, Lewis publishers Inc. p. 201-205.

20. Ball J. P. - Mrsh Management by waterlevel manipula-tions or other natural techniques - A community approach, coastal wetlands, lewis publishers Inc. P. 263-273.

21. Soley H.T. - The former Province of Sind (including Khairpur state). The Gazetteer of west Pakistan Chapter-II, geology of Sind, p. 19-24.

The Application of the General Energy Theory to the Studies of Chemical Engineering: Some Fundamental Aspects of the Behavior and Characteristics of the Colloids

K. PARTHASARATHY
Department of Electrical Engineering
Annamalai University
Annamalai Nagar 608 002, India

ABSTRACT :

In this paper, the basic aspects of the colloids are being investigated through the principles of the General Energy Theory (GET). The electrical double layer, the mechanism of the colloidal formation and some aspects of colloidal stability are also discussed. The paper sets forth the various physical reasons that go to form the electrical double layer, the slip plane and the significance of the zeta potential. The method of computing the location of the slip plane and the magnitude of the zeta potential are also presented. The nature of formation of the colloids and their accretionary details are presented vividly. Some of the aspects of the colloidal stability are also discussed from the creational point of view of the atoms enabling better perspective into the phenomena of the colloids. Correlation to the experimental data are also made.

1. INTRODUCTION :

In view of the great industrial importance, colloidal systems have been extensively investigated in the past through the conventional scientific concepts. However, many of the associated phenomena of the colloids would demand a greater comprehension and detailed analyses so as to bring out an effective unification of the various inter related information available. To this end, the principles of the General Energy Theory (GET) can be used to study colloids from the creational aspects of the atoms. In this way, we will be able to visualise many of the intricate phenomena of the colloidal chemistry.

A great amount of information on the concepts of the GET have been already brought out(9-45). These will enable us to understand and analyse the basic properties of the colloids, for example, like the formation and dynamics of the electrical double layer, the coagulation and stabilization aspects of the colloids, the influence of the various parameters like the temperature or stabilizer effects etc.. truly in their entirety. In this paper, many of these are investigated. In the companion papers (on sedimentation and the aerosols) inter related information will be discussed more fully.

2. THE ELECTRICAL DOUBLE LAYER :

As the studies on the colloidal systems involve the detailed investigation on the adsorptive properties of the solids, liquids and gaseous atoms, it becomes necessary to study the surface phenomena of the various types of the atoms in contact with each other. In this regard, in the conventional literature, the electrical double layer that would form at typical interface (between solid-solution solid-gas and solid-liquid) have been extensively studied. But many of the assumptions involved in the various theories would require greater perception and analyses, which can be made only through the principles of the GET.

On this score, we shall first consider the theories of Helmholtz-Perrin, Gouy-Chapman and Stern critically from the physical point of view. This approach enables us to visualise not only the actual mechanism of the double layer formation, but also deduce more information on the nature of the slipping plane, the zeta potential and the like phenomena. The analyses can be made to include the ideas of the triple layer model and the water-dipole model of the electric double layer.

The short comings of the earliest Helmholtz-Perrin theory and the subsequent improvement made upon the same are well-known. As it were, the failure of the earliest thin concentrated layer assumed by Helmholtz and Perrin to account for the electrokinetic phenomena, the zeta potential etc., had led to the inclusion of the additional diffuse layer of sufficient width(to the Helmholtz layer)by Gouy and Chapman. However, the latter theory also suffered considerable difficulties. Primarily, there is the difficulty of correctly fixing up the slip plane in the solid-liquid system. The incorrect double layer thickness so deduced gave capacitance values for the model to be very much in excess of the observed values. This was due to the theory not taking into account the ionic radii although it freely incorporated the concept of the ionic environmental clouds for the potential determining(or surface)ions and the counter ions- a concept which was supposed to be introduced by Debye and Hückel many years later. Furthermore, the reversal of the zeta potential due to the introduction of the polyvalent electrolytes into the system and the influence of the ionic radii on the size of the electric double layer etc., were problems that could bot be answered satisfactorily. To tide over these, the later introduction of the Stern's theory presupposed two main assumptions viz., i) the definite ionic radii that preclude the potential forming ions of the solution from coming into closer distances to the surface beyond a certain minimum(corresponding to the radius of the ion and ii) the introduction of the additional non electrical forces due to adsorption effect etc. Although these improved the electric double layer model considerably, there remained the basic aspect of the nature of the ionic field associated with the ions to be still expounded clearly. It must be emphasised that the inclusion of the ionic field had been the underlying assumption in the Gouy-Chapman and Stern's theories on the electric double layer and the Debye-Hückel theory on the strong electrolytes. Although the surrounding environments influencing the given ion is

guessed by the Debye-Hückel theory through the manifestation of
the ionic field, it will be seen through the GET that the situation
is not very correctly brought out in any case. This has already
been pointed out by the GET(34). This will also become clear subsequently when the double layer is more fully analysed. As it should,
some of the basic assumptions like the nature of the adsorption
potential remaining constant irrespective of the concentration of
the solution etc., that are employed in the Stern's theory also
are questionable. Again the slipping plane associated in the theory
happens to differ considerably from the assumed or observed one
and on this score, the zeta potential should be different in both
theoritical and observed results. Attempt to correctly strucutre the
Stern's double layer led to the triple layer model which was first
pointed out by Essin and Markov and later fully developed by
Graham in which the identification of the Helmholtz layer(of hydrated ions) in contact with the metal and the outer Helmholtz layer
(of solvated ions) were postulated. A further refinement to the
above model became necessary by considering the electric dipole
moment that were introduced by the neutral molecules adsorbed to
the surface. This was originally pointed out by Lange and Mischenko
but later fully developed by Bockris, Devanathan and Muller(1).
Many of the drawbacks of the various theories given above have
been greatly reduced in the last one.

But in all the above, considerable developmental work is based
on the concept of the equivalent electrical capacitance effect of
the electric double layer. Excepting for the fact that a capacitance property exists between two charged plates separated by a
suitable dielectric, the current concepts of the electrostatic
theory cannot bring out any true insight into the very phenomena
of the electrostatic field and its dynamics. On this score, the
nature of the charge build up at the interface (leading to the
formation of the double layer) the exact phenomenon of the slipping plane and its potential will lack proper physical understanding etc.,. However, these can be greatly expounded by the concepts of the GET.

In the discussion on the generation of the electrical double
layer, it is presumed that the surface layer of the solid at the
interface usually has adsorption(both physical and chemical)of
ions which are selective. This feature is often considered to be
the consequence of the polarizability and ionic hydration effects.
However, there is the basic aspect of the cause for these effects
to be explained in the first place. For instance, while we understand that hydration is merely the process of permitting the given
ion(anion or cation) be surrounded partly or completely by the
water molecules as to why the complex structure must be effective
in adsorption only for one particular type of charge on the ion
(+ve or -ve)is not known. Obviously, these difficulties arise
because of our inability to perceive the various phenomena at sub-
atomic level. However by applying the principles of the GET, we
could overcome many of these difficulties.

3. THE DOUBLE LAYER AS VISUALISED BY THE GET :

It is in order that we deduced the electric double layer through the principles of the GET. To this end, the information on the GET made available (9-45) must be referred.

As already explained by the GET, a given atom has a certain amount of free buffer cloud surrounding it. The nature of this buffer is determined from the constitution of the nucleus of the given atom. Thus a p+ nuclear core would tend to have a large amount of p- buffer and a p- core will have a high p+ buffer. Reference to the periodic table will show that most of the atoms(of virgin elements) will have a high order of p+ absorption (spectral)character so that the buffers involved are generally p- in character. Thus, it also goes in harmony with the terrestrial surroundings where the earth's immediate atmosphere is dominated by the p- cloud flow(16).

The buffer constitution of the molecule will be determined basically from that of the atoms. For instance, when we consider the absorption of the oxygen atom, the maximum for the atom occurs at $\lambda = 7775.5$ Å. This will correspond to an energy level of 2.5×10^{-12} erg($E = h\nu$), where h is the Planck's constant and ν is the frequency). In the computations of the GET, the unit of energy is defined as 1 unit= 10^{-12} erg. Hence, oxygen has typically 2.5 units of p- absorption(in as much as the p- must correspond to the IR end of the spectrum, while the p+ corresponds to UV end), or simply 2 5p-. Likewise hydrogen will have 3p-, carbon 8p+ etc.,. Then a CH_2 radical will have a net energy of 8-2x3 = 2p+. In as much as the basic unit of all the organic compounds are invariably CH_2, it becomes evident that all organic molecules will be highly p+ in character and so their buffer will be essentially p- in character. Finally it should be rememberedthat the buffer clouds of a given body will be highly mobile (although firmly coupled to the parent mass)so as to permit interaction with the buffers of other bodies. The free buffer clouds will always tend to generate spherical objects in free space wherever permitted and where otherwise it is not possible, the body tends to become tetrahedral or cylindrical.

On the above basis, we can now investigate into the environment of the interface generated by typical liquid(of organic character) in contact with a solid surface. The situation will become apparent from the figure-1. We find that individually the liquid and the solid will have their buffers predominantly of p- character(figure-1a). When the two are brought into contact forcibly(figure-1b) the buffers will tend to merge into a common envelope. However, the inherent tendency of the buffers of the two parent bodies (the solid surface on the one hand and the liquid in bulk acting as a body on the other hand)is always to mutually exist exclusively independent of one another, especially under circumstances of one of the bodies (liquid)being highly mobile. Thus the tendency of this separation which is against the forced coalescence brought in by setting up the liquid in contact with the surface will lead to the common zone of the buffer interaction to thin out and be specifically located at considerable distance (when compared to the radius of the atoms involved)from the surface. The buffer of the solid will be highly oriented (high density)at the surface zone due to the inherent tendency of the A-circles of the individual atoms constituting the surface coalescing and generating a common contour(of p+) that would run parallel to the true solid surface and at close distance (of about 1-2 A°) from it. The buffer of all the combined atoms will therefore be governed by their property of the A-circle coalescence and will therefore be very dense in the immediate vicinity of the surface and it will gradually wane towards the interior of the liquid zone. The liquid atoms, on the other hand, are highly mobile and hence an unique coalescence of their A-circles into an appropriate plane surface is not quite possible(and cannot also be stable due to the Brownian movement of the liquid atoms)and hence the overall liquid buffer clouds will tend to be only marginally spread towards the metal surface . The obvious coalescence of these two separate buffers will become evident and at the common zone of the coalescence (figure-1c)the p- density will be marginally enhanced. This plane will therefore promote the possibility of greater free liquid cations being located in it as otherwise such a free layer would distort locally the charge value at the coalescence zone and would therefore lead to unstable conditions(tendency to push back both the original buffer layers) . It will also become apparent that this layer will permit an easy shear plane through it. Any tendency for the bulk liquid (along with its interacting buffer whose terminal periphery is at the shear plane)to pull itself from the solid surface will therefore be aided by the easy shear along coalescence plane and on this account this plane becomes the slip plane which is identified otherwise in conventional theories .

At the instant when the liquid is forced to be in contact with the solid surface , the establishment of the slip plane and the disposition of the p- buffers of the two parent bodies (solid plane and the liquid) will preclude any free molecule or atom be located in the zone between the slip plane and the actual solid surface proper. But soon this condition will be passed over due to the density gradation effect of physical systems in free space. The solid surface is at higher density level when compared with the liquid region and hence as we try to plot the density variation with distance measured from the surface , we will observe that there is a sharp discontinuity in the graph corresponding to the zone lying between the surface and the slip plane (figure-2a)(as it is made of only the buffer clouds and hence at very low density level when compared with the solid surface or of the liquid column). This sudden dent in the density distribution curve cannot be stable in free space . Hence there is a spontaneous drift of some of the molecules of the general liquid column But in this act. the nature of charge of the drifting molecule will become important .

The water molecules have a net 8.5p- and hence it will freely possible for them to drift into the double layer zone and adsorb on the metal surface (figure-1c). This drift of the water molecules is greatly enhanced by the repulsion generated by the p- of the bulk liquid . On the same token, the high p- character of the anions of the liquid will permit their penetration (against the p- of the buffer of the metal surface) and adsorption on to the surface.Nextly the cations , which by their part or total hydration will become overall p- in character and hence could drift into the solid surface-slip plane zone. Every such solvated ion by itself would form an independent system and hence its density distribution curve would always position it at sufficient distance from the actual metal surface (figure-2c). It will become evident from figures 2-b and 2-c that the density distribution curve of the cations and anions together with the free water molecules would generate the stable density gradation effect and the entire double layer zone. It should be borne that the anions would easily get adsorbed at the metal surface owing to their buffer being p+ while that of the metal surface being p- and this is one of purely electrostatic in character. But in as much as the high p- character of the anion would generally discourage its entry close to the surface which is already highly p- and hence any such ion sticking to the surface would largely depend on the effectiveness with which its buffer (p+) will be able to couple itself with the p- of the buffer of the metal surface. On this score, only a few anions may find favourable adsorption to the solid surface Lastly there can be the drifting of the neutral solvent molecules also into the surface-slip plane zone. This occurs because the water molecules being highly p- in character, would tend always to stick to the solvent molecules which are generally p+ in character. This sticking is considerable in true solutions. Thus the complex water-solvent molecules will drift towards the surface just as any cation or the anion did. In this respect, the neutral solvent molecules behave like the solvated cation(with an overall p- charge on them)but can travel and settle closer to the surface than is possible for the cation. The details of such an adsorption can be considered in a subsequent section.

The above logic of the GET would immediately permit us to deduce the location of the boundary of the double layer. In this respect, we notice that the atoms constituting the solid surface , would by virtue of the λ-circle coalescence , be elevated to a density level of 10^{45} p/cc (particles per cc). But the free interstellar substratum density is 3.3×10^{38} p/cc(which will be realised in an undisturbed zone between two sufficiently displaced solvent atoms. However,owing to the Brownian movement, the density at such a location would change periodically and attain as maximum the value of 1.24×10^{42} p/cccorresponding to the radiation level. Such a fluctuating density variation will then lead to a geometric mean density level of $\sqrt{1.24 \times 10^{42} \times 3.3 \times 10^{38}}$ $= 2 \times 10^{40}$ p/cc , which is characteristic of the free space(between the atoms of the solvent). Such a density must obviously exist at the boundary plane of the double layer.

Since the double layer must be graded densitywise , each surface atom of the solid at the interface with its radius r will tend to generate the contour of the boundary plane of the double layer at R= r x $(10^{45}/2 \times 10^{40})^{1/3}$ = 36.8 r. Thus with typical r=1.5 A°. the double layer must have its limiting plane at 36.8x1.5 = 55.2 A° . Experimental data would would show that for a uni-bivalent and bi-univalent type of electrolyte with a molar concentration of 0.001 M(very dilute) the thickness of the double layer (reciprocal of the thickness of the ionic atmosphere) as 55.5 A°. The variation of this thickness with the concentration and type of the ions involved can also be discussed.

Experimental values for the double layer thickness quoted in the literature are as follows (the values are given in the order of molar concentration of 0.1M , 0.01M , and 0.001M respectively): For uni-univalent type of electrolyte:9.62, 30.4, 96.2 ; for uni-bivalent and bi-univalent electrolytes 5.56, 17.6, 55.6 ; for bi-divalent electrolytes 4.81, 15.2, 48.1 ; for uni-tervalent and ter-univalent 3.92, 12.4, 39.2. (all values are in Å)(6).

We observe from the above that for a given concentration of the solution, the thickness of the double layer tends to decrease as the valency of the constituent atoms of the electrolyte increases.This factor is in accordance with the logic of the GET given above . The overall shape of the given set of the atoms making up the molecules will be the least spherical for the uni-univalent com-bination. In this case, the combining atoms form a dumb-bell shaped molecule with a large overall radius r for the buffer and hence the double layer thickness given by 36.8 r rule will be a maximum.As we consider the polyvalent ionic combinations, the structure of the molecule tends to become more spherical or tetrhedral in formation so as to have the decreasing values of r with increasing valencies. The numerical values quoted can be justified when we have more infor-mation on the on the actual type of ions involved and the radii of the corresponding neutral atoms.

The variation of the double layer thickness with the molar con-centration for a given electrolyte can also be deduced. From the GET point of view, we will notice that the increasing concentration of a given electrolyte is to bring in more electrolyte molecules in a given volume and so the interaction between energy particles on the outer surface of the buffers of the adjacent molecules will increase. Since the number of such interacting particles on the sur-face of each buffer will increase with the radius of the buffer and since the increase in the concentration of the electrolyte brings about an increasing interaction (due to the reduction in the molecular distances) , it is apparent that we could arrive at $c \propto R^2$, where c-the concentration of of the electrolyte and R- the buffer of the molecule(figure-3a).

We had already seen that the actual value of the intermolecu-lar space density will depend upon the geometric mean of the true space density and the electromagnetic radiation level. Now, in the changing concentration of the electrolytes, we will notice that the density δ_m will also change and this variation can be easily compu-ted with refernce to figure- 3b . Thus, let us consider the number of particles contained in a given molecule as N so that the surface density at the buffer becomes $\delta = N/(4\pi R^2 t)$ where R-the radius of the bufferand t-thickness of the thin shell of particles(that actually interact). Since the value of t is usually constant (corresponding to the monolayer of the energy particles), we see that $\delta \propto 1/R^2$ and hence the charges on small masses A and B shown in the figure-3a with a volume dV can each be expressed as δ dV (with $m_A = m_B$). Then the force of repulsion existing between them by Coulomb's law becomes $f \propto 1/(R^4 d^2)$. Under stable conditions, let us presume that the molecules are separated by the constant distance $d(= 2/\sqrt{R}$ - see below). Then the force becomes:

$$f \propto 1 / (R^4 d^2) \quad _{d = 2/\sqrt{R}} \propto 1 / R^3 \tag{1}$$

If the molecules tended to move under the action of this force, we would then have for any mass , say for mass A ,

$$f = m_A \ a \propto dV/dt \propto V \tag{2}$$

m_A and m_B remain constant and for the infinitely small dR, when the radius R also remains constant, then $R \gtrless dR$. We thus have from equations (1) and (2) ,

$$v \propto 1 / R^3 \tag{3}$$

But from the basic force-energy equation of the GET (21), we have already known that

$$E = f \ dx = dV_p \ (\delta \ lv/2) \text{ and } dV_p = 2 \ A \ v \tag{4}$$

where E-energy , dV - volume of the energy cloud displaced , A- area of cross section of the moving body , δ - density of the body , l- length of the body , v - velocity of movement. Thus rearranging the equation (4), we have,

$$\delta \propto E/ (Alv^2) \propto 1/ v^2 \propto R^6 \tag{5}$$

In equation (5), we observe that for the energy particle (E/Al) would remain constant under the given conditions. The mass density δ obtained in eq(5) will also correspond to the energy density in the GET and now define the factor of equivalent enhancement of the molecular density (within the buffer) and so the δ_m will now become $\delta_m = [(3.3 \times 10^{38})(1.24 \times 10^{42}) \delta]^{\frac{1}{2}} = 2 \times 10^{40} R^3$. From these, we can now connect up the concentration c of the of the electrolyte with the energy density as ,

$$\delta \propto R^6 \propto (R^3)^2 \propto c^3 \quad \text{ since } c \propto R^2 \tag{6}$$

equation (6) will therefore enable us to directly compute δ without the need for the determination of the R of the molecule . Then we can easily compute the double layer thickness using eq (6) As an example, extending the previous computations for the uni-bi' valent and bi-univalent type of electrolyte, when the concentration changes from 0.001 to 0,01 and 0.1 M , concentration factors of 10 and 100 are involved. Hence δ must go up byfactors of 10^3 and 100^3 respectively. These will give finally the corresponding δ_m as 6.4×10^{41} and 2×10^{43} . From these, we obtain the thickness of the double layer as 17.4 A° and 5.52 A° respectively (compare with the data given above).

Incidentally, these above computations will also permit us to deduce the famous London equation (49), connecting the molecular energy of attraction and their distances in the colloidal solutions, which is $u = -a / \gamma^6$ where γ is the distance between the molecules and a is a constant.

Referring to figure-3b, which is the same as the earlier situation(figure-3a)we notice that the distance d between adjacent molecules could be defined as $d/2 = 1/\sqrt{R}$. This follows from the fact that when we consider the surface density δy that the shell of the molecule will develop would correspond to $(4\pi/3)R^3\delta_x =$
$4\pi R^2(d/2)\delta y=$ that is, the shell has an equivalent number of particles . This will lead to $\delta x/\delta y = 3d/2R$. Under circumstances, this shell thickness of $d/2$ of the buffer will be seen to be sufficient to balance all internal particles contained in the molecule along any radial line upto a thickness of $R/3$. In this situation,we will notice that, due to the nucleii of the molecules proper, a considerable zone internally within the atom(or molecule) will not be available for counter balancing the thrust generated by the buffer clouds along any line and on this basis, it is reasonable to assume that about $R/3$ at the outer region(electronic zone)is only effective (figure-3c). Thus along any radial line, considering a small volume of area of cross section dA, we have,

$(R/3)dA)(\delta_x)=(d/2)\ dA(\delta y)$ from which, $(\delta x/\delta y)$ $= 3d/2R$ ..(7)

When the shell of buffer of thickness $d/2$ forms on the molecules adjacent to each other , the crowding of the buffer energy particles in the interspace (between the molecules)will decide the δy value. We note that the overall buffer clouds for the two molecules being $8\pi/3\ R^3$, the cylindrical volume that can form between the two diametrical planes of the molecules(perpendicular to the axis connecting the molecules) will be $\pi R^2 x\ 2R = 2\pi R^3$. Clearly, when this intermolecular space is filled up , the deficit will be $8\pi/3\ R^3 - 2\pi R^3 = 2/3\pi R^3$. This half spherical volume will now be shared by both the molecules so that the overall volume engaged by each molecule will be $3/4(4\pi R^3/3)$ .Thus the buffers so collected will exhibit the density relation δy to the density δz of the nuclear particles contained within the molecule as $(4\pi/3)\ R^3\delta z=(3/4)\ \dot{x}$ $(4\pi R^2/R^2)x\ 1X(\delta y)$. It should be noted that the thin shell of unit thickness on the spherical surface of radius R will produce at the center an equivalent effect of q/R , where q is the surface charge given by $4\pi R^2\ (\delta y)$. This will also imply that the distribution is such as to be independent of the radius and will directly correspond to the solid angle subtended at the center.(in the previous case, the solid angle is $3/4 x\ 4\pi$). From the above, we can compute the δ_x , the geometric mean density of δy and δ_z as $\delta_x =\sqrt{\delta_y \delta_z}$ so that finally we have the relation,

$\delta x/\delta y$ $= 3/2\ (1/R^{3/2})$ (8)

It should be seen from eq(8) above, that the desired result of $d = 1/\sqrt{R}$ will be obtained by substituting eq(7) in it . This would lead to the earlier discussions on the relationship between the concentration and the double layer thickness to be fully established. On the same lines, we notice that from eq(5), that $E\propto v^2$. In as much as the intermolecular distance r = 2 R , we can write the above identity as $E\propto 1/(r^6)$ or $E = a/(r^6)$., which is the London equation. The negative sign in the London equation indicating a reduction in the molecular energy as the molecules tend to come closer will be otherwise evident from the logic of the GET , when we notice that the size of the buffer envelope decreases(leading to decrease in the density) in the intermolecular space, as the molecules move closer. As a passing mention, it should be made clear that the GET identifies the same circumstances under which the London equation is   deduced-viz., that (as presumed by the conventional theories of quantum mechanics), the orbiting electrons of atoms which are far away from the nucleus proper(corresponding to the shell zone as identified by the GET)so that the energy of the atom is essentially concentrated in this region. However, the Quantum mechanics does not visualise this energy field and its distribution in its entirety in the intermolecular region as is known obviously otherwise.

The location of the slipping plane within the electric double layer can be easily determined . This follows freom the fact that this plane is essentially generated by the coalescence of the buffer of the solid and the liquid media. We have already fixed up the nature of these buffer fields. The buffer field of the solid surface has a density of 10^{45} p/cc while that of the liquid medium in the bulk will correspond to 2×10^{46} p/cc . The coalescence of these fields will obviously generate at the slip plane, the mean geometric density that would be $\sqrt{10^{45} \times 2 \times 10^{46}} = 4.46 \times 10^{42}$ p/cc. Thus with the surface atoms of radius $r = 1.5$ A°, (for the uni-bivalent and bi-univalent type of electrolyte with c=0.001 M considered earlier), we now have $R = r \times (10^{45} / 4.46 \times 10^{42})^{1/3} = 9.1$ A°-. Similar computations can be made for various other types of ions and ionic concentrations.

Reference to the literature (1, 5, 6, 46, 49)would show that there is considerable speculation as to the deduction of the slipping plane. In as much as the characteristic distribution of ions(in the Stern and water dipole models)in the Helmholtz region is known, this aspect of the location of the slip plane can be easily fixed. Thus referring to figure-2b, we notice that the central plane of the adsorbent layer on the solid surface including the anions would be about 1.5 to 2 A° from the surface constituting the inner Helmholtz plane. The solvated cations will form the complex, of about 6 to 7 A° in diameter from the inner Helmholtz plane. Thus the slipping plane would be almost gracing the outer Helmholtz plane and hence can be presumed to be at the boundary of the Helmholtz and the Gouy layers for very weak electrolytes.However, its location from the solid surface will be almost fixed due to the characteristic density values of the surface layer of the solid and the bulk liquid. This will mean that as the concentration of the electrolyte increases, the size of the diffuse(Gouy)layer will decrease with the constant size for the Helmholtz layer. This will automatically result in the decrease in the zeta potential of the double layer.

The potential jump at the electrical double layer can be easily computed. This follows from the fact that the charged anions of the electrolytes, by virtue of their acquiring the free electrons in their outer periphery (of buffer), will determine the number of the energy particles collected on the ionic surface that will go to give the potential of the double layer. For example, an anion of about 3 A° in diameter adsorbed on the surface atom of about 1.5 A°

in radius would generate a density value of $10^{45} \times (1.5/4.5)^3 =$
3.7×10^{43} p/cc at its surface that faces the bulk of the electrolyte.
When we consider of about 6.6×10^{-11} cm (electronic diameter) thick
shell where the electronic and free particles charges can accumulate,
the shell volume will become $4\pi (3 \times 10^{-8})^2 \times 6.6 \times 10^{-11} = 7.48 \times 10^{-25}$ cc
which will now be made available for the adsorbing anions. The shell
will therefore accumulate a total of about $(7.48 \times 10^{-25}) \times (3.7 \times 10^{43}) =$
2.7×10^{19} particles. In as much as 1.1×10^{19} of the p+ particles
of the buffer of the electron would constitute one volt , this accumulation of the p+ in the entire sheath would correspond to 2.5
volts. Since each anion may house only about 2 to 4 electrons in its
shell, we could allow a weighting factor of about 6 to 8 to account
forthe actual energy particles accumulating in the shell of the
anion which thereby would give a potential of about 300 to 400 mV
as usually found in the various experiments. This weighting factor
would give about 45-60° (steradians) of exposed surface of the anion
for the electronic attachment which is highly reasonable.

The zeta potential can be deduced from the fact that the potential distribution curve for the double layer would follow almost
the cubic gradation (density)law for weak electrolytes. Thus for
an electrolyte of 0.001 M , of the uni-bivalent and bi-univalent
type considered earlier, the zeta potential can be obtained as
equal to $400 \times (4.5/9.1)^3 = 50$ mV which is a typical experimental
value. It will also be evident that for the increasing concentration the potential curve would become steeper and steeper with in
the Helmholtz region so thst the potential variation should be
represented by $\xi = f(x)^{-n}$ where $n = 4,5,6..$ etc., and x- the distance from the surface.

We can finally consider the concept of the ionic atmosphere
that is visualised by the conventional theories. In the Debye and
Hückel model, the ionic atmosphere is considered by presuming say
a cation like X , to be surrounded by anionic charges like A,B etc.,
The Boltzmann's statistics is then evoked to determine the probability of the given cation being counteracted by the anionic field.
On this basis, the ionic atmosphere is visualised , with the inherent difficulty of exactly confining the radius of this influence
zone, still present. A little consideration would show that a typical anion like A so considered as above, in itself must be compen-
sated by the surrounding cations which obviously is not incorporated
in the model. We have only the cation X and anions like B,C etc.,
surrounding A and clearly the symmetry sought earlier for the cation
X is now lost for the anion A. But still, the Debye model successfully accounted for the development of the theory of the double layer
thickness because it tacitly assumed the existence of a negative
charge zone around the cation and a positive charge zone around the
anions. In other words, it will now be seen through the GET that
this amounted to accounting for the buffer in typical ions, although
the visualisation of the situation is not carried out in the Debye
model as it should be evident from the logic of the GET.

4. FACTORS INFLUENCING THE ZETA POTENTIAL :

As we had seen above, the zeta potential can be influenced

typically by a number of factors. When an indifferent electrolyte is introduced into a colloidal system, it has the effect of reducing the thickness of the double layer and hence the zeta potential. In as much as the ions of the indifferent electrolytes do not dissociate, the ionic concentration of the initial solvent electrolyte will still be increased as the neutral molecules of the former is identical with that of the latter. This increasing concentration of the electrolyte leading to the reduction of the double layer has already been analysed. This situation is typical for the introduced indifferent electrolyte when it has one of its ions as the counter ions of the initial electrolyte. However, when the indifferent electrolyte has no common ion with that of the initial electrolyte exchange of ions in solution can occur. This exchange of the counter ions in the bulk solution between the indifferent and stabilizer electrolytes is logical from the buffer considerations of the GET. Also, the ratio of the concentration of these two types of counter ions in the double layer zone must be the same as for the bulk solution. In as much as a given number of electrons removed would generate almost the same void zones in different atoms, the buffer involved in every case will be the same. This implies that for equivalent (but different types of) counter ions, we could expect a 1:1 exchange of counterions as logical. However, a polyvalent atom will lose more buffer and become more +ve or -ve (but always charged opposite) with respect to the solid surface than is possible for an univalent atom. On that score, the equilibrium must be shifted obviously towards the higher valency atoms as is found otherwise . Such polyvalent atoms by virtue of their greater buffer strength, when selectively adsorbed on the solid phase, can even completely reverse the polarity of the surface charge and hence that of the zeta potential which is also observed in practice.

The influence of the non indifferent electrolytes will also become obvious. The dissociated ions of these electrolytes can adsorb more effectively and precisely in the same way as that of the ions of the stabilizer electrolyte . The buffers of both groups of ions are identical and hence this would increase the potential of the double layer (by improving the surface density of the solid phase and permitting greater particle absorption by the electronic sheath as explained earlier). The surface selective adsorption can prevail even to the extent of reversing the charge of the surface. On this score, the zeta potential can even exhibit change of polarity as indicated in the experiments.

In these above discussions and also in the general case of double layer formation , the conventional literature would presume the crystal forming tendency of the potential determining ions of the stabilizer electrolyte with that of the solid surface (49). From the GET point of view we will notice that this assumption is introduced only to find a cause for the selective adsorption of the potential determining ion on to the surface. While the detailed investigation on the adsorption features of the various interfaces are considered in the companion papers, we presently note that the buffer polarity of the potential determining ions will always be opposite to that of the surface generated buffer in that the two will always lock with each other(electrostatic attraction as per conventional parlance).

The pH of the electrolyte will influence the zeta potential in a characteristic manner. In this regard, as per the GET, we
will notice that the acid highly p- in character and the base
is highly p+ in character. This also goes in harmony with the
Lewis and Brönted models of acids and bases. The acid, by donating the proton (or by loving an electron) becomes more p- in its
character while the base by accepting the proton(or by donating
an electron) becomes highly p+ in character. Then for example,
if the stabilizer electrolyte like $Al(OH)_3$ in a weakly acidic medium is considered, it becomes apparent that this breaks into a
more positive $Al(OH)_2^+$ and a negative $(OH)^-$ ion. It is apparent
that the positive $Al(OH)_2^+$ ions will be the potential determining
ions while the $(OH)^-$ will be the counterions. The weakly acidic
electrolyte is feably p- or strongly p+ and obviously the OH^-
acting as counterions will be easily permeated into the electrolyte. On the same token , the $Al(OH)_2^+$ will be repelled by the
bulk electrolyte and so would be deposited on the solid surface
more as a potential determining ion. However, when the electrolyte becomes highly acidic(p-), then the counterions should be highly p+ in character. This is ensured by the $Al(OH)_3$ breaking
up as $Al(OH)_2 O^- + H^+$. Clearly the H^+ will be the counterion and
the $Al(OH)_2 O^-$ will be the potential determining ion as is known
otherwise.

The existence of the isoelectric point becomes obvious, since
as the medium goes from acidic to basic, the reversal of polarity
of the surface charges would change the polarities of the associ -
ated potential jump and the zeta potential taking the latter through the zero(isoelectric state) point.

The effect of increasing the temperature will be reckoned with
an increasing of the size of the electric double layer with attended increase in the zeta potential. In the conventional literature
this is presumed otherwise as due to the Brownian motion being enhanced by the increased temperature. From the GET point of view,we
notice that the effect of increasing the temperature is to inject
more of p+ particles into the system. Such clouds can be easily
absorbed more by the anions in their buffer and by the cations in
their nuclear space. Thus when the anions, when acting as the potential determing ions would strongly adsorb to the solid surface
increasing the total potential jump of the double layer. Also,due
to the greater adsorption, the inner Helmholtz layer is much stren-
gthened leading to the spread out of the potential distribution
curve. At the same time, the cations which became the counterions
could be more effectively solvated and therefore would then tend
to   increase in number within the double layer zone. This would
obviously broaden the double layer. Coupled with the earlier effect
on the potential distribution we find that the zeta potential
should also increase.

There are many other associated phenomena of the electric
double layer which may merit detailed discussions. However, these
are not presently attempted for space considerations. Investigation
on the various electrokinetic phenomena are made more fully in the

companion paper on sedimentation theory.

5. STRUCTURE OF COLLOIDAL SYSTEMS AND COLLOIDAL MICELLES :

Colloidal solutions are generally obtained through the processes of condensation, dispersion and peptisation. The studies on these would bring out a number of very interesting phenomena associated with the colloids.

Condensation techniques of colloidal formation is often presumed to be the process of crystallization (the particles forming minute crystals) The crystals are generally formed by first the nucleation being initiated by a supersaturated solution and subsequently followed by the growth of the nuclei. In this connection, it was assumed , for long, that nucleation occured spontaneously. On this score, the concentration differences between the supersaturated state (C_{sup})and the normal saturated state(C_s) was con - sidered to be a measure of the nucleation rate. Such a difference was believed to initiate spontaneously the nucleation process in a given condition. However, it was experimentally found that, as a rule, nucleation centers were initiated only around any minute extraneous dust particles that got into the solution. In such instances, the crystallization was instantaneous.On the otherhand, in extremely purified and supersaturated solutions, the nucleation did not take place even on prolonged standing.These led to the conclusion that the nucleation could occur only at the already formed interface (between the solid and the mother liquid).

The growth of the crystals is the result of the deposition of the substance from the supersaturated solution. Gibbs etal., proceeded to study the phenomenon from the connection between the crystal shape and its surface energy . The diffusion theory postulates that the crystal face grows at infinitely high rate and hence depends on the rate of supply of substance from the solution-i.e., on the diffusion rate. Volmer proposed peculiar two dimensional adsorption layer to be responsible for the crystal growth. However, the connection between the crystal growth and supersaturation is very complicated and still many details are not made clear by the above theories.

The nucleation rate u_1 must be large when compared with the crystal growth rate u_2 for a large number of crystallization centers to be initiated(mono dispersed sols). If $u_2 > u_1$, it will lead to only a few number of large crystalline growth centers(poly dispersed solutions.). The concentration of the reacting solutions is important in obtaining colloidal systems. In a chemical reation yielding slightly soluble substances, sols are obtained for low concentrations of the reactants. By increasing the concentration, precipitates and gels are only obtained. Usually such precipitates are highly stable and do not yield sols even under high stirring.

The introduction of the inhibitors and retarding agents into the solution would highly either prevent or retard the formation of the colloidal solutions. Examples of the former are $K_3(Fe(CN)_6)$ and $K_4(Fe(CN)_6)$ while for the latter are KBr and KI . The former arrest the crystal growth until such times when a dust particle is introduced into the system which would initiate immediate nucleation. It is presumed that such substances would form a thin layer on the crystal surface and prevent the deposition of the substance from the solution. Special additives would tend to change the rate of growth of the crystal faces.

The formation of the amorphous or crystalline state of the colloidal systems depends largely on the rate of ordering and rate of aggregating of the molecules. It was believed that colloidal systems having amorphous particles which only later become crystalline may be obtained by quick precipitation. Haber etal., assumed that if the rate of ordering was higher crystalline particles always result, while if the rate of aggregating is high , it would lead to the amorphous state. Kargin etal., had shown that the formation of the amorphous state , as a rule, always preceeded in the colloidal formation. They found, by using electron diffraction techniques, that amorphous particles always appear first when sols of Titanium dioxide, silicon dioxide, etc., are obtained. These particles are 1000-8000 A in size and aggregate in chains. But soon these amorphous particles disintegrate into crystalline structures. Rings of point of reflexes simultaneously appearon electron diffraction patterns. The life time for these amorphous particles depends upon the type of the colloidal system. For example, particles of gold sol crystallise in 3-5 minutes after preparation; vanadium pentoxide sol after one hour; titanium dioxide sol after 1-2 hours; aluminium hydroxide sol after one day; silicic acid sol after 2 years. Temperature seems to strongly affect the rate of crystallization . For example, amorphous particles of aluminium hydroxide and titanium dioxide at 80-90°C , give immediately the crystalline structures(there is no incubation period given above)Aging of these colloidal systems often give chains or reticular structures suggesting surface sticking active regions on them.

The application of the principles of the GET would enable us to investigate into all the above phenomena . In this context, we observe that in a supersaturated solution, the number of molecules of the solution is so high that the buffers of them would be tightly engaged(under normal state, the number of molecules in the solution per unit volume would only correspond to the Avagadro number and hence the buffers are fully relaxed). Therefore, buffer coalescence can easily occur. However, such a coalescence will not generally be spontaneous since in a homogeneous solution the density of the substratum is same everywhere and hence any given molecule is in no way better than any of its neighbour to act as a central core or nucleation center so as to generate the successive accumulation of more energy particles to simulate the cubic density grdation law that governs the existence of a body in space (in this case the solution). Any tendency for the coalescence of two adjacent molecules is automatically counteracted by the empty space

created around into which the original (combining)molecules will tend to get back. Thus, mere supersaturation alone cannot generate the nucleation centers. On the otherhand, by the greater crowding of the molecules and the consequent buffer stresses involved, the molecules now acquire an inherent tendency to deposit themselves

on any density wise graded solid that may be present in the solution. In this regard, the very minute dust particles act as typical nucleation center. It must be seen that typical dust particles
have sizes ranging from 500 A°(ZnO smoke) to 1000,000 A°(Flue
gas). Thus, even in the extreme, a dust particle of 500 A°in size
would correspond to a collection of about $(500/3)^3$ =1.4x10^6 molecules (of about 3 A°in dia).From atomic considerations, this dust
particle is therefore a macroscopic body and hence it   must establish the density gradation law. Therefore, to this parent body ,
further successive additions of solution molecules is imperative,
especially when the molecules surround the dust particle and tend
to compress it. The addition of every molecule to the parent(dust)
body would tend to increase the central (equivalent)core density
of the dust particle considered as a compound structure. Then this
will promote the addition of more and more molecules of the solution. In other words, the initial nucleation is automatically generated and subsequently deposition(crystallization) promoted.

In this connection, some of the basic aspects of the crystal
formation can also be considered. The crystal is in essence, the
particular combination of a given number of atoms arranged in a
definite pattern. It can be seen that this feature is the outcome
of the density gradation law coupled with the interaction of the
A-circles of the basic atoms involved. We should observe that basically the A-circle of the given atom(21)will determine its ability
to form the gaseous , liquid or solid body. When the A-circle is
displaced sufficiently farther off the central nuclear zone of the
atom(at a given temperature), interaction of such A-circles of
neighbouring atoms will be pronounced. This mutual repulsive action
exhibited by the A-circles will preclude the common coalescence
of the various atoms and hence the given number of atoms freely
exist as a gas. If the A-circles are brought closer to the nuclei
of the atoms by abstracting the p+ particles from within the atom
(cooling effect), then the given atoms can come more closer simulating the liquid wherein partial buffer coalescence is made possible. In solids, the atoms are brought to the maximum closeness due
to the shrinking of the A-circles to the fullest extent(consistant
with the temperature). In this case, a greater buffer interaction
will exist and such common buffer cannot easily be brokenup—in
other words,the atoms constituting the solid are very difficult
to be separated or the body exhibits considerable cohesivity and
hardness.

The crystalline state of the body is this aggregation of the
given group of atoms. In this aspect, the group of atoms also under
go the density gradation effect . Specifically,consider a given
atom A in the group of the given atoms. This atom has its own graded density distribution when it is considered all by itself in free
space. If now a group of atoms surround this atom A all around in
the three dimensions and if an attempt is made to compress these
surrounding atoms, their buffers will tend to coalesce with that
of the atom A. In this process,for the complex structure so formed,
the central(nuclear)density must be enhanced. In as much as the
central density attainable for the atom A is limited to the maximum
of 4.08x10^{63} p/cc, there will automatically be a limitation imposed

in the accumulation process. For example, considering the limits of accretion density for physical bodies which varies from 2.69×10^{59} p/cc (corresponding to the relative density of 1.0 in the conventional parlance) to 4.08×10^{63} p/cc (relative density 24.8) we notice, say, an iron atom (density of 7.8) will correspond to 1.25×10^{62} p/cc. Hence as the macroscopic body is generated out of the aggregation of more Fe atoms surrounding this central one, we find that the maximum radius to which such an accumulation can occur is given by $R(4.08 \times 10^{63}/1.25 \times 10^{62})^{1/3}$ where R is the radius of the Fe atom. With R= 1.1 A', this amounts to about 3.52 A'. This would show that to the central Fe atom we may just add only one layer (spherical distribution) of Fe atoms (so that the radius of the complex will be about $1.1 + 2 \times 1.1 = 3.3$ A'). Thus it becomes clear that the most favourable set up will correspond to the b.c.c. and the f.c.c. that are more characteristically observed in practice. Therefore as the aggregates of the atoms of Fe tend to form a macroscopic body , it can always occur only in this well-ordered complex (crystal) groups. Obviously, in as much as two neighbouring complex groups will each exhibit the density gradation law, the buffer for each such complex will be independent of any other. Any interaction brought in between the two buffers (adhesion) would largely be physical in character enabling the cleavage at the buffer contact when so attempted. Then we observe that the macroscopic Fe lump cleaves along the crystal lattices easily. This is the general scheme of crystal formation and in the light of this, the colloids can be studied further.

When aggregation of the crystalline masses are considered, it can be seen that the effectiveness of the successive buffer coalescence would again be the deciding factor. This in turn, will depend upon environmental factors like temperature and pressure to which the system is subjected. The deposition rate on the given parent crystal surface will also be dependent upon the concentration of the solution and hence on this basis, we could define the depletion rate of crystalline complexes in a given solution (in tending to form bigger units) by the usual relation

$$d N / dt = - k N \qquad (9)$$

where N-the number of units present at a given time t. This will lead to the characteristic result

$$N = N_0 \ e^{-kt} \qquad (10)$$

where N - will be the initial number of units present with the time constant of the system being specified as τ (crystallization time for the depletion process) and where $\tau = (1/k)$.

The value of τ can be defined in the usual way as applicable to any exponentially decaying phenomena (like heat conduction) as the time taken for the system to reach 63.2 % of the final steady state value. The usual final value will be presumed to be attained (for such phenomena) at a time $\tau' = 4\tau$ (when the value reached will be within ± 3 % of the final 100 % -value). The constant k can be easily determined through the logic of the GET. Thus for the series of the experimental data reported above we shall now try to obtain the crystallization times . In this respect, we shall first consider the gold sol in as much as it corresponds to purely elemental particles.

The radius of the Au atom is 1.34 A°. Hence, it offers a cross sectional area of $\pi \times (1.34 \times 10^{-8})^2$ cm^2 for absorption of incident light. The light photons can be shown to have a typical radius of influence of 6.6×10^{-11} cm (electronic diameter) so that the number of beams of light that can strike a given atom , at any instant, will be $\pi \times (1.34 \times 10^{-8})^2 / \pi \times (6.6 \times 10^{-11})^2 = 0.41 \times 10^5$. Thus when light of $(4000° + 7000°)/2 = 5500$ A° is involved , in as much as each light photon corresponding to the energy of $E = h\nu$ would give 0.88×10^5 energy particles (1 erg = 6.67×10^{30} particles), the total particles intercepted will , per atom per second, be equal to $(0.41 \times 10^5) \times (3 \times 10^{18} / 5500) \times (0.88 \times 10^5) = 1.98 \times 10^{24}$.

When the amorphous aggregates are formed, they are typically 1000-8000 A° in size. Hence, on the average, let a nucleation zone correspond to 4000A° . If the complex is made up of 3370 centers of nucleation, then the number of the Au atoms that will be present in each nucleation zone will be $(4\pi/3)(4000 \times 10^{-8})^3 / (4\pi/3)(1.34 \times 10^{-8})^3$ x 3370 = 8 x 10^6 . (this value of 3370 is directly obtained from the density law otherwise). As the spectral absorption of Au will correspond to 8.3 p+ (2428 A°), the energy particles that will be absorbed by the complex will be $(8 \times 10^6)(8.3 \times 10^{-12})(6.67 \times 10^{30}) = 4.4 \times 10^{26}$. Hence, as the complex is subjected to the light adsorption at 5500 A°, (characteristic median value for the visible light spectrum) the time taken to supply the above quantity of particles will be given as $(4.4 \times 10^{26}) / (1.98 \times 10^{24}) = 220$ seconds or 3.73 minutes. This agrees well with the reported value of 3-5 minutes for the Au sol. In the above computation, it must be noted that the complex of 8x10^6 atoms were supplied essentially through a single central atom. This amounts to the zone of influence of a given nucleating Au atom of radius R as $R(10^{49} / 1.24 \times 10^{42}) = 200$ R. Therefore a buffer shear can be expected for the complex at this radius (see below). Then the number of atoms involved in the zone 200^3 = 8x10^6 which is used above.

It now remains to be seen as to how the other types of sols can be dealt with . This follows easily from the fact that when compared with the gold sol particles , the volume and matter density of the various sol atoms will vary and using these we can compute the crystallization times as shown in table-I. In the table, it should be noted $\tau' = 4/k$ (4τ)= (volume of the given sol particle / volume of the gold sol particle) x (density of gold/density of the parent atom of the sol)x(τ' of gold) x (energy factor).

In the table-I the energy factors introduced would need explanation. With V_2O_5 , althigh the net spectral energy value is 0.3p+ in as much as 5.0-atoms surround the V-atoms, the net effect of the O-atom will be dominant. The energy value for the O-atom being 2.5p- it bears an energy factor of 2.5/8.3 = 1/3 with that of the Au atom. The TiO$_2$ being an open molecule, its net 0.5p+ has no effect in the surrounding and hence has an energy factor of 1.0 . The Al(OH)$_3$ has by virtue of its basic character and larger size could attract only 3 water molecules around it taking the total energy value for the complex as 8.5x3 +11.4 = 36.9 p- . This will then give an energy

284

factor of 36.9/8.3 = 4.45 . Finally, the H_2SiO_3 will have a greater
energy factor of about 74.4/8.3 = 8.95 since in this case, the silicic acid being a compact molecule would attract about 8 H_2O molecu-
les(b.c.c.configuration).In addition, energy particles will tend
to collect at the surface of the molecule (2.85 A°) to a depth of
$6.6x10^{-11}$ cm (electronic diameter) . This will give a ratio of
$(2.85x10^{-8}/ 6.6x10^{-11})$=435 so that collectively the energy factor
would become 8.95x435 = 3900. It should be seen that the last feature is characteristic of the acid molecule and does not apply to
others in the table. The correlation with the practical data will
become apparent.

In view of the above logic, it now becomes immediately apparent
that for low concentrations only sols can exist as in this case the
cleavage planes can develop at appropriate areas . In the above we
had seen that typically zones of about 200R are generated as isolated regions and their initial state must only be amorphous. This is
due to the inherent tendency of the molecules existing initially
as separate entities. Buffer coalescence follows due to the excitation which can obviously be the light photons or a more effective
heat photons. In as much as the heat photons are short stubs of
energy particles(81 particles), they can be packed to a greater extent in a given cross section and since they do not have to obey
the wavelength-frequency ($\lambda = c/\vartheta$) relation, it is readily seen
that their contribution of energy to the nucleation zones will be
of high order so as to reduce the crystallisation time to a bare minimum. On this score, the conversion to crystals from amorphous state
may even be instantaneous depending upon the temperature employed.
As the concentration of the solution is increased, it is obvious
that the buffers of the individual crystal zones would easily interact and can generate particles or complete coalescence leading to
the formation of larger aggregates which in turn could either exist
individually (precipitate) or again be connected with one another
with a certain overall flexibility (gel). Thus when the concentration corresponds to the formation of precipitate, mere agitation
cannot disengage the coalescence of buffers any longer and hence
no sol can be obtained by such a process.

The role of the inhibitors and retarders must now be considered.
The spectral energy for the inhibitors , $K_3Fe(CN)_6$ and $K_4Fe(CN)_6$
are 75p+ and 67.2p+ respectively while for the retarding agents
KBr and KI the values are 7.1p+ and 1.6p+ respectively. These energy values when compared with that of the table values for the sols
(table-I) will indicate the nature of their activities. Highly p+
substances when added to the solutionwill absorb all the incident
p+ (preferentially with respect to the sol matter)and also prevent
the entry of p+ into the nucleation centers. This would mean that
the crystallization would be delayed indefinitely until when a dust
particle is introduced . In this case, the larger accumulation of
the atoms making up the dust particles has already introduced density gradation effect which is now beyond the control of the inhibitors. Hence crystallization becomes effective. On the same token,
the less effective role of the retarding agents will become obvious
when their p+ values are much lower. Sol formation proceeds when
the p+ content is lowest or the p- content is   enhanced as

for the substances listed in the table-I. This will also clarify
the role of the special additives that can moderate the nucleation
times on the lines discussed for $Al(OH)_3$ and H_2SiO_3 .

As a passing mention, we could note that the typical aging of
sols developing crystalline chains and reticules will also be a
logical conclusion based on the above analyses. In as much as the liquid molecules are always at a higher energy density level during
nucleation and crystallization when compared with the free atmosph-
ere(in contact with the solution), it is logical that some amount
of energy clouds will gradually pass out in due course from the liquid into the atmosphere. This will greatly reduce the buffer strength and size so that the adjacent crystal zones could be brought
nearer to each other. In this manner the coalescence of a number of
crystalline or amorphous units forming chains will be apparent.

The formation of the colloidal solutions by dispersion through
disintegration process is also very important. Usually an inert
medium would be employed for the dispersion. Unlike the dissolution,
dispersion does not occur spontaneously. Hence efforts must be taken
to achieve this. Studies have shown that a solid disintegrates primarily due to a large number of micro cracks contained in it. The
failure of a solid is attributed to cleavage at these microcracks
which widen on the application of external load. Such cracks are
found to be spread out , on the average, at 100R every where on the
surface of the given body(where R is the intermolecular distance)
According to Rehbinder,etal., these microcracks disappear or heal
on the removal of the load(49). The presence of various substances
adsorbed on the micro cracks would greatly alter the development
of the cleavage along these regions. It is believed that such surfactants would generate a large amount of force in aiding the split-
ting of the body along the micro cracks.

The principles of the GET will enable us to visualise the mechanism of formation and dynamics of the microcracks . The details for
this have already been given above. We notice that while the given
atom in a group receives the input energy (particles)in the form of
a force, the density enhancement will force it to act as the core
of the nucleation center. The transfer of the energy particle to the
surrounding neighbours will be from this central atom(in a plane
perpendicular to the line of force). This action will soon set up
in the present instance, the buffer zone around the core atom which
typically would extend to about 36.8R This value is deduced from
the fact that in the free atmosphere, the application of force to
the core atom would raise the density to 1045 p/cc while the substratum value will be $2x10^{40}$ p/cc . Hence the buffer zone will extend
to about R($10\,45\ /2x10^{40})^{1/2}$ =36.8R . A large number of such core atoms
would then generate corresponding isolated zones with the cleavage
planes at the boundaries of the buffers. These cleavage planes along
which the material can shear are the so-called microcracks. Thus
for typical atoms, which have R=1.1 to 1.5 A°, the size of each
microzone , on the average, would be about 36.8x1.3= 48°A. The distance between adjacent microcracks would then be on the average of
about 96 A°, which is close to the quoted value.

The mechanism of formation of the microcracks will be evident

from the figure-4. We find that in a group of atoms of a zone ,
any atom like A can act as the receiver of energy particles. Then
the neighbours like B,C etc., cannot act as the cores . Then another
such core atom like X is displaced from A far away at a mean distance of about 36.8R x1.3x2 = 96 A`so that an identical group of
atoms as for the first zone would occur. Then the cumulative effort
of all all core atoms in sharing the load will become evident.

It can be seen that the nature of the microcracks is purely
physical in character. As long as the load exists$d$, the core atoms
maintain their density level promoting the cleavage at the crack-
lines.and the removal of this load would automatically restore the
original buffer coalescence. In this manner, the microcracks will
disappear on the removal of the load. It must be noted that this
generation and destruction of the microcracks will be the basic property of any solid surface. The extent to which the adjacent surface
regions will retain adhesion during the development of the micrcracks
(for loaded surface) will depend also on the spectral characteristics of the core atom. This will become lear from the fact that
atoms have different types of absorption (p+ and p- to different intensity levels) This would mean that the composition of the
buffer would vary considerably. It can then lead to a situation of
adjacent buffers tending to retain contact by more of electrostatic
attraction than by one of purely physical means. For instance, in
an alloy, atoms of p+ and p- will be generally making up the
material molecules. When such a surface is loaded, the first core
atom may be a p+ atom while the neighbour (more decided locationa-
lly by the density gradation law) may be p- . This means that the
buffers (now opposite in character) will tend to hold together more
firmly than what they might do with a buffer of a polarity like
their own.On this basis, eutectic (mechanical mixtures)alloys can
show a great variation in strength . The role of the specific adsorbed surface atoms will be exactly the same. They tend to either
greatly neutralise or reinforce the buffer of the zones especially
surrounding the microcrack regions. Then the property of the material can be greatly affected by these surfactants during the loading and promote or retard cleavage on lines given above.

6. STRUCTURE OF THE COLLOIDS :

We must now consider the details of the physical and chemical
view points of the structure of the colloids. Jurdis was able to
establish that impurities in the preparatory solutions were always
present and that they were non indifferent to the colloidal systems.
Thus colloids were complex substances and Duclaux was able to establish the more general form of these as micelles. He identified the
active part of the system(a small part of the stabilizer) as the
one which is responsible for interaction with an external applied
electric field. Pauli believed that the micelle contained an inert
nucleus and an active part that could be ionised. (ionogenic complex.). However, these schemes proposed by him did not explicitly
account for the binding of the inert nucleus and ionogenic complex.
Further, the observation of double layer and the zeta potential
with the colloidal system could not be explained by this model.

Paneth assumed the physical adsorption as a causative factor . He showed that   some water insoluble substances adsorb vigorously to ions which form slightly soluble substances with ions of opposite character in crystals. Fajans presumed the colloids as ultramicroscopic crystals . Summing up, both the physical and chemical concepts of colloidal formations lead to the same conclusion , viz., that the ions of the stabilizer electrolyte largely hinders crystal growth in a colloidal particle and impart charges to it thereby improving the aggregate stability of the system. Based on the above and the Paneth-Fajans rule , Peskov defined the structure of the micelle and the colloids . Accordingly, a central region made of the aggregates of atoms , molecules or ions of the dispersed phase together with the potential determining ions of the electrolytes would constitute the nucleus. Surrounding this would be the layer of counterions (part only when compared with the bulk electrolyte) that are so firmly attached to the nucleus that the complex moves always as a whole in an electric field. This is referred to as the colloidal particle proper and will be seen to contain the electric double layer and exhibit a charged state. The bulk liquid surrounding will provide a diffuse zone of counterions and an envelop of the micelle could be considered upto the radius where the condition for electroneutrality is exhibited . On the application of an electric field , the central colloidal particle shears itself from the outer zone of the diffused counterions along the shear plane (of the double layer of the system).

We will find that the above deductions and description of the colloidal particle and micelle are in harmony with the principles of the GET. We had already established that the crystallisation process involved in the building up of a density gradation effect . In as much as the central core would be composed of neutral atoms or molecules or ions , a net charge of either p+ or p- could always result , this charge effect being the manifestation of the spectral absorption. Then the buffer will likewise be polarised and oppositely charged to that of the core. To this many of the potential determining ions (same charge as that of the buffer) could attach themselves and this combined system will always attract some of the counterions of opposite charge. This coupled assembly must behave as an integral unit in as much as it could be identified as the colloidal particle proper. The introduction of the counterions would naturally tend to generate the shear plane as had already been considered in detail earlier. Therefore the presence of the diffused counterion zone beyond the shear plane is also obvious. The micelle can then be defined by the condition of electroneutrality by including the diffused zone of appropriate dimensions.

7. STABILITY OF COLLOIDAL SYSTEMS AND THE KINETICS OF COAGULATION :

Stability of the colloidal systems will be greatly influenced depending upon whether the system is lyophilic or lyophobic. In lyophobic systems, the coagulation rate sharply increases as electrolytes are added. Typically, after passing the critical concentration, the coagulation rate reaches a maximum . Lyophilic systems coagulate if the concentration of the electrolyte added is very high. Unlike in the lyophillic system, in lyophobic system the critical concentration sharply reduces as the charge concentration of the counterion increases. The rapidity of coagulation is decided by the sticking of the particles during the Browninan motion. Smoluchowski found that the force of repulsion between adjacent molecules could be greatly reduced by adding electrolytes of high concentration. Experimental work connected with this had shown that molecular aggregation commences for a given concentration of the electrolyte , when the intermolecular distance D is related to the radius of the molecule as $D/r =2.3$. Muller had shown that ionised particles and particles of different dimensions always aggregate better than smooth and uniform particles. The coagulation rate also increases as the molecules are elongated.

When the principles of the GET are applied to the above, we find that the stabilizing action on the colloidal particles will be pririly dependent upon the buffer characteristics. Since lyophilic systems tend to reduce the number of water molecules adsorbed and would therefore correspond to basically enriched p- central cores and corresponding p+ buffers . Hence if the coagulation rate must be increased, it must be necessary that the electrolyte should have a p- character in order that it reduced the strength of the adjacent buffers(p+) and promoted coalescence. In other words, the elctrolyte will have the same character as that of the counterion or opposite to that of the potential determining ions. This had been amply confirmed by Hardy etal., (see companion paper on sedimentation). It is also obvious that greater is the concentration of the electrolyte, the greater would be the coagulation effect. On this same basis, lyophilic systems (p+ core) must go well with the corresponding electrolyte (p+).

The sharp reduction of the electrolytic concentration in lyophobic systems with increasing counterions will be logical. In as much as the electrolyte and counterions have the same character(p+ or p-) it becomes obvious that increasing the concentration of either one of them will have a sharp effect on buffer neutralisation. Therefore increasing the counterion concentration would be attended by the reduction in the concentration of the electrolyte and on this basis, this feature must appear to be the same for both the lyophilic and lyophobic syatems. However, for the lyophobic systems , the p- charge introduced on the colloidal particles will be relatively very low due to the small number of water molecules adsorbed to the colloid particles , when compared with the lyophilic systems. Hence, on a one to one basis, the reduction in the charge of the buffer will be more effective for the lyophobic system than for the lyophilic system(that tends to build a powerful p- buffer sheath.) .

The influence of the Brownian motion on coagulation can be understood. Although more details of this phenomenon are included in the companion papers, in essence, the motion will obviously lead to the molecules or particles of the colloids approaching one another closer together. Under these conditions, buffer interaction and coalescence (which are interpreted as the molecular attraction in the conventional parlance) will become automatic. In this respect, the minimum distance of approach of the molecular aggregates for coalescence can be easily deduced . Thus, when two such units came closer and separated by a distance D , with the internal density being 10^{45} p/cc, we find that(for the radius of the unit, r) the density at D /2 is given by

$10^{45}x(r/(D/2))^3$, due to each unit considered separately. Thus the
the combined effect on the density by the two units will be $2x10^{45}x$
$(r/(D/2))^3$. If this value reached 10^{45} p/cc, then the buffers in
the zone between the two molecular units being at the same density
level they will coalesce promoting aggregation. Then , $2x10^{45}x(r/(D/2))^3$
= 10^{45} from which D=2.5r . The measured value of D=2.3 r will then
agree well with this.

The physical phenomenon involved will become clear by referring
to figure-5b. It has already been shown by the GET (21) that the
A-circle location for the protonic mass will be at 4.0 A° and that
for any atom of atomic mass m this will be $4x(m)^{1/3}$ A°. Thus for example,
if an element had m=27 amu, its A-circle radius will be at about
12 A° while the actual radius of the atom may be about 1 A°. Hence
when two such atoms are brought closer to each other (figure-5b;A),
to an inter atomic distance of 2.5 r=2.5 A° it is easy to see that
the A-circles are compressed heavily (figure-5b;B).The density at
this face for the A-circle will peak up by a factor of $(12/1.25)^3$=
1000 times or reach $10^{45}x2x1000$ = $2x10^{48}$ p/cc. This is almost the
threshold of the light formation and hence the energy particles at
the interface will tend to lump and produce the light photon which
can be ejected out. This ejection is automatic as the size of the
light photon is very small(although densitywise quite high) when compared with the atoms(which have a lower density). The atoms tend to
remain in place in space and establish density gradation effect.
Alternatively if the light photon got off, then the vacuum left behind would pull the atoms together. This will involve the gradual
coalescence of the outer regions of the A-circles as shown (figure-
5b;C)until the closest approach is made possible(figure-5b;D). This
corresponds to the case of just 10^{45} p/cc being established at the
mid point of the system which corresponds to the condition of D=2.5r
(figure-5 a).

The aggregation of the colloidal particles depending upon the
coalescence effect of the buffers will be evident from figure-5c.
When particles of smooth surface(spherical surface)of same dimensions tend to coalesce it will become obvious that that the gap X among
the units will be more, calling for greater coalescence effect or
additional energy clouds -i.e., the coagulation is more difficult.
But when particles of unequal sizes are involved, the gap space Y
is minimum and hence easy building of the aggregates will be possible. On the same token, rough surfaces will generate protruding buffers Z in some directions where attachment of an incoming unit will
be relatively be relatively easy(guffer coalescence). This will mean
that rough surfaces will promote easy aggregation as known experimentally.

There are many other properties of colloids that would require
investigations. These are covered in the companion papers on aerosols
and sedimentation theory.

REFERENCES :

The references(being common to all), are listed in the companion
paper on the sedimentation theory.

TABLE - I ., CALCULATION OF CRYSTALLISATION TIME :

Type of Sol	overall molecular size	(Vol of molecule / Vol of Au particle)	Spectral energy ($\times 10^{-12}$ erg)	density (δs)*	$\delta s/\delta g$	energy factor	τ'_{cal}	τ' exptl.
Au	1.34A°		8.3p+	19.3(δg)			0.0623	Hr
V_2O_5	3.74	21.6	0.3p+	5.9	3.28	0.33	1.5Hrs	1Hr
TiO_2	2.70	8.15	0.5p+	4.54	4.26	1	2.18Hrs	1-2Hr
Al(OH)$_3$	3.26	14.4	11.4p-	2.7	7.15	4.45	27.8Hr	1 day
H_2SiO_3	2.72	8.35	6.4p-	2.4	8.05	3900	685 days	720days
* Values correspond to			parent	atoms	V, Ti, Al, Si			

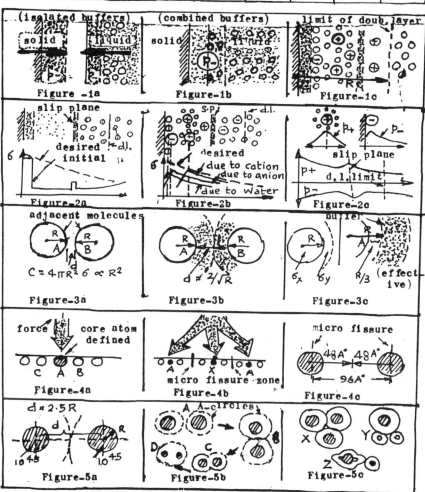

(isolated buffers)

Figure -1a

(combined buffers)

Figure-1b

limit of doub. layer

Figure-1c

slip plane · desired · initial

Figure-2a

desired · due to cation · due to anion · due to water

Figure-2b

slip plane · d.l.limit

Figure-2c

adjacent molecules

$C \approx 4\pi R^2 \quad 6 \propto R^2$

Figure-3a

$d \approx 2/\sqrt{R}$

Figure-3b

buffer · R/3 (effective)

Figure-3c

force · core atom defined

Figure-4a

micro fissure zone

Figure-4b

micro fissure · 48A° · 48A° · 96A°

Figure-4c

$d \approx 2.5R$

Figure-5a

A A-circles

Figure-5b

Figure-5c

291

FLUIDIZED BEDS

Mathematical Modeling of Spouted Bed Hydrodynamics and Numerical Solution

B. KILKIŞ
Middle East Technical University
İnönü Bulvarı 06531
Ankara, Turkey

K. DÜNDAR
Gazi University, Tech. Ed. Fac.
Beşevler, Ankara, Turkey

Abstract

Spouted beds are used specifically for drying, coating, combustion and heat exchanging purposes. Determination of the spoutability characteristics, particle migration and spout shape is crucial at the design stage for an efficient and satisfactory operation. In this paper a mathematical model which gives the exact solution in two dimensions for the flow pattern and the pressure distribution of the fluid in the annulus is developed by using only the mass and momentum equations. For the spout; spout diameter, fluid and the solid flow patterns are solved numerically along the bed height. Bubble growth is also determined which occurs at low fluid velocities. An iterative computer program was developed which tabulates and graphically demonstrates the spouted bed hydrodynamics in terms of fluid, and particle flow, spout shape and bubble growth.

Paper gives relevant theory, algorithm and sample solutions.

1. INTRODUCTION

Some of the basic concerns in the analysis and design of spouted beds are the determination of the flow pattern and the pressure field of the fluid and spout shape. These are closely related with other design considerations such as spoutability, minimum spouting velocity and maximum spoutable bed depth. In order to analyse the hydrodynamics, two distinct regions can be distinguished. These are the spout region and annulus region (see Figure 1). Following this distinguisment, a two zone model has been developed.

Figure 2 shows the annulus element variables. Figure 3 shows the centrally located spout element variables.

U_r is the horizontal component of the superficial fluid velocity across the annulus boundary. U_a is the superficial velocity in the annulus. Another important parameter is the pressure P in the annulus. Both p, U_r and U_a depend upon the height (z) and radius r. The void fraction ε_a in the annulus element is taken to be constant because there is no fluidization.

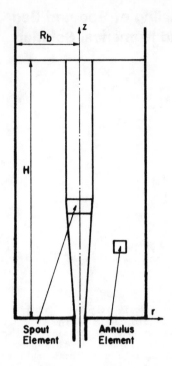

Fig. 1. - Two Region Model of a Cylindrical Spouted Bed.

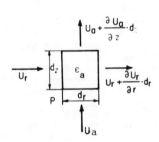

Fig. 2. - Annulus Element.

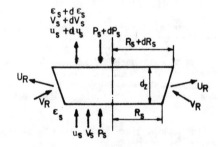

Fig. 3. - Spout Element.

In the spout element, R_s is the variable spout radius, P_s is the spout pressure, ε_s the void fraction in the spout at a given level. u_s is the interstital velocity of fluid in the spout and V_s is the av.velocity of solid particles in the spout U_R is superficial fluid velocity towards the annulus and V_R is the incoming particle velocity to the spout. These are variables of height (z).

Volpicelli [1] used Darcy's equation to formulate the pressure drop around an incipient spout, still in the form of a bubble. He assumed that spout pressure is constant in two dimensions and further took the fluid density to be constant. Using Laplace equation and applying boundary conditions, he obtained the pressure field;

$$\nabla p = - KU \tag{1}$$

$$\nabla^2 p = 0 \tag{2}$$

Mathur [2] assumed that pressure drop around the spout is one dimensional and modified Darcy's law (Equation 1) for the annulus:

$$\frac{dp}{dz} = -KU_a^{\,n} \tag{3}$$

Here, K and n are experimentally determined constants. Lefroy and Davidson [3] assumed a constant spout radius, R_s, and expressed the conservations of mass and momentum for fluid and solid particles in the spout as follows:

$$\pi R_s^{\,2} \frac{d(\varepsilon_s u_s)}{dz} + 2\pi R_s U_R = 0 \tag{4}$$

$$\pi R_s^{\,2} \frac{d[(1-\varepsilon_s)V_s]}{dz} - 2\pi R_s(1-\varepsilon_s)V_R = 0 \tag{5}$$

$$\rho_f \frac{d(\varepsilon_s u_s^{\,2})}{dz} = - \varepsilon_s \frac{dP_s}{dz} - \beta(u_s - V_s)^2 \tag{6}$$

$$\rho_s \frac{d[(1-\varepsilon_s)V_s^{\,2}]}{dz} = - (1-\varepsilon_s)\frac{dP_s}{dz} - (\rho_s-\rho_f)(1-\varepsilon_s)g + \beta(u_s - V_s)^2 \tag{7}$$

β is the interaction factor and is expressed at high Reynolds numbers where the drag coefficient C_D is constant:

$$\beta = \frac{3C_D\rho_f(1-\varepsilon_s)}{4d_p\varepsilon_s^{\,1.78}} \tag{8}$$

At low Reynolds numbers and high solidity regions, C_D is not constant. Mathur [2], and Ergun [4] discussed the following empirical equation for a wide range of void fractions:

$$C_D = \frac{200(1-\varepsilon)\mu}{d_v \rho_f U_m \varepsilon^3} + \frac{7}{3\varepsilon^3} \tag{9}$$

U_R is proportioned by Mathur [2] to the net downward pressure for solids in the annulus, P_b as defined by Hattori [5]:

$$P_b = (\rho_s - \rho_f)g(1-\varepsilon_a)(H-z) - P_a \tag{10}$$

$$P_b = k\rho_f \frac{U_R^2}{2} \tag{11}$$

2. MATHEMATICAL MODELLING

The following assumptions were made:
- Fluid density is constant
- Void fraction in the annulus (ε_a) is constant.

2.1. Annulus Region

For the element shown in Figure 2, continuity equation:

$$[U_r + \frac{\partial U_r}{\partial r} dr]dz - U_r dz + [U_z + \frac{\partial U_z}{\partial z} dz] \, 2rdr - U_z 2rdr = 0 \tag{12}$$

or shortly:

$$\frac{\partial U_z}{\partial z} + \frac{\partial U_r}{\partial r} + \frac{U_r}{r} = 0 \tag{13}$$

Pressure and velocity relations are obtained from Darcy's law:

$$\frac{\partial p}{\partial z} = - KU_a \tag{14}$$

$$\frac{\partial p}{\partial r} = - KU_r \tag{15}$$

Upon insertion of Equations (14) and (15) into Equation (13), one obtains the Laplace Equation (Eq.2):

$$\frac{\partial^2 p}{\partial z^2} + \frac{\partial^2 p}{\partial r^2} + \frac{\partial p}{r \partial r} = 0 \tag{16}$$

where the boundary conditions are:

For $\quad z = H \qquad p = 0 \tag{17}$

For $\quad z = 0 \qquad U_z = 0 \quad$ or $\quad \dfrac{\partial p}{\partial z} \Big]_{z=H} = 0 \tag{18}$

For $r = R_b$ $\qquad U_r = 0$ or $\dfrac{\partial p}{\partial r}\Big]_{r=R_b} = 0$ $\hfill$ (19)

For $z = H$ $\qquad U_z = U_{mf}$ or $\dfrac{\partial p}{\partial z}\Big]_{z=H} = -KU_{mf}$ $\hfill$ (20)

With the separation of variables:

$$p = \Gamma \cdot \xi \hfill (21)$$

Thus Laplace Equation becomes:

$$\frac{d^2\xi}{dz^2} + \xi\frac{d^2\Gamma}{dr^2} + \frac{\xi}{r}\frac{d\Gamma}{dr} = 0 \hfill (22)$$

Dviding it into Γ and ξ product and changing places:

$$\frac{1}{\xi}\frac{d^2\xi}{dz^2} = -\frac{1}{\Gamma}\frac{d^2\Gamma}{dr^2} - \frac{1}{\Gamma r}\frac{d\Gamma}{dr} = \text{a constant} = \pm\,\lambda^2 \hfill (23)$$

$-\lambda$ is compatible with the boundary conditions:

$$\frac{d^2\xi}{dz^2} + \xi\lambda^2 = 0 \hfill (24)$$

$$\Gamma\frac{d^2\Gamma}{dr^2} + \frac{d\Gamma}{dr} - \lambda^2 r\Gamma = 0 \hfill (25)$$

With the general solutions:

$$\xi = C_1\sin(\lambda z) + C_2\cos(\lambda z) \hfill (26)$$

$$\Gamma = C_3 I_o(\lambda r) + C_4 K_o(\lambda r) \hfill (27)$$

Here I_o and K_o is the p.th order Bessel function's zero order series:

$$I_p(\lambda r) = \sum_{k=0}^{\infty} \frac{(\frac{\lambda r}{2})^{2k+p}}{k!(k+p)!} \hfill (28)$$

$$K_p(\lambda r) = (-1)^{n+1}I_p(\lambda r)\,\ln(\frac{\lambda r}{2}) + \frac{1}{2}\sum_{k=0}^{p-1}\frac{(p-k-1)!(-1)^k}{2^{-p+2k}\,r}(\lambda r)^{-p+2k}$$

$$+ (-1)^p\frac{1}{2}\sum_{\ell=0}^{\infty}\frac{(\lambda r)^{p+2k}}{2^{p+2\ell}\ell!(p+\ell)!}[\psi(\ell+p)+\psi(\ell)] \hfill (29)$$

$$\psi(m) = 0.5772157 + 1 + \frac{1}{2} + \frac{1}{3} + \ldots\ldots + \frac{1}{m} \tag{30}$$

Only the zeroth and first order functions are used. When it is necessary, the following differentiation properties were employed:

$$\frac{d}{dr} [rPI_p(\lambda r)] = \lambda rPI_{p-1}(\lambda r) \tag{31}$$

$$\frac{d}{dr} [r^P K_p(\lambda r)] = -\lambda r^P K_{p-1}(\lambda r) \tag{32}$$

$$\frac{d}{dr} I_p(\lambda r) = \lambda I_{p-1}(\lambda r) - \frac{P}{r} I_p(\lambda r) \tag{33}$$

$$\frac{d}{dr} K_p(\lambda r) = -\lambda K_{p-1}(\lambda r) - \frac{P}{r} K_p(\lambda r) \tag{34}$$

$$\frac{d}{dr} I_p(\lambda r) = +\lambda I_{p+1}(\lambda r) + \frac{P}{r} I_p(\lambda r) \tag{35}$$

$$\frac{d}{dr} K_p(\lambda r) = -\lambda K_{p+1}(\lambda r) + \frac{P}{r} K_p(\lambda r) \tag{36}$$

For small λr values the following approximations can be made in order to reduce computation time, by Hildebrand [6] :

$$I_0(\lambda r) \sim 1 \tag{37}$$

$$I_1(\lambda r) \sim \frac{\lambda r}{2} \tag{38}$$

$$K_0(\lambda r) \sim - \ln(\lambda r) \tag{39}$$

$$K_1(\lambda r) \sim \frac{1}{\lambda r} \tag{40}$$

Applying boundary condition (18) to Equation (26):

$$C_1 = 0 \tag{41}$$

Applying boundary condition (19) to Equation (27):

$$C_4 = C_3 \frac{I_1(\lambda R_b)}{K_1(\lambda R_b)} \tag{42}$$

From boundary condition (17) and Equation (26):

$$\lambda = \frac{\pi}{2H} (2n-1) \qquad n=1,2,3 \tag{43}$$

For the most suitable smallest χ value for b. condition 20, Equation (43) is interpreted:

$$\chi = \frac{\pi}{2H} \tag{44}$$

In this case a second condition is imposed to b. condition 20 :

for $\quad z = H \quad$ and $\quad r = R_{st} \qquad U_z = U_{mf} \quad$ or $\quad \dfrac{\partial p}{\partial z} \bigg|_{\substack{z=H \\ r=R_{st}}} = -KU_{mf}$ $\hfill (45)$

Here R_{st} is the spout radius at the top (see Figure 4). Using b. values in Equations (41) to (45) in Equation (26) and (27), Equation (21) can be solved:

$$p = aKcos(\lambda z) \ [\ I_0(\lambda r) + bK_0(\lambda r)\ ] \tag{46}$$

$$\lambda = \frac{\pi}{2H} \tag{47}$$

$$b = \frac{I_1(\lambda R_b)}{K_1(\lambda R_b)} \tag{48}$$

$$a = \frac{U_{mf}}{\lambda[I_0(\lambda R_{st})+bK_0(\lambda R_{st})]} \tag{49}$$

Using Equations (14) and (15) superficial velocities in the annulus can be found:

$$U_r = - \lambda acos(\lambda z) \ [I_1(\lambda r) - bK_1(\lambda r)] \tag{50}$$

$$U_a = \lambda a \ sin(\lambda z) \ [I_0(\lambda r) + bK_0(\lambda r)] \tag{51}$$

K can be found from the following condition:

$$\frac{dP}{dz} = -KU_{mf} = -(\rho_s-\rho_f) \ g \ (1-\varepsilon_a)$$

Thus:

$$K = \frac{(\rho_s-\rho_f) \ g \ (1-\varepsilon_a)}{U_{mf}} \tag{52}$$

U_{mf} is given in Equation (79).

Streamline function ψ is determined in following manner:

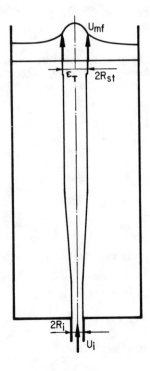

Fig. 4. - Spout Velocity Profile at the Top.

$$\frac{d_z}{d_r} = \frac{U_z}{U_r} = \frac{\lambda a \sin(\lambda z) \ [I_o(\lambda r) + bK_o(\lambda r)]}{-\lambda a \cos(\lambda z) \ [I_1(\lambda r) - bK_1(\lambda r)]} \tag{53}$$

Rearranging and upon integration:

$$\int \frac{\cos(\lambda z)}{\sin(\lambda z)} \, dz = - \int \frac{I_o(\lambda r) + bK_o(\lambda r)}{I_1(\lambda r) - bK_1(\lambda r)} \, dr$$

Using definitions in Equations (33) and (34),

$$X = I_1(\lambda r) - bK_1(\lambda r)$$

$$\frac{dX}{dr} = I_o(\lambda r) + bK_o(\lambda r) - \frac{I_1(\lambda r)}{r} + \frac{bK_1(\lambda r)}{r}$$

$$\int \frac{\cos(\lambda z)}{\sin(\lambda z)} \, dz = - \int \frac{\frac{dX}{dr} - \frac{X}{r}}{X} \, dr = - \int \frac{dX}{X} + \int \frac{dr}{r}$$

$$\ell n(\sin(\lambda z)) = - \ell nX + \ell nr + \ell n \ C_5 = \ell n \ C_5 \ \frac{r}{X}$$

$$\sin(\lambda z) = C_5 \ \frac{r}{X}$$

$$C_5 = \frac{\sin(\lambda z)}{r} \ [I_1(\lambda r) - bK_1(\lambda r)] \tag{54}$$

C_5 is replaced by streamline function ψ:

$$\psi = \frac{\sin(\lambda z)}{r} \ [I_1(\lambda r) - bK_1(\lambda r)] \tag{55}$$

2.2. Spout Region

A modelling was presented earlier by Dündar and Kılkış [7]. Using the element shown in Figure 3 and writing the mass conservation:

$$\pi d(\rho_f \varepsilon_s u_s R_s^2) + 2\pi R_s \rho_f U_R dz = 0 \tag{56}$$

$$U_R = \sqrt{U_{Rr}^2 + U_{Rz}^2} \tag{57}$$

$$U_{Rr} = -\lambda a \cos(\lambda z) \ [I_1(\lambda R_s) - bK_o(\lambda R_s)] \quad \text{(see Eqns.50,51)} \tag{58}$$

$$U_{Rz} = -\lambda a \sin(\lambda z) \ [I_o(\lambda R_s) + bK_o(\lambda R_s)] \tag{59}$$

From momentum conservation:

303

$$\pi d(\rho_f \varepsilon_s u_s^2 R_s^2) + 2\pi R_s \rho_f u_s U_R dz = -\pi \varepsilon_s R_s^2 dP_s + F_a + F_s + F_v \tag{60}$$

Multiplying Equation (56) by u_s and the resulting term is subtracted from Equation (57):

$$\frac{\pi \rho_f \varepsilon_s R_s^2}{2} d u_s^2 = -\pi \varepsilon_s R_s^2 dP_s + F_a + F_s + F_v \tag{61}$$

where:

$$F_a = -\pi \varepsilon_s R_s^2 \rho_f g \, dz \quad \text{(weight)} \tag{62}$$

and:

$$F_s = -C_D A \frac{\rho_f}{2} U_m^2 = -\pi R_s^2 \beta (u_s - V_s)^2 \quad \text{(fluid-solid friction), A:Area} \tag{63}$$

For high velocities, C_D is unity [3] and for β, Eq. (8) is used. For F_v (viscosity effect). The term u_s is average fluid velocity in the spout.

$$\text{For} \quad r = 0 \qquad \frac{du_{sr}}{dr} = 0 \quad , \quad u_{sr} \text{ is the radial component} \tag{64}$$

$$\text{For} \quad r = R_s \qquad u_{sr} = \frac{U_{Rz}}{\varepsilon_a} \tag{65}$$

$$u_s \pi R_s^2 = \int_0^{R_s} u_{sr} \, 2\pi r dr \tag{66}$$

$$u_{sr} = 2 \frac{r^2}{R_s^2} \left(\frac{U_{Rz}}{\varepsilon_a} - u_s \right) + 2u_s - \frac{U_{Rz}}{\varepsilon_a} \tag{67}$$

$$F_v = \tau . A = \mu \frac{du_{sr}}{r} \Big]_{r=R_s} \varepsilon_s 2\pi R_s dz = \pi 8 \mu \varepsilon_s \left(\frac{U_{Rz}}{\varepsilon_a} - u_s \right) dz \tag{68}$$

Upon insertion of these terms and simplification, Equation (61) becomes:

$$\frac{\rho_f}{2} \frac{du_s^2}{dz} = -\frac{dP_s}{dz} - \rho_f g - \frac{\beta}{\varepsilon_s} (u_s - v_s)^2 + \frac{8\mu}{R_s^2} \left(\frac{U_{Rz}}{\varepsilon_a} - u_s \right) \tag{69}$$

Here P_s is the pressure at annulus boundary, R_i, through Equation (46). For radii smaller than R_i:

$$P_s = aK\cos(\lambda z) [I_o(\lambda R_i) + bK_o(\lambda R_i)] \tag{70}$$

304

From mass conservation and momentum conservations for solids:

$$\pi d[\rho_s(1-\varepsilon_s)R_s^2 V_s] - 2\pi R_s \rho_s (1-\varepsilon_s) V_R \ dz = 0 \tag{71}$$

$$\frac{1}{(1-\varepsilon_s)R_s^2} \frac{d[\rho_s(1-\varepsilon_s)R_s^2 V_s^2]}{dz} + \frac{F_m}{dz} = -\frac{dP_s}{dz} - (\rho_s-\rho_f)g$$

$$+ \frac{\beta}{(1-\varepsilon_s)}(u_s-V_s)^2 - \frac{8\mu}{(1-\varepsilon_s)R_s^2}\left(\frac{U_{Rz}}{\varepsilon_a} - u_s\right) \tag{72}$$

Here F_m is the momentum function of solid particles penetrating to annulus:

For $\quad V_R \geq 0 \qquad F_m = 0 \tag{73}$

For $\quad V_R < 0 \qquad F_m = \rho_s(1-\varepsilon_s) \ 2R_s V_R V_s \ dz \tag{74}$

Equations (56), (69), (71), (72) are the four out of five equations necessary to solve ε_s, R_s, u_s, V_s and U_R. Assuming a linear dependence of ε_s along z, gives the fifth equation:

For $\quad z = H, \qquad \varepsilon_s = \varepsilon_T \tag{75}$

For $\quad z = 0 \qquad \varepsilon_s = 1 \tag{76}$

$$\varepsilon_s = 1 - \varepsilon_T \frac{z}{H} \ , \quad \varepsilon_T \text{ is the voidance at the top.} \tag{77}$$

U_{mf} is found by solving Equations (69) and (70) simultaneously:

$$u_{if} = \frac{-\dfrac{150\mu d_p(1-\varepsilon_a)}{d_v^3 \varepsilon_a} + \sqrt{\left(\dfrac{150\mu d_p(1-\varepsilon_a)^2}{d_v^3 \varepsilon_a}\right) + 4(\rho_s-\rho_f)g \ \varepsilon_a^2 \ \dfrac{1.75\rho_f d_p^2}{d_v^3}}}{2\dfrac{1.75\rho_f d_p^2}{d_v^3}} \tag{78}$$

u_{if} is minimum fluidization superficial fluid velocity.

$$U_{mf} = \varepsilon_a u_{if} \ , \quad U_{mf} \text{ is min.flu.superficial velocity.} \tag{79}$$

3. NUMERICAL SOLUTIONS

For the solutions of equations elaborated in the previous section, an interactive computer program was developed. It was written in BASIC Language and for APRICOT PC. The details of the algorithm was given by Kürşat [8]. The program is mainly capable of outputting 4 solutions. These are:

- Spout shape
- Equal pressure lines
- Stream lines
- Pre-spouting regime geometry.

The typical outputs are shown in Figure 5.

It is also possible to elaborate on these results like the determination of the fluid percentage transferring to annulus. A typical graph for wheat is shown in Figure 6. Experimental results in the literature show an agreement within 20 % with the findings.

It is also possible to determine the minimum spouting conditions. A typical output for wheat grains spouted by air is given in Figure 7 for the following inputs:

inlet radius=0.64 cm.
$$R_b = 7.6 \text{ cm} \qquad \varepsilon_a = 0.40$$
$$H = 30 \text{ cm.}$$

wheat diameter: 3.20 mm

Minimum spouting conditions thus found were compared with literature for several cases. A comparison is shown in Figure 8, where the minimum spouting velocity is plotted against bed diameter.

4. CONCLUSIONS

A computer program was developed with extensive analytical solutions both for the spout and the annulus region. Computer program idealizes the spouted bed in horizontal finite elements. By imputing the bed geometry, particle and fluid inlet properties feed rates, it is quite possible to accurately predict the spouting hydrodynamics this will enable designers to choose the optimum bed geometry and operational characteristics. Sample solutions gave agreeable results with previous approaches and experiments available in the literature.

5. NOMENCLATURE

a,b,c	Constants
C_D	Drag coefficient
D_b	Bed diameter
D_i	Inlet nozzle diameter
d	Particle diameter , (d_v = volume equ. particle diameter)
F^p	Fluid weight function
F^a	Momentum function for particles transferring to annulus
F^m	Fluid-particle friction function
F^s	Viscosity function
g	Gravitational acceleration
H	Bed height
p,P	Pressure , $p = p(r,z)$, $P = P(z)$
P	Annulus pressure
p^a	Solid particle weight pressure
p^b_o	Pressure at the top of bed

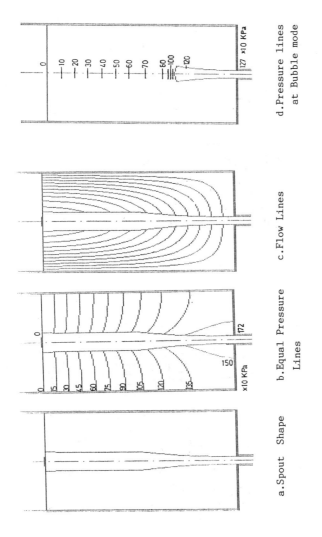

a.Spout Shape   b.Equal Pressure   c.Flow Lines   d.Pressure lines
                Lines   at Bubble mode

Fig. 5. – Typical Outputs of the Program.

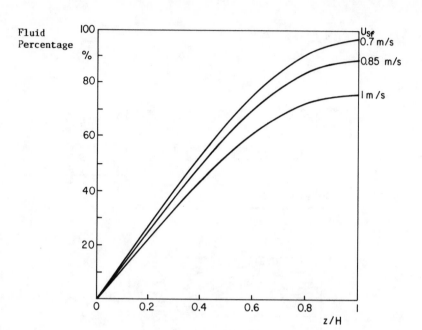

Fig. 6. - Fluid Percentage Passing to Annulus.Material: Wheat.

U_{sf}:Inlet Fluid Velocity

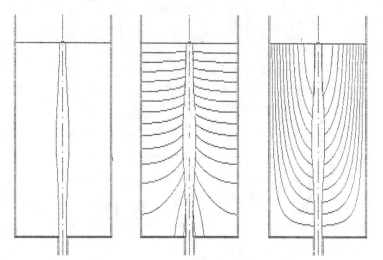

Fig. 7. - Solution at Minimum Spouting Conditions.

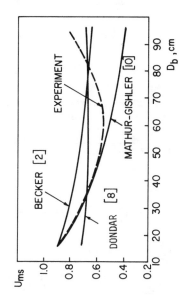

Fig. 8. – Dependence of U_{ms} on Vessel Diameter.

U_{ms}:Inlet Fluid velocity corresponding to minimum spouting conditions.

P_s	Spout pressure
r	Radial dimension
R_b	Bed radius
R_s	Spout radius
z	Vertical axis
β	Interaction factor
ε	Void fraction
ψ	Streamline function
ρ_f	Fluid density
ρ_s	Solid density
μ	Viscosity

REFERENCES

1. Volpicelli, G., "Gas-Solid Systems with Pulsating Feed", Chem. Ind., Milan 45, 1362 (1963).

2. Mathur, K.B., and Epstein, N., Spouted Beds, Academic Press, London (1974).

3. Lefroy, G.A., and Davidson, J.F., "Mechanics of Spouted Beds", Trans. Inst. Chem. Eng., 47, T120 (1969).

4. Ergun, S., "Fluid Flow Through Packed Columns", Chem. Eng. Prog., 48, 2, 89 (1952).

5. Mamuro, T., and Hattari, H., "Flow Patterns of Fluid in Spouted Beds", J. Chem. Eng. Jap., 1, 1 (1968).

6. Hildebrand, F.B., Advanced Calculus for Applications, 3rd. ed., Prentice Hall, Inc., New Jersey (1962).

7. Kürşat, D., Kılkış, B., "Numerical Analysis of Spouted-Bed Hydrodynamics", The Can. J. of Chem. Eng., 61, 293 (1983).

8. Kürşat, D., Mathematical Modelling of Spouted Bed Hydrodynamics, (in Turkish), Ph.D. Thesis, Gazi University, Ankara (1988).

9. Becker, H., "An Investigation of Laws Governing the Spouting of Coarse Particles", Chem. Eng. Sci., 13, 245 (1961).

10. Mathur, K.B., and Gishler, P.E., "A Technique for Contacting Gases with Coarse Solid Particles", A.I.Ch.E.J., 1, 157 (1955).

Modeling of Liquid-Solid Jets from Fluidized Beds

R. B. H. TAN and J. F. DAVIDSON
University of Cambridge
Department of Chemical Engineering
Pembroke Street, Cambridge CB2 3RA, UK

Abstract

A simple model for the axi-symmetric flow of a liquid-solid suspension in a converging nozzle has been developed. A theoretical nozzle profile is obtained which eliminates interparticle stresses. An improved method of accounting for the liquid-solid slip velocity based on a realistic expression for the drag in a concentrated suspension is presented. Experiments on the generation of dense solid-liquid jets from a fluidized bed at high pressure are briefly described. Data for solid and liquid flowrates at various bed liquid fractions and the effect of pressure on solid and liquid discharge agree well with predictions of the proposed model.

1. INTRODUCTION

It has been known for some time that fluidized solids will flow horizontally through an orifice in a manner very much like a liquid jet. Massimilla, *et al* [1], Stemerding, *et al* [2], Jones and Davidson [3] and Burkett, *et al* [4] have shown experimentally that the flowrate of solids from a gas-fluidized bed through an orifice in the side of the containing vessel obeys Bernoulli's equation. This result has been confirmed for flow from moderately pressurized air-fluidized beds of up to 0.4 bar overpressure by Martin and Davidson [5].

These investigations have also shown that the volume fraction of fluid in the jet is higher than in the fluidized bed. This effect was explained by considering percolation of fluid through the interstices between the moving particles [3]. Various theoretical models have been developed to predict the flow of particles and fluid through the orifice. Some of these have been reviewed by Massimilla [6]. Jones and Davidson [3] proposed a solution of the one-dimensional equations of motion for the separate phases: Bernoulli's equation was used to predict the particle velocity and Darcy's Law for the percolating flow. Their theory also proposed a nozzle shape for which the theoretical interparticle stress was specified to be zero.

The development of theory for the axi-symmetric case is considered here. Also, the validity of Darcy's Law in predicting the fluid-particle slip velocity is critically examined, and an alternative approach based on an estimation of drag in a concentrated suspension is proposed. Experiments are reported which give data for solid and liquid flowrates of water-sand jets from a particulately fluidized bed; these data are compared with the theory. In particular, the effects of varying the supply pressure and the bed voidage on the flowrates of solid and liquid will be examined.

2.   EQUATIONS OF MOTION

The approach of Jones and Davidson [3] for one-dimensional gas-solid flow is here extended to model the flow of a fluidized liquid-solid suspension through an axisymmetric nozzle.   Figure 1 shows the coordinate system.   The nozzle has a circular cross-section, with exit radius r_e and length   L.   The r-coordinate is measured radially from the nozzle centreline and the z-coordinate measured axially from the nozzle exit.

It is assumed that fluid and particles enter the nozzle from a particulately fluidized bed, at pressure P_0,   voidage fraction ε_0,   and that the nozzle discharges to atmosphere.   Gravity effects are neglected.

Under these conditions, the equations of continuity for the separate phases are

$$\text{div} \, [\, \underline{v} \, (\, 1 - \varepsilon \,) \,] = 0 \qquad , \qquad\qquad\qquad\qquad (1)$$

and

$$\text{div} \, [\, \underline{u} \, \varepsilon \,] = 0 \qquad , \qquad\qquad\qquad\qquad (2)$$

where $\underline{v}$ and $\underline{u}$ represent the absolute particle and liquid velocities respectively. Following the two-phase theory as expounded by Davidson and Harrison [7], assuming constant voidage fraction ε_0, $\underline{v}$ and $\underline{u}$ and the fluid interstitial pressure P_f are related by the equation

$$(\, \underline{u} - \underline{v} \,) = -K \, \text{grad} \, P_f \qquad , \qquad\qquad\qquad\qquad (3)$$

where   K   is the constant for Darcy's Law relating interstitial velocity and the pressure gradient for the bed of particles.   Further, we assume that the bulk flow of particles and fluid under an applied total pressure gradient is governed by Bernoulli's equation

$$\text{grad} \, (\, P + \frac{1}{2} \rho_0 \, | \, \underline{v} \, |^2 \,) = 0 \qquad , \qquad\qquad\qquad\qquad (4)$$

where ρ_0 is the bulk density of the suspension,

$$\rho_0 = \rho_s \, (\, 1 - \varepsilon_0 \,) + \rho_f \, \varepsilon_0 \qquad . \qquad\qquad\qquad\qquad (5)$$

The total pressure,   P , is assumed to be the sum of P_f , the interstitial fluid pressure and P_p , an interparticle pressure. P_p may be interpreted physically as arising from the stresses between solid particles in contact in the fluid stream.   Then, following Jones and Davidson [3], we seek a solution to equations (1) to (5) subject to the condition that $P_p = 0$, ie the particles may approach true inviscid flow.   This requires

$$P = P_f \qquad . \qquad\qquad\qquad\qquad (6)$$

312

This condition is consistent with the above-mentioned assumption that the voidage ε remains constant at ε_0, since the particles have no tendency to compact or move apart. Hence,

$$\varepsilon = \varepsilon_0 \qquad . \qquad\qquad\qquad\qquad (7)$$

An analytical solution of equations (1) to (7) in axi-symmetric coordinates has been obtained by Tan [8]. A nozzle profile based on a theoretical streamline is shown in Figure 2 for a nozzle with $r_e = 1.0$ mm, $L = 25$ mm. The corresponding one-dimensional solution of Jones and Davidson [3] is also plotted for comparison.

3. NUMERICAL SOLUTION

Defining a velocity potential function ϕ by $\underline{v} = \text{grad } \phi$, equations (1) and (7) yield

$$\nabla^2 \phi = 0 \qquad . \qquad\qquad\qquad\qquad (8)$$

Also, substituting equations (1), (2), (6) and (7) into (3),

$$\nabla^2 P = 0 \qquad . \qquad\qquad\qquad\qquad (9)$$

Equations (8) and (9) are Laplace's equation, and are readily solved by a numerical finite-difference technique with realistic boundary cobnditions for the region defined in Figure 3 as follows:

a. At the vertical vessel wall, $\dfrac{\partial P}{\partial z} = 0$; $\dfrac{\partial \phi}{\partial z} = 0$.

b. Along the theoretical nozzle profile, $\dfrac{\partial P}{\partial n} = 0$; $\dfrac{\partial \phi}{\partial n} = 0$.

where n is the direction of the normal to the nozzle profile.

c. At the free jet surface, $P = 0$; $\dfrac{\partial \phi}{\partial r} = 0$.

d. At a cross-section of the jet downstream from the nozzle,

$P = 0$; $\dfrac{\partial \phi}{\partial z} = v_0$, where $v_0 = \sqrt{\dfrac{2P_0}{\rho_0}}$.

e. Along the central axis of the nozzle, $\dfrac{\partial P}{\partial r} = 0$; $\dfrac{\partial \phi}{\partial r} = 0$.

by symmetry.

313

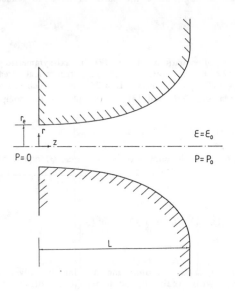

Fig. 1. – Coordinate System for Flow in an
Axi-symmetric Converging Nozzle.

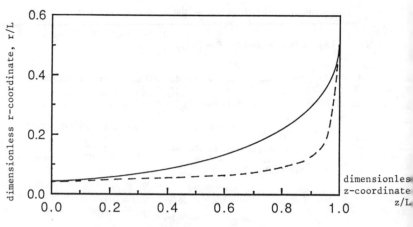

Fig. 2. – Theoretical Nozzle Profiles for
Zero Particle Stresses.
———— 3-dimensional solution (Tan 8)
-----1-dimensional solution (Jones & Davidson 3)

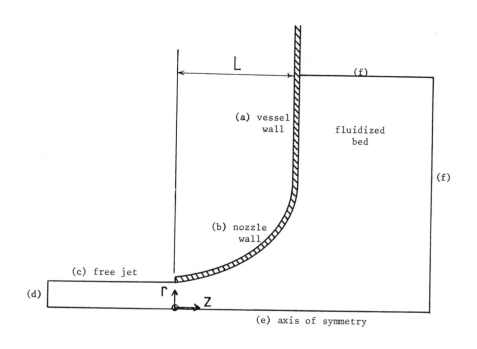

Fig. 3. – Domain for Numerical Solution of
Pressure P, and Velocity Potential ϕ.

(a) to (f) correspond with (a) to (f)
in Section 3.

f.   In the fluidized bed, sufficiently far from the nozzle, take
$P = P_0$; $\phi = 0$.

A numerical solution for the axi-symmetric nozzle profile with $r_e = 1$ mm, and $L = 25$ mm was obtained by iterative solution of the finite-difference equations. A variable grid-size procedure was developed in order to create a finer mesh of points in the region of greatest change in P or ϕ , ie near the nozzle exit.

Numerical values of pressure, pressure drop, velocity potential and gradient of velocity potential in the z-direction at $z = 0$ (the nozzle exit) are tabulated in Table 1 for different radial positions.

The results in Table 1 show that the flow at the nozzle exit is essentially one-dimentional. Further, the magnitude of the mean slip velocity between the liquid and solid phases may be obtained from equation (3),

$$(u-v)_e = \frac{1}{\pi r_e^2} \int_0^{r_e} 2\pi r K \left.\frac{\partial P}{\partial z}\right|_{z=0} dr \tag{10}$$

By simple quadrature, we find, for the nozzle in question, that $(u-v)_e = 3.714 \ (KP_0/L)$. Note that (KP_0/L) is the theoretical percolating velocity in a porous slab of thickness L , permeability K , under a total pressure drop P_0.

The magnitude of K may be estimated by the Carman-Kozeny equation, giving

$$K = \frac{d_p^2 \ \varepsilon^2}{180 \ \mu_f \ (1-\varepsilon)^2} \tag{11}$$

With $\mu_f = 1 \times 10^{-3}$ Ns/m^2 for water, $d_p = 0.12$ mm for spherical particles and taking a typical voidage fraction $\varepsilon_0 = 0.5$, the slip velocity is

$$(u-v)_e = 1.2 \times 10^{-5} \ P_0 \ \text{m/s}, \tag{12}$$

the units of P_0 being N/m^2. Defining an effective Reynolds number (Kay and Nedderman [9])

$$Re = \frac{2 \ \varepsilon_0 \ \rho_f \ (u-v)_e \ d_p}{3 \ (1-\varepsilon_0) \ \mu_f} \ , \tag{13}$$

we find that $Re = 9.6 \times 10^{-4} \ P_0$. Therefore, for all but the lowest pressures ($P_0 < 0.1 \times 10^5$ N/m^2, say), Re will be greater than 10 ; note that $Re = 10$ is about the upper limit of validity of Darcy's Law.

316

TABLE 1: CONDITIONS AT NOZZLE EXIT (z=0) FROM NUMERICAL SOLUTION

r	$\dfrac{P}{P_0}$	$\dfrac{\partial(P/P_0)}{\partial z}$	$\dfrac{\phi}{v_0}$	$\dfrac{1}{v_0}\dfrac{\partial \phi}{\partial z}$
(mm)	($\times 10^{-2}$)	(mm^{-1})	(mm)	(-)
0.0	0.895	0.1433	0.297	1.00
0.2	0.870	0.1445	0.297	1.00
0.4	0.767	0.1450	0.297	1.00
0.6	0.570	0.1470	0.297	1.00
0.8	0.315	0.1497	0.297	1.00
1.0	0.000	0.1530	0.297	1.01

4.   ALTERNATIVE THEORY FOR PREDICTING THE SOLID-LIQUID SLIP VELOCITY

The simple linear form of Darcy's Law is convenient for obtaining approximate analytical and numerical solutions of two-phase flow in a nozzle.   In reality, expressions such as Darcy's Law, or the more general Ergun equation are a representation of viscous or viscous and inertial energy loss in flow through a packed bed of particles in contact with one another.   These equations do not necessarily apply to the case of a flowing suspension of particles and fluid, except where the suspension is very close to the minimum voidage fraction.   Further, it is clear from the preceding section that the validity of Darcy's Law at high pressure is doubtful.

An alternative approach for estimating the drag on a concentrated suspension by means of well-tested empirical correlations is proposed.   The voidage ε_0 is assumed constant as before.

The Richardson and Zaki [10] equation relating the relative superficial velocity um in a fluidized bed of voidage ε, and u_i, the effective terminal settling velocity of a single particle is

$$\frac{u_m}{u_i} = \varepsilon_0^n \qquad , \qquad\qquad\qquad (12)$$

where the exponent n is a function of $Re_i = \dfrac{\rho_f u_i d_p}{\mu_f}$.   Equation (12) is known to represent liquid-particle systems particularly well.   More importantly, it expresses the link betweens two relative velocities which produce the same effective drag force on a particle: u_m in a suspension of voidage ε and u_i for a single particle in isolation.

This view of equation (12) may then be coupled with an experimental correlation for the drag force f on a single particle in isolation.   This takes the form

$$f = \text{function } (Re_i) \qquad\qquad\qquad (13)$$

for a particular liquid.   Clift, et al [11] review twelve such correlations, the most reliable apparently being one due to Clift and Gauvin [12] for $Re < 3 \times 10^5$.   It is convenient to define F , the drag force per unit volume of suspension, as

$$F = f \times \frac{6(1-\varepsilon_0)}{\pi d_p^3} \qquad\qquad\qquad (14)$$

Finally, for the one-dimensional case and neglecting wall effects, F can be equated to the pressure gradient:

$$\frac{dP_f}{dz} = F \qquad\qquad\qquad (15)$$

which may be integrated subject to the conditions $P_f(z=1) = P_0$ and $P_f(z=0) = 0$ to yield

$$P_0 = \int_0^L F \, dz \quad . \tag{16}$$

For a specified nozzle shape, voidage ε_0, and exit superficial relative velocity $u_{m,e}$, the bed pressure P_0 may then be computed by a straight-forward numerical integration. At each value of z, u_i is obtained from u_m, using equation (12) and then u_i is substituted into equation (13) to give f and thence F from equation (14); at each section, u_m is related by continuity to the exit velocity $u_{m,e}$.

The quantity $u_{m,e}$ is related to $(u-v)_e$ from the previous section by

$$u_{m,e} = (u-v)_e \, \varepsilon_0 \quad . \tag{17}$$

5. EXPERIMENTAL RESULTS AND DISCUSSION

Horizontal jets of sand (mean diameter 0.12 mm) and water issuing from the side of a particulately fluidized bed were studied experimentally. The dimensions of the fluidized bed column were : 10 cm ID and 2 m over-all height. The nozzle was constructed of stainless steel according to the three-dimensional theoretical profile (Figure 2), with $r_e = 1$ mm, $L = 25$ mm.

A steady pressure in the bed during discharge was achieved by a downflow of water within the column to replace the outflow of fluidized suspension. A centrifugal pump supplied fluidized bed pressures of up to 20 bar above atmospheric, corresponding to jet velocities in excess of 40 m/s.

Extremely coherent and steady jets were observed, even at these high velocities. The jet material was collected over a measured time interval. Its weight and volume were measured. From the known densities of sand and water, the volume flow rates of solid and liquid, Q_s and Q_f were calculated. Solids fractions in the jet (ie $Q_s / (Q_s + Q_f)$) as high as 50% were achieved.

The erosion of the steel nozzle was quite significant. The exit diameter of the nozzle was re-measured after every four runs - it was found to have increased from 2 mm to 2.3 mm during an equivalent of about two hours of continuous running.

Figure 4 is a plot of experimental solids flowrate Q_s against the theoretical solids flowrate $\pi r_e^2 (1-\varepsilon_0) \sqrt{2P_0/\rho_0}$, allowing for nozzle erosion. The results are seen to lie approximately on a single straight line, the gradient of which may be viewed as an effective mass discharge coefficient, C_d. From Figure 4, this value is 0.79. Martin and Davidson [5] found a mean value of C_d of 0.7, while Jones and Davidson [3] reported a range of values of 0.62 to 0.87 for flow of air-fluidized particles through several shaped nozzles with the Jones and Davidson profile. The data from the present work covers a much larger range of pressures than the two earlier studies.

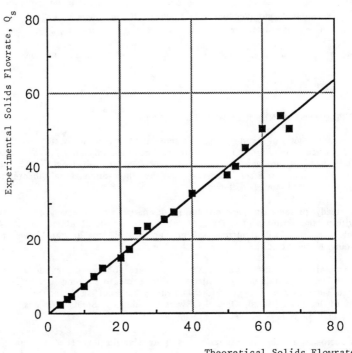

Fig. 4. – Sand-Water Flow through a Shaped Nozzle,
exit diameter 2-2.3 mm: Experimental
solids flowrate plotted against
theoretical solids flowrate.

■ Experimental data

— Best fit straight line, gradient 0.79

Close-up photographic measurements of the jet diameter as it leaves the nozzle show that no contraction of the jet occurs at the nozzle. This has been confirmed by jet velocity measurements from photographs of jet trajectory. In view of the fact that jet contraction at the nozzle exit is negligible, the reasons for C_d being somewhat less than unity are not clear. It has been suggested [5] that bed viscosity may give rise to significant shear stresses at the nozzle walls. Values of apparent viscosity of 0.1 to 1 Ns/m^2 for dense liquid-solid suspensions have been quoted [13]; these viscosities would, if the particle jet behaved as a viscous liquid with no wall slip, give values of C_d significantly less than unity.

The "extra" liquid flowrate Q_a may also be deduced from the experimental data. This is the additional flow of liquid due to percolation between the moving particles. Following Jones and Davidson [3],

$$Q_a = Q_f - \frac{Q_s \varepsilon_0}{(1-\varepsilon_0)} \qquad , \qquad (18)$$

where Q_f is the total liquid flowrate, and $\dfrac{Q_s \varepsilon_0}{(1-\varepsilon_0)}$ is an equivalent liquid flowrate due to entrainment. Experimentally determined values of Q_a for different bed liquid fractions ε_0 at $P_0 = 1$ bar and $P_0 = 10$ bar are shown in Figure 5. These are compared with the theoretical predictions from Darcy's Law (from equation (10)) and from the Richardson-Zaki approach (equations (12) to (16)). These results show that both the theoretical predictions are in quite good agreement with the data at the lower pressure, but at 10 bar, the Darcy equation significantly overpredicts the extra flowrate of liquid, while the Richardson-Zaki method remains in good agreement with the experimental data.

This trend is more clearly seen in Figure 6, where the experimental values of apparent jet voidage ε_j are plotted against discharge pressure from a fluidized bed at constant voidage fraction, $\varepsilon_0 = 0.5$. Again the Richardson-Zaki approach gives a much better fit to the data. Note that ε_j is an apparent jet voidage, assuming no slip between phases, so that

$$\varepsilon_j = \frac{Q_f}{(Q_f + Q_s)} \qquad . \qquad (19)$$

At high pressures, the improved prediction of the Richardson-Zaki method is because the interstitial liquid flow is proportional to the square root of pressure, like the particle flow; it follows that at high pressure, ε_j tends to a constant as indicated in Figure 6.

6.   CONCLUSIONS

A model for the flow of liquid-particle jets through a nozzle from a fluidized bed at elevated pressure has been developed. Simple experimental measurements of liquid and particle flowrates at discharge pressures up to 20 barg have yielded sound evidence regarding the validity of the basic assumptions of the theoretical model, as follows:

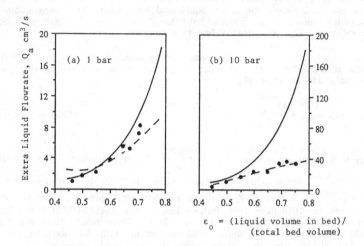

Fig. 5. - Variation of Q_a with Bed Voidage at Bed Pressures
(a) 1 bar, (b) 10 bar above atmospheric.
——— Darcy Law (eqn. 10)
----Richardson-Zaki (eqns. 12 - 16)
• Experimental points

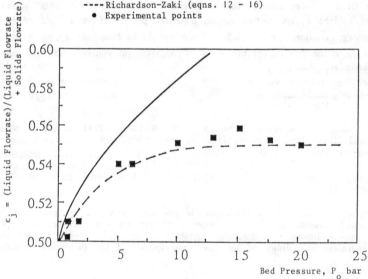

Fig. 6. - Variation of Jet Liquid Fraction with Bed Pressure
from a Steel Shaped Nozzle, exit diameter 2-2.3 mm.

——— Darcy Law (eqn.10) ; --- Richardson-Zaki (eqns.12-16)

■ Experimental data

322

a.   The flowrate of solids, Q_s is approximately proportional to $\sqrt{P_0}$ as predicted by Bernoulli's theorem.   A mean discharge coefficient of 0.79 was found for the stainless steel shaped nozzle.

b.   The Darcy's Law approach seriously overpredicts the liquid-solid slip velocity in the nozzle at pressures above about 1 barg.   An improved method based on the Richardson-Zaki drag term for the concentrated suspension gave good agreement with the experimental data.

7.   NOMENCLATURE

d_p	particle diameter	$\underline{v}$	absolute particle velocity
f	drag force on a single particle	v_0	dimensionless velocity, $\sqrt{2P_0/\rho_0}$
F	drag force per unit volume of suspension	z	axial coordinate
K	Darcy Law constant		
L	nozzle length	ε	voidage (liquid fraction)
P	total pressure	ε_0	bed voidage
P_0	pressure in fludised bed	ε_j	jet voidage fraction, Eqn. (19)
P_f	fluid interstitial pressure	ρ_0	bulk density, equation (5)
P_p	interparticle pressure	ρ_f	fluid density
Q_f	liquid flowrate	ρ_s	solid density
Q_s	solids flowrate	ϕ	solids velocity potential
Q_a	extra liquid flowrate	μ_f	liquid viscosity
r	radial coordinate		
r_e	nozzle exit radius		
Re	Reynolds number, $\rho_f u d_p / \mu_f$	Subscripts	
$\underline{u}$	absolute liquid velocity		
u_m	superficial relative velocity	e	condition at nozzle exit
u_i	terminal velocity of particle	o	condition in fluidized bed

Acknowledgement

     Financial assistance provided by the ICI Joint Research Scheme is gratefully acknowledged.   We are also grateful to staff from ICI Billingham who constructed the high pressure apparatus.

REFERENCES

1. Massimilla, L. Betta, V. and Della Rocca, C., 1961, A study of streams of solids flowing from solid-gas fludized beds. *A.I.Ch.E. J.* 7, 502-508.

2. Stemerding, S. de Groot, J.H. and Kuypers, N.G.M.,1964, Engineering aspects of fluidisation. *Proc. of Symp. on Fluidisation* (London, 11-12 Mar, 1963), Soc. Chem. Ind., 35-46.

3. Jones, D.R.M. and Davidson, J.F.,1965, The flow of particles from a fluidised bed through an orifice. *Rheologica Acta,* 4(3), 180-192.

4. Burkett,R.J., Chalmers-Dixon, P., Morris, P.J. and Pyle, D.L., 1971, On the flow of fluidised solids through orifices. *Chem. Engng Sci.* 26, 405-417.

5. Martin, P.D. and Davidson, J.F., 1983, Flow of powder through an orifice from a fluidised bed. *Chem. Engng Res. Des.* 61, 162-166.

6. Massimilla, L., 1971, Flow properties of the fluidized dense phase. In *Fluidization* (Davidson, J.F. and Harrison, D., eds.), Chap. 15, Academic Press, London.

7. Davidson, J.F. and Harrison, D., 1963, *Fluidised Particles.* Cambridge University Press.

8. Tan, R.B.H., 1986, *Particle Jets from Fluidised Beds.* CPGS dissertation, Univ. of Cambridge.

9. Kay, J.M. and Nedderman, R.M.,1983 *Fluid Mechanics and Transfer Processes.* Cambridge Univiversity Press, Cambridge, U.K.

10. Richardson, J.F. and Zaki, W.N., 1954, Sedimentation and fluidisation, Part I. *Trans. Instn Chem. Engrs* 32, 35-53.

11. Clift, R., Grace, J.R. and Weber, M.E., 1978, *Bubbles, Drops, and Particles,* p.111. Academic Press, New York.

12 Clift, R. and Gauvin, W.H., 1970, The motion of particles in turbulent gas streams. *Proc. Chemeca '70* (Melbourne and Sydney, 19-26 Aug., 1970), Session 1, 14-28.

13. Darton, R.C., 1985, The physical behaviour of three-phase fluidised beds. In *Fluidization, 2nd Edn.,* (Davidson, J.F., Clift, R. and Harrison, D. eds.), Academic Press, New York, 495-528.

Computation of Dynamics and Erosion for Small Tube Arrays in Fluidized Beds

R. W. LYCZKOWSKI, J. X. BOUILLARD, S. L. CHANG, and S. FOLGA

Argonne National Laboratory
9700 South Cass Avenue
Argonne, Illinois 60439, USA

ABSTRACT

The erosion issue in both atmospheric and pressurized fluidized bed combustors (AFBC's and PFBC's) has received increased attention in the last several years. The problem of erosion (more generally referred to as metal wastage) is so severe that in some cases the lifetime of tubes is as little as several thousand hours. The problem is international in scope and is hampering the commercialization of AFBC's and PFBC's for burning coal in a clean manner. The purpose of this paper is to discuss Argonne National Laboratory's (ANL's) current state-of-the-art computational capabilities in predicting erosion of fluidized bed tube arrays.

Hydrodynamic calculations were performed for generic few tube approximations of the Coal Research Establishment's cold model erosion experiment. Predicted solids-gas velocities and void fractions, using FLUFIX computer code, were used as inputs to three erosion models - The monolayer energy dissipation erosion model, the extended Finnie impaction erosion model, and the combined ductile-brittle Neilson-Gilchrist erosion model - All erosion models predict order of magnitude agreement with each other and with limited erosion data. However, the monolayer energy dissipation model produces more reliable trends of overall erosion rates.

1. INTRODUCTION

The erosion issue in atmospheric and pressurized fluidized bed combustors (AFBC's and PFBC's) was reviewed in depth at the 1987 International Conference on Fluidized Bed Combustion.[1] At that time, Lyczkowski et al.[2] compared time-averaged transient predictions for the preliminary monolayer energy dissipation (MED) and Finnie erosion models for a rectangular obstacle in a thin "two-dimensional" fluidized bed. Since that time, a cooperative research and development effort between U.S. DOE, Argonne National Laboratory, EPRI, State of Illinois CRSC, TVA, ASEA Babcock PFBC, CE and Foster Wheeler has been established to investigate erosion in bubbling AFBC's.[3] The purpose of this paper is to discuss Argonne National Laboratory's (ANL's) current state-of-the-art computational capabilities for bed dynamics and erosion in fluidized bed tube arrays.

The hydrodynamic model in the FLUFIX computer code, presented by Lyczkowski et al.[2], was improved and viscous stress terms were added to both the gas and solids phases. The preliminary monolayer energy dissipation erosion model was also improved by adding the solids viscous stress terms. The Finnie impaction erosion model was extended to account for abrasive (zero impaction angle) erosion through the inclusion of an applied normal force and sliding friction. The extended model unifies the concepts of abrasion and cutting wear. The combined ductile-brittle Neilson-Gilchrist erosion model was implemented. Hydrodynamic calculations were performed for two generic fewtube approximations of the Coal Research Establishment (CRE) cold model erosion

325

experiment.[4-5] Several major bed oscillation frequencies have been identified in the range of 1-4 Hz. For various fluidizing velocities, the erosion predictions from these erosion models are discussed and compared.

2. HYDRODYNAMIC MODEL

Hydrodynamic models of fluidization use the principles of conservation of mass, momentum and energy. The general mass conservation equations and the separate phase momentum equations for transient and isothermal fluid-solids non-reactive multiphase flow (in vector notation) are written in conservation-law form for two hydrodynamic models as follows (refer to the Nomenclature section):

Continuity

$$\frac{\partial}{\partial t}\left(\epsilon_k \rho_k\right) + \nabla \cdot \left(\epsilon_k \rho_k \vec{v}_k\right) = 0 \tag{1}$$

Momentum

$$\underbrace{\frac{\partial}{\partial t}\left(\epsilon_k \rho_k \vec{v}_k\right) + \nabla \cdot \left(\epsilon_k \rho_k \vec{v}_k \vec{v}_k\right)}_{\text{Acceleration}} = \underbrace{\nabla \cdot \bar{\bar{\sigma}}_{ke}}_{\text{Stress}} + \underbrace{\epsilon_k \rho_k \vec{g}}_{\text{Gravity}}$$

$$+ \underbrace{\sum_{i=1}^{n} \bar{\bar{\beta}}_{ik} \cdot \left(\vec{v}_i - \vec{v}_k\right)}_{\text{Interphase Drag}} \tag{2}$$

Each phase is denoted by the subscript k and the total number of phases is n.

The effective stress tensor, $\bar{\bar{\sigma}}_{ke}$, contains pressure, viscous and coulombic components according to the convention

$$\nabla \cdot \bar{\bar{\sigma}}_{ke} = \begin{cases} - \epsilon_k \nabla \cdot \left(P\bar{\bar{I}}\right) + \nabla \cdot \left(\epsilon_k \bar{\bar{\tau}}_{kv}\right) - \delta_{ks_i} \nabla \cdot \bar{\bar{\tau}}_{kc} & \text{(3a)} \\ \quad \text{(Hydrodynamic Model A)} \\ - \left(1 - \delta_{ks_i}\right) \nabla \cdot \left(P\bar{\bar{I}}\right) + \nabla \cdot \left(\epsilon_k \bar{\bar{\tau}}_{kv}\right) - \delta_{ks_i} \nabla \cdot \bar{\bar{\tau}}_{kc} & \\ \quad \text{(Hydrodynamic Model B)} & \text{(3b)} \end{cases}$$

The Kronecker delta function δ_{ks_i} is given by:

$$\delta_{ks_i} = \begin{cases} 1 \text{ if } k = s_i \\ 0 \text{ if } k \neq s_i \end{cases} \tag{4}$$

where $k = s_i$ is a solids phase. The set of momentum equations given by Equations (2) and (3a) are sometimes referred as the annular flow model or the basic equations set[6-7] which we refer to as "hydrodynamic model A." The treatment of the pressure gradient term in the fluid- and solids-phase momentum equations results in an initial-value

326

problem that is ill-posed (i.e., one possessing complex characteristics, as discussed in detail by Lyczkowski et al.[8]). This situation leads to a conditionally stable numerical solution.

The set of momentum equations given Equations (2) and (3b) is well-posed (i.e., one possessing all real characteristics), was first given by Rudinger and Chang[9], and was extended by Lyczkowski.[10] We refer to this set of momentum equations as "hydrodynamic model B." Either model may be selected by the user with the latest version of FLUFIX/MOD2.

The solids elastic modulus, $G(\varepsilon_k)$, is defined as the normal component of the solids coulombic stress through the following relationship:

$$\nabla \cdot \bar{\bar{\tau}}_{kc} = G(\varepsilon_k) \bar{\bar{I}} \cdot \nabla \varepsilon_k \tag{5}$$

where $\bar{\bar{I}}$ is the unit tensor. $G(\varepsilon_k)$ is the only component of the solids coulombic stress presently used. For a two-phase fluid/solids mixture, Equation (5) becomes:

$$\nabla \cdot \bar{\bar{\tau}}_{sc} = G(\varepsilon_s) \bar{\bar{I}} \cdot \nabla \varepsilon_s = - G(\varepsilon_s) \bar{\bar{I}} \cdot \nabla \varepsilon \tag{6}$$

Simple semi-empirical theory, used to develop a generic form of the solids elastic modulus, may be written as:

$$G(\varepsilon_s)/G_o = \exp \left[c(\varepsilon_s - \varepsilon_s^*) \right] = \exp \left[-c(\varepsilon - \varepsilon^*) \right] \tag{7}$$

In this study, we use c = 600, ε^* = 0.376, and G_o = 1.0 Pa, unless otherwise noted.

Hydrodynamic models A and B with the phase viscous stress term, $\bar{\bar{\tau}}_{kv}$, set to zero form the basis of the two-dimensional transient computer code FLUFIX/MOD1[11] which is used to compute two-phase fluid-solids motion in fluidized beds.

Both hydrodynamic models A and B given by Equations (2), (3a) and (3b), prior to the addition of the viscous stress terms, together with accepted expressions for the interphase drag, $\bar{\bar{\beta}}$[12], have been evaluated by comparison with experimental data taken in a thin, "two-dimensional" fluidized bed containing a rectangular obstacle.[2] Cartesian coordinates were used for all computations. A detailed description of the initial and boundary conditions may be found in Ref. 2.

The viscous stress terms, $\bar{\bar{\tau}}_{kv}$, were added to the hydrodynamic models. The form of these stress is the same as that in the original K-FIX computer program.[13] They may be written vector form as:

$$\bar{\bar{\tau}}_{kv} = 2\mu_k \bar{\bar{D}}_k - \frac{2}{3} \mu_k \nabla \cdot \vec{v}_k \bar{\bar{I}} \tag{8}$$

where the deformation tensor $\bar{\bar{D}}_k$ is given by:

$$\bar{D}_k = \frac{1}{2} \left[\nabla \vec{v}_k + (\nabla \vec{v}_k)^T \right] \qquad (9)$$

This formulation is identical to that for a Newtonian fluid and follows earlier work[14,15]. The microscopic phase viscosity is μ_k and the symbol T stands for the transpose of a tensor. These terms were added to both the fluid and solids phases in the FLUFIX/MOD1[11] computer program. We refer to the modified computer program as FLUFIX/MOD2.

3. IMPLEMENTATION OF THE NEILSON-GILCHRIST EROSION MODEL

The Neilson-Gilchrist erosion model[16] was chosen to be added to the post-processor erosion computer program EROS because it has the capability of computing erosion at 90° angle of particle incidence relative to the eroding surface. It also has normal and parallel velocity threshold components below which erosion does not take place. In place of a single value for hardness, two material properties are needed to characterize cutting and deformation energy absorption.

The Neilson-Gilchrist erosion model may be written as:

$$\dot{E} = \dot{E}_d + \dot{E}_b = \begin{cases} \dfrac{\dot{m}_s \vec{v}_s^{\,2} \cos^2(\alpha)\sin(n\alpha)}{2\phi} + \dfrac{\dot{m}_s(|\vec{v}_s|\sin(\alpha) - V_{el})^2}{2\,\varepsilon_b} & , \; \alpha < \alpha_o \qquad (10) \\[3mm] \dfrac{\dot{m}_s \vec{v}_s^{\,2} \cos^2(\alpha)}{2\phi} + \dfrac{\dot{m}_s(|\vec{v}_s|\sin(\alpha) - V_{el})^2}{2\varepsilon_b} & , \; \alpha > \alpha_o \qquad (11) \end{cases}$$

with the proviso that $\dot{E} = 0$ when $|\vec{v}_s|\sin(\alpha) < V_{el}$, $\dot{m}_s$ is the solids mass flux given by $\dot{m}_s = \rho_s \varepsilon_s |\vec{v}_s|$, n is an empirical constant, and $\alpha_o = \pi/(2n)$.

The ductile and brittle wear factors, ϕ_b and ε_b published by Neilson and Gilchrist[16] for 210 μm aluminum oxide particles and aluminum plate target were curve fit in dimensionless form as:

and

$$\phi(\tau/\rho_s) = 54.21 \left[1 + \exp(0.06793 \, (115.27 - V_{ref}))\right] \qquad (12)$$

$$\varepsilon_b(\sigma/\rho_s) = 82.52 \left[1 + \exp(0.07942 \, (110.13 - V_{ref}))\right] \qquad (13)$$

where τ is the yield shear stress and σ is the yield tensile stress of the target.

With α_o chosen to be 18.43°, the same angle used in the Finnie erosion model[17], n becomes 4.89 from the relationship $\alpha_o = \pi/(2n)$. The wear factors calculated from the above empirical formula are in excellent agreement with the experimental data in a velocity range from 100 to 150 m/s.[16]

Neilson and Gilchrist neglected the threshold velocity normal to the target, V_{el}, in their analysis of aluminum plates because they stated it is usually small relative to the particle velocity, therefore, they did not report its value. It can be estimated from the theoretical expression from the Hertz contact theory given by Engel[18] to be 0.056 m/s for glass beads and aluminum target material.

4. EXTENDED FINNIE ABRASION/CUTTING EROSION MODEL

Finnie[17] introduced a two-dimensional theory of erosive wear to explain the erosion of ductile materials. The theory assumes that a hard particle impinging upon a smooth surface at an incident angle will cut into the surface and cause erosion. This theory predicts zero erosion at both 0 and 90 degree incident angles, and a maximum erosion rate at about 18 degrees. Finnie's predictions are fairly accurate when compared with the experimental data up to 45 degrees.

A new concept was developed by Chang[19] to predict the erosion rates near zero incident angle. This erosive model takes into account the applied normal force, N, and a friction force, fN, between the colliding particle and the target surface and used Finnie's erosive cutting methodology to predict wear rates. The application of the normal force results in a combined abrasion/cutting erosion mechanism.

The governing equations of the erosive cutting in the z direction (tangential) or the x direction (normal) are given by (refer to the Nomenclature):

$$m \frac{d^2x}{dt^2} + \psi bpX = fN \tag{14}$$

$$m \frac{d^2z}{dt^2} + \psi bpkZ = N \tag{15}$$

$$I \frac{d^2\theta}{dt^2} + \psi bprz = -rfN \tag{16}$$

By solving Equations (14) through (16) using the same initial conditions as used by Finnie, the vertical displacement of the particle (or the erosion path) becomes a sinusoidal function of time as:

$$Z(t) = \left[N_1 \sin \left[\beta t - \gamma \right] + N \right] / m\beta^2 \tag{17}$$

where

$$\beta^2 = \psi bpk/m \tag{18a}$$

$$N_1{}^2 = N^2 + \left[m\beta V \sin(\alpha) \right]^2 \tag{18b}$$

and

$$\gamma = \tan^{-1} \left[N/\left(m\beta V \sin(\alpha) \right) \right] \tag{18c}$$

The volume of target material removed, W, is obtained as

$$W = \int_0^{t^*} bZ(t)V_t(t) \, dt \tag{19}$$

where V_t is the particle cutting velocity which is given by

$$V_t = \frac{dX}{dt} + r \frac{d\theta}{dt} \tag{20}$$

and t* is the cutting duration.

By integrating Equation (19) over the cutting period t*, the volume of target material removed, W, is given by

$$W = \frac{mV^2}{k\psi p} F\left(\alpha, \ \tau^*, \ \bar{N}, \ \bar{N}_1, \ k, \ \gamma\right) \tag{21a}$$

where $F\left(\alpha, \ \tau^*, \ \bar{N}, \ \bar{N}_1, \ k, \ \gamma\right)$

$$= \ \bar{W} = \bar{N}\left[\cos(\alpha) - \frac{3\bar{N}_1}{k} \cos(\gamma)\right] \tau^*$$

$$- \frac{3\bar{N}^2}{2k} (1 + fk) \ \tau^{*2}$$

$$+ \frac{3\bar{N} \ \bar{N}_1}{k} \left[\sin(\tau^* - \gamma) + \sin(\gamma)\right]$$

$$+ \bar{N}_1\left[\cos(\alpha) - \frac{3\bar{N}_1}{k} \cos(\alpha)\right] \left[\cos(\gamma) - \cos(\tau^* - \gamma)\right]$$

$$+ \frac{3\bar{N} \ \bar{N}_1}{k} (1 + fk) \left[\tau^* \cos(\tau^* - \gamma) - \sin(\tau^* - \gamma) - \sin(\gamma)\right]$$

$$+ \frac{3 \ \bar{N}_1^2}{4k} \left[\cos(2\tau^*) - \cos(2\tau^* - 2\gamma)\right] \tag{21b}$$

where $\bar{N}_1 = N_1/(m\beta V)$ $\bar{N} = N/(m\beta V)$ and $\tau^* = \beta t^*$.

In order to apply the extended Finnie model to compute erosion rates in fluidized beds, the total particle mass m is replaced by the mass flux of solids, $\dot{m} = (1 - \epsilon)\rho_s|\vec{v}_s|$, where $(1 - \epsilon) = \epsilon_s$ is the solids volume fraction, ρ_s is the particle density, and $|\vec{v}_s|$ is the magnitude of the velocity of the solid phase. The mass flux $\dot{m}_s$ is assumed to be positive toward the eroding surface and the particle incident velocity V is replaced by $\vec{v}_s$ to obtain

$$\dot{E} = C \frac{(1 - \epsilon)\rho_s|\vec{v}_s|(\vec{v}_s)^2}{p} F\left(\alpha, \ \tau^*, \ \bar{N}, \ \bar{N}_1, \ k, \ \gamma\right) \tag{21c}$$

where $\dot{E}$ is the erosion rate (in m/s) and $C = c/\psi k$. As in the N-G erosion model, the erosion rate is positive if the solids velocity vector points toward the eroding surface; otherwise, it is zero.

Finnie took k = 2, where k is the ratio of vertical to horizontal (frictional) force, and ψ = 1, where ψ is the ratio of the depth of contact to the depth of the cut. The constant c allows for the fact that many particles will not be as effective as the idealized model particle; Finnie arbitrarily took c = 1/2. These same values are used in the extended Finnie model.

Several wear mechanisms are possible: 1) the particle after gouging the target bounces back off the target surface, 2) the particle is stopped before it leaves the target and 3) the particle cuts the target material at zero impaction angle, α, (abrasion). A transition incident angle, α_o, is introduced as a criterion to differentiate between mechanism (1) and (2). The normalized cutting period, τ^*, is determined for each mechanism as follows:

mechanism (1) $\qquad$ $\tau^* = \pi + 2\gamma$ $\qquad\qquad \left(\alpha \leq \alpha_o\right)$ $\qquad$ (22a)

mechanism (2) $\qquad$ τ^* is solution of

$$\frac{3}{k} \sin(\alpha) \cos(\tau^*) + \frac{3\bar{N}}{k} \sin(\tau^*) - \frac{3\bar{N}}{k} (1 + fk)\tau^*$$

$$= \frac{3}{k} \sin(\alpha) - \cos(\alpha) , \qquad \left(\alpha_o > \alpha\right) \qquad (22b)$$

mechanism (3) $\qquad$ τ^* is solution of

$$\frac{3\bar{N}}{k} \sin(\tau^*) - \frac{3\bar{N}}{k} (1 + fk)\tau^* + 1 = 0 \qquad \left(\alpha = 0\right) \qquad (23)$$

The transition incident angle, α_o, in Equations (22a) and (22b) is given by

$$\cos\left(\alpha_o\right) = \frac{6\bar{N}_1}{k} \cos(\alpha) + \frac{3\bar{N}}{k} (1 + fk) (\pi + 2\gamma) \qquad (24)$$

If the incident angle, α, is zero, the particle follows a sinusoidal trajectory until its horizontal cutting velocity is exhausted. In other words, the erosion rate at zero incident angle depends on the particle incident velocity and is not zero. As shown in Figure 1, which plots the dimensionless wear volume, $\bar{W}$, as a function of the dimensionless applied normal force, $\bar{N}$, for a sliding friction coefficient f = 0.1, the erosion at zero incidence angle can be of the same order as the maximum erosion rate calculated by Finnie's original model. When both the applied force and sliding friction coefficient are set to zero, the original Finnie erosion model results.

Further development of this extended Finnie erosion model is continuing in order to compute the erosion rate of an obstacle surface in FBCs using the values of porosity, solids velocity, incident angle, and normal exerting force computed from the FLUFIX code. This model is useful at low particle incident angles because substantial abrasive wear (about the same order as Finnie's maximum erosion) is predicted.

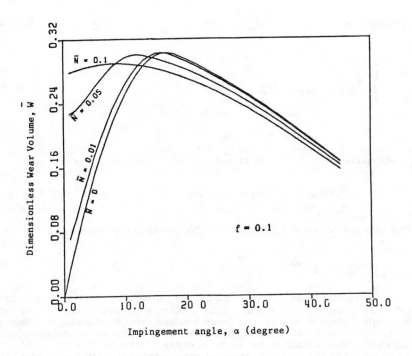

FIGURE 1 Effect of Force Number on Erosive/Abrasive Wears.

5.   IMPROVEMENTS IN THE MONOLAYER ENERGY DISSIPATION EROSION MODEL

A monolayer dissipation erosion model was developed by Lyczkowski et al.,[2] by extending the so-called power dissipation erosion model for low-angle erosion caused by slurries proposed by Ushimaru et al.[20] This model was based on hydrodynamic model A and is herein extended to include solids viscous stress and generalized to allow for hydrodynamic model B.

A standard analysis which combines the mixture kinetic energy equation with the mixture total energy equation written in terms of entropy production rate shows that the total rate of kinetic energy dissipation, E_{vTOT} (which is related to the entropy production rate), is given by:[21-22]

$$E_{vTOT} = \left(\varepsilon_g \bar{\bar{\tau}}_{gv} \right) : \nabla \vec{v}_g + \left(\varepsilon_s \bar{\bar{\tau}}_{sv} \right) : \nabla \vec{v}_s$$

$$+ \left(\vec{v}_s - \vec{v}_g \right) \cdot \bar{\bar{\beta}} \cdot \left(\vec{v}_s - \vec{v}_g \right) , \geq 0 \tag{25}$$

Note that E_{vTOT}, given by Equation (25), is always positive and therefore represents the irreversible degradation of mechanical energy as explained by Bird et al.[23]

The first and second terms of Equation (25) extend the expression given by Bird et al.[23] for single phase flow to multiphase flow; note that the sign convention of the viscous terms used is opposite to that of Bird et al.[23] and agrees with that used in the K-FIX computer code[13]. The third term represents the rate of kinetic energy dissipation due to interfacial drag.

As shown in Figure 2, the rate of kinetic energy dissipation given by Equation (25) in a monolayer of particles in the vicinity of stationary surfaces (obstacles), may be converted into: a) heat which is transferred between the gas and solid particles, between the gas and obstacles, and between the solids and obstacles, b) erosion of obstacles, and c) attrition of solid particles.

Clearly, the rate of kinetic energy dissipation which can result in erosion is, thus, always only a fraction of the total energy dissipation rate given by Equation (25) since heat transfer and particle attrition will always occur.

Lacking definitive erosion experiments, uncertainties in the proper boundary conditions for evaluating the kinetic energy dissipation in the monolayer of particles in the vicinity of the stationary surfaces which can result in erosion, lead to several approximate formulations of this term and, therefore, to several alternative erosion models. For all these models, the following major assumptions are made: a) negligible particle attrition, b) negligible gas phase contribution to erosion of obstacles and c) negligible heat generation.

One of these alternatives is extended from Bouillard[24] who, in a companion paper, derived an approximate closed form solution of energy dissipation which can result in erosion by integrating the solids mechanical energy equation over a virtual quasi-static transformation. This expression when including the solids viscous stress is given by:

$$E_{vCF} = 1/2 \left[\left(\varepsilon_s \bar{\bar{\tau}}_{sv} \right) : \nabla \vec{v}_s + \vec{v}_s \cdot \bar{\bar{\beta}} \cdot \vec{v}_s \right] \tag{26}$$

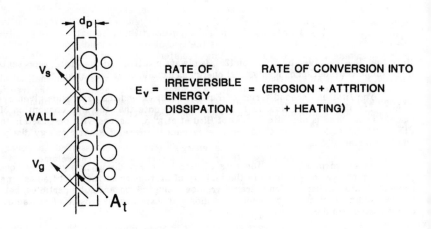

FIGURE 2 Conceptual Picture of Monolayer Energy Dissipation Erosion Model.

Equation (26) will be referred to as the closed form solution. In this formulation, the energy dissipation rate is not directly dependent upon the gas velocity.

In order to complete the derivation of the monolayer energy dissipation (MED) erosion model, a linkage must be made between the rate of kinetic energy dissipation per unit volume and the rate of removal of the eroding surface.

Based upon thermodynamic considerations, a general relationship between the rate of kinetic energy dissipation and the volumetric rate of the eroding surface removal, V_t, referred to as the target is given by

$$C \, E_{v\alpha} \, V_f = P_t \dot{V}_t \tag{27}$$

Equation (27) states that the total rate of kinetic energy dissipation, $E_{v\alpha} \, V_f$, is proportional to the thermodynamic work required to remove material from the target. $E_{v\alpha}$ represents E_{vCF} if Equation (26) is used. Several other alternatives are possible and are explained by Bouillard et al.[25] V_f is the two-phase fluid-solids volume in which the energy dissipates, P_t is the eroding target material pressure which is related to the hardness, p, or as Ushimaru et al.[20] refer to it, the specific energy of the eroding surface, E_{sp}. C is an empirical factor which accounts for the observation that not all the dissipation results in material removal. A similar factor to account for the fraction of particles cutting in an idealized manner was used by Pourahmadi and Humphrey[26] in their continuum approach to erosion modeling of curved and straight two-dimensional channels.

With respect to Figure 2, the energy dissipation is assumed to occur in a monolayer of particles in contact with the eroding surface having the volume

$$V_f = A_t \cdot d_p \tag{28}$$

where A_t is the area of the target and d_p is the average diameter of the particle which is of the order of the particle mean free path.[27] The volumetric rate of target removal may be expressed as:

$$\dot{V}_t = A_t \dot{E} \tag{29}$$

where $\dot{E}$ is the target erosion rate (m/s).

Combining Equations (27) through (29), we obtain the MED erosion model as

$$\dot{E}_{ED\alpha} = \dot{E} = C\left(E_{v\alpha} \, d_p / E_{sp}\right) \tag{30}$$

where E_{sp} is the specific energy of the eroding surface, $\dot{E}_{ED\alpha}$ is the erosion rate in m/s, and $E_{v\alpha}$ represents E_{vCF} given by Eq. (26) or any of the other alternatives.[25]

In calculating the erosion rates for the monolayer energy dissipation (MED) erosion model, we applied the empiricism that when metal is removed, fewer than 10% of the grains in contact with the surface actually remove metal (Ushimaru[20]). Therefore, the factor C in Equation (3) is taken to be 0.10. The remaining particles may only cause elastic deformation, not resulting in material wear.

6. COMPARISON OF EROSION PREDICTIONS FOR TWO GENERIC FEWTUBE FBC GEOMETRIES

The erosion models are compared in this section for two two-dimensional idealizations of the Coal Research Establishment (CRE) 0.3 m x 0.3 m cold model of the IEA Grimethorpe tube bank "C1" configuration Parkinson et al.[5,28] shown in Figure 3. In the following simulations, the bed consists of glass beads (500 μm diameter) and has a solids viscosity of 0.1 Pa s. Such a value is typical of that deduced from bubble shapes in fluidized beds consisting of 550 μm diameter glass beads.[29] The actual cold model consisted of 17 PVC tubes of 33 mm outside diameter, three columns wide and eight rows high. The vertical distance between tube row centers was 80 mm and the horizontal distance between tube column centers was 89 mm.

The FLUFIX models shown in Figure 3 consist of only two and three rows of tubes. The grid size chosen is $\Delta x = \Delta y = 1.7$ cm with the assumption of symmetry. The number of computational cells in the x-direction is 9 with 48 in the axial direction, for a total of 432. With this nodalization, the round tubes are approximated as square obstacles consisting of four nodes each. The perimeter of the square tube is therefore 13.6 cm as compared with the actual round tube perimeter of 10.4 cm. The vertical distance between the obstacle centers is 8.5 cm.

The CRE cold model air distributor had 16 standpipes on a 75 mm square pitch. Each standpipe had 20 round holes, 10 of which were 3.2 mm in diameter and 10 of which had a diameter of 4 mm for a total open area of 2.06 cm^2. The FLUFIX model standpipes are modeled resembling bubble caps as shown in Fig. 3. These bubble caps are comprised of five obstacle cells in each bed half, each consisting of only one computational cell. The two rows of obstacles are arranged so that the upper row is situated above the two inlets as shown, deflecting the air flow sideways. If the depth of the bed is chosen to be 4.85 cm, the effective flow area of each simulated standpipe is the same as in the CRE cold model. The average distance between the simulated standpipes is 6.3 cm (3.4 cm on each side and 11.9 cm across the centerline).

For both beds, the superficial gas velocity, U, was maintained at 23.5 cm/s, just above minimum fluidization velocity, U_{mf}, equal to 20.9 cm/s (U/U_{mf} = 1.12) for two seconds to obtain a reasonable initial condition for subsequent runs for a bed near minimum fluidization ($\varepsilon \sim 0.4$). The compaction gas volume fraction, ε^*, was increased to 0.42 to accomplish this. Considering that the total open area of the standpipes is 2/9 of the total bed area, this means the jet velocity (V_{jet}) into each of the standpipes shown in Figure 3 is 105.81 cm/s. The jet velocities near the centerline were set 5 percent higher at 111.10 cm/s and the jets near the sidewalls were set 5 percent lower at 100.52 cm/s in order to induce sweeping of solids off the obstacles near the side walls.

Several simulations of these fluidized-bed models were undertaken. The simulation for U/U_{mf} = 1.12 was used as the initial condition for the case U/U_{mf} = 1.7 (average V_{jet} = 158.7 cm/s). The jets near the centerline were set five percent higher at 166.64 cm/s and the jets near the sidewalls were set five percent lower at 150.76 cm/s. The transient was run to 4.5 s (2.5 s beyond the case of U/U_{mf} = 1.12).

The simulation for the case of U/U_{mf} = 1.7 at 4.5 s was used as the initial condition for the case of a still higher jet velocity, 2.3 x U_{mf} (V_{jet} = 216.21 cm/s). This time, the jets near the centerline were set to 5 percent higher at 227.02 cm/s and the

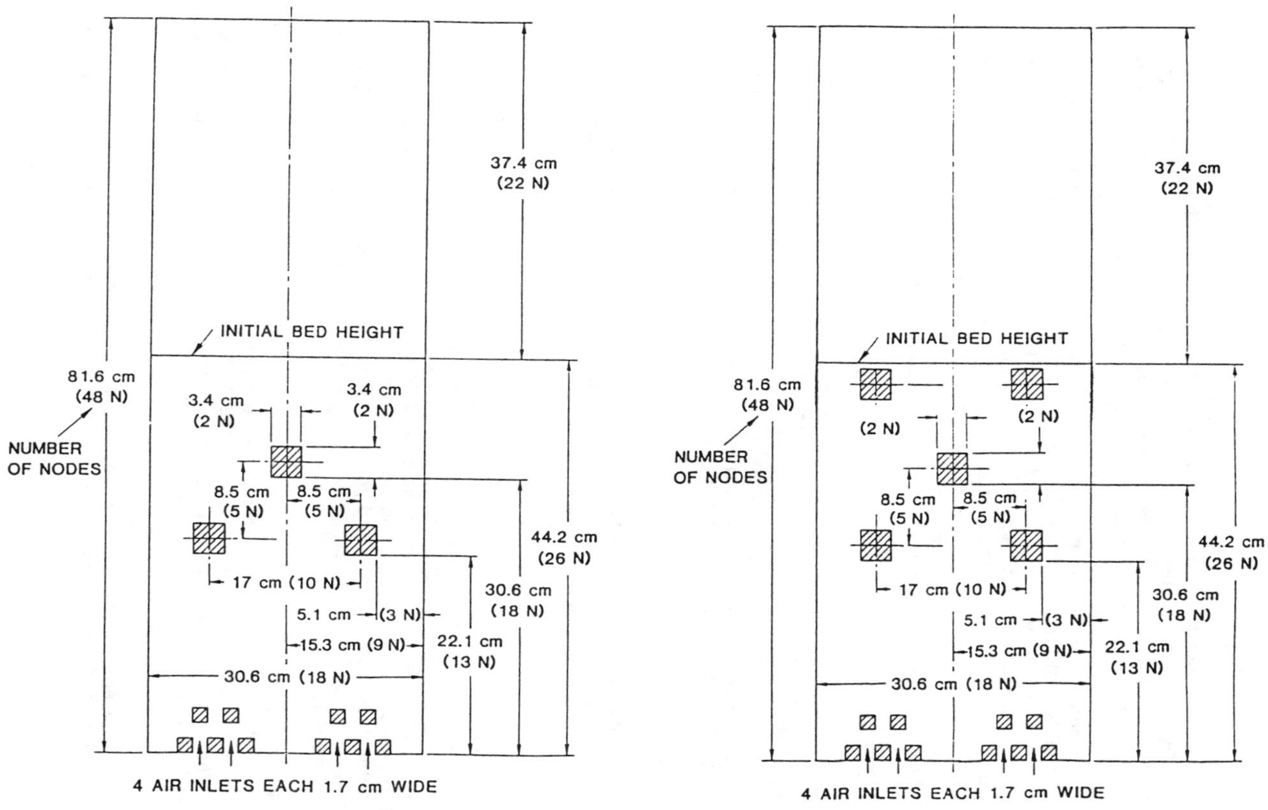

FIGURE 3 Two Generic Fewtube FBC Geometry FLUFIX Models.

337

jets near the sidewalls were set 5 percent lower at 205.39 cm/s. The simulation was run to 7.0 s (2.5 s beyond the case of U/U_{mf} = 1.7).

6.1 Hydrodynamic Results

Interesting time averaged solids flow patterns are revealed for U/U_{mf} = 1.2, 1.7 and 2.3 in Figures 4 and 5. Solids flow patterns are similar for the two geometries, however, the five-tube bed appears to be more quiescent and thus will be subject to less erosion as discussed later. The solids are entrained by the gas at the air distributor standpipes, converge toward the center of the bed beneath the upper tube, deflect on the side of this tube, and fall down the walls of the fluidized bed toward the distributor. For both beds, the increase of fluidizing velocity increases the bed expansion as well as the strength of solids motion. The time-averaged porosity contours indicate the solids building up on the top of the center tube. The reason for the buildup on the center tube is the assumption of symmetry which does not allow transverse flow across the centerline to sweep solids off the top of the center tube. It was also noted that the bubble formation near the off center tubes alternated from one side to the other. This observation leads to the conjecture that if the symmetry assumption were removed, solids would not build up on top of the center tube to such a large extent as computed.

6.2 Bed Dynamics

In another companion paper[30], Chang et al. describe the dynamics of the bed containing three tubes by performing a spectral analysis of the transient computed results generated by the FLUFIX computer code. A basic bed frequency of 1 Hz was found as well as frequencies of 3 Hz localized underneath the obstacles. Typical vertical void fraction propagation velocities ranging between 0.5 m/s to 1.0 m/s were predicted. For further details, the reader should refer to Reference 30.

6.3 Erosion Predictions

Local computed time-averaged erosion rates of the immersed tubes using the monolayer energy dissipation (MED), Neilson-Gilchrist (N-G) and the original Finnie erosion models are summarized in Figures 6 through 9 for U/U_{mf} = 1.7 and 2.3. The square tubes are assumed to be made of aluminum having a hardness of 30 Kgf/mm^2 (294 mn/m^2). As can be seen, the erosion rates are increasing more rapidly on the sides of the two lower and upper tubes facing each other (Figures 6-9) as the fluidizing velocity increases. A large erosion rate of the top of the center tube is predicted and is probably too high due to the solids buildup on top of the center tube artificially caused by the symmetry assumption, thereby preventing any flow across the line of symmetry. The erosion rate on the right sides of the tube closest to the walls is almost independent of fluidizing velocity since it corresponds to the downcomer area in which the solids descend at almost constant velocity and porosity to the distributor. On the average, the erosion rates predicted for the five-tube geometry are lower than that predicted for the three-tube geometry.

The performance of the N-G and Finnie models is somewhat inconsistent in contrast to the MED model. For the three-tube geometry, the erosion on the sides of the two tubes facing each other computed from the N-G model decreases with fluidizing velocity. The Finnie model exhibits the same trend. This is because the solids impingement angle is becoming closer to 0° (See Figure 4) and both models predict zero erosion at this angle.

The time-averaged transient erosion rates for all three erosion models were arithmetically averaged and are plotted in Figures 10 and 11 as a function of fluidizing

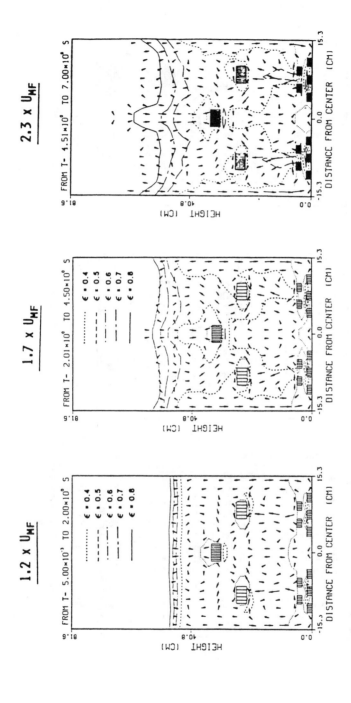

FIGURE 4 CRE Fewtube (3 Tubes) Generic FBC Geometry Time-Averaged Porosity and Solids Velocity.

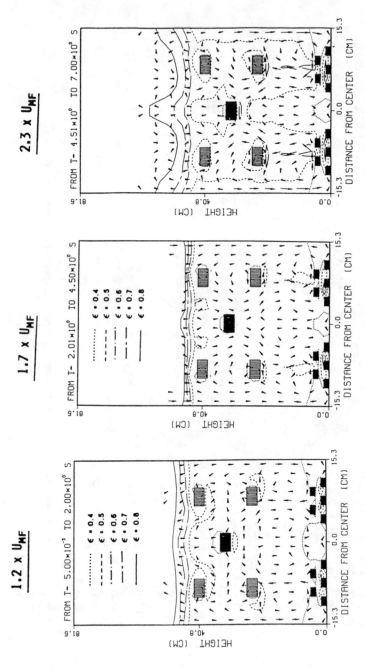

FIGURE 5 CRE Fewtube (5 Tubes) Generic FBC Geometry Time-Averaged Porosity and Solids Velocity.

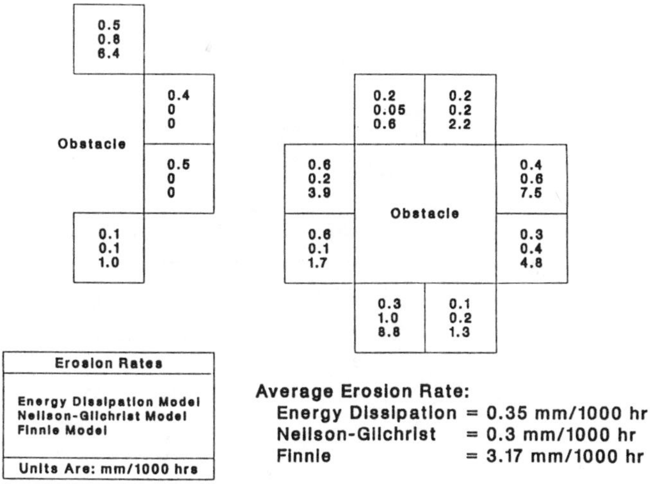

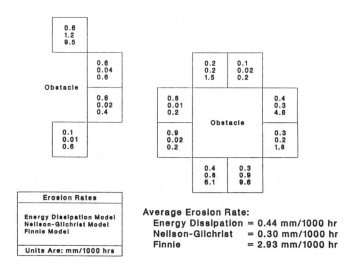

FIGURE 6 Computed Time-Averaged (over 2.5 s) Transient Erosion Rates for Aluminum, for the Monolayer Energy Dissipation Model E_{EDCF}, Neilson–Gilchrist and Finnie Erosion Models (1.7 x U_{mf}, d_p = 500 µm).

FIGURE 7 Computed Time-Averaged (over 2.5 s) Transient Erosion Rates for Aluminum, for the Monolayer Energy Dissipation Model (E_{EDCF}, Neilson–Gilchrist, and Finnie Erosion Models (2.3 x U_{mf}, d_p = 500 µm).

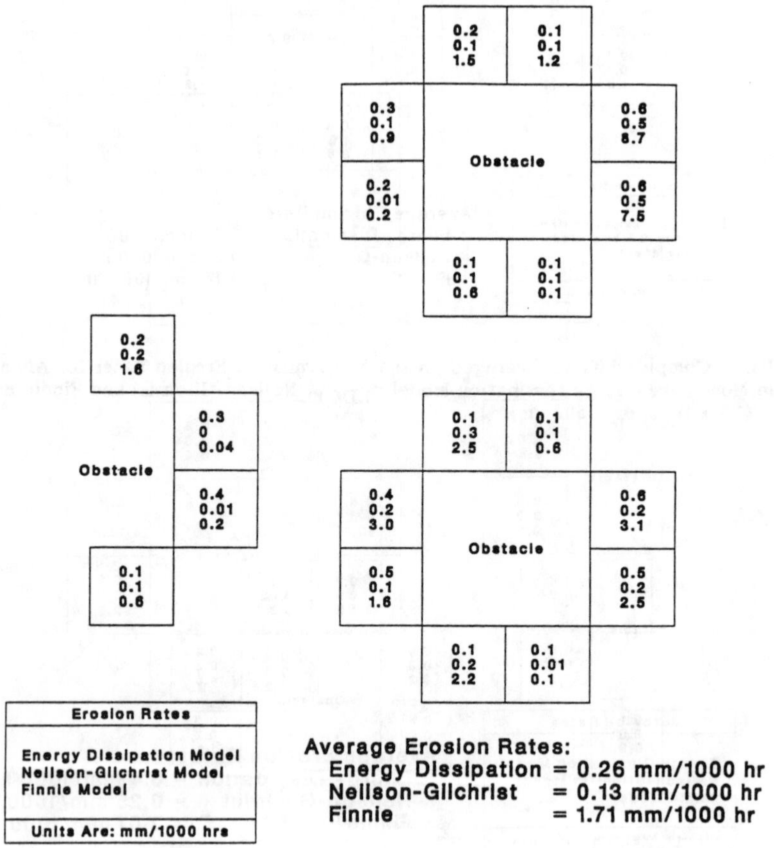

FIGURE 8 Time-Averaged Transient Erosion Rates for Aluminum (d_p = 500 μm, 1.7 x U_{mf}).

342

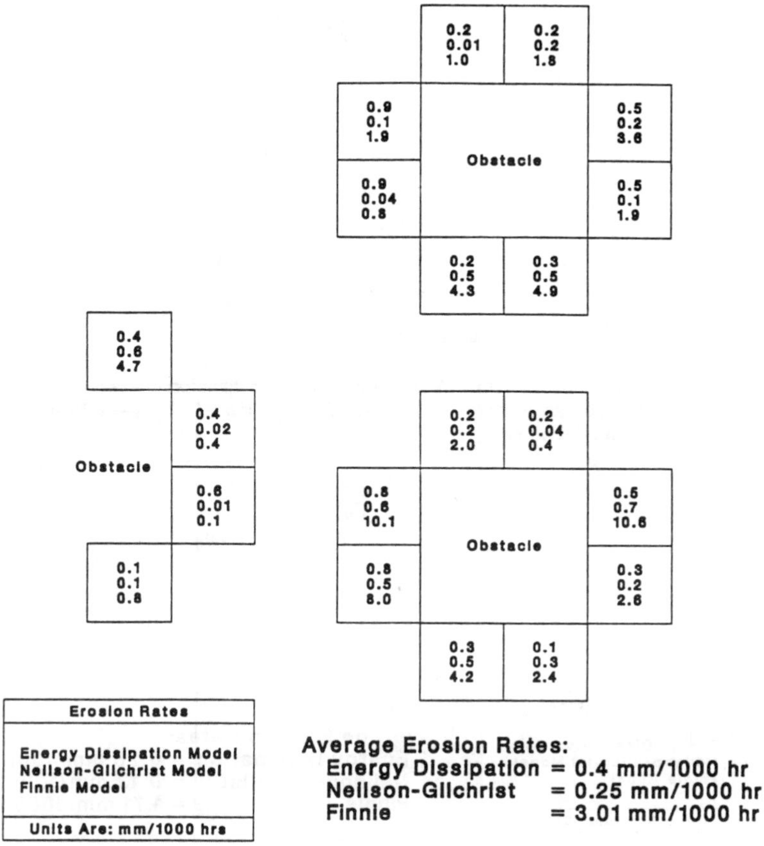

FIGURE 9   Time-Averaged Transient Erosion Rates for Aluminum (d_p = 500 μm, 2.3 x U_{mf}).

343

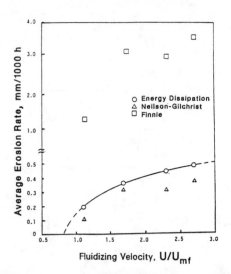

FIGURE 10 Comparison of Time-Averaged Transient Erosion Rates for Aluminum (3 Tubes) for the Monolayer Energy Dissipation Erosion Model, $\hat{E}_{EDCF}$, Neilson–Gilchrist and Finnie Erosion Models.

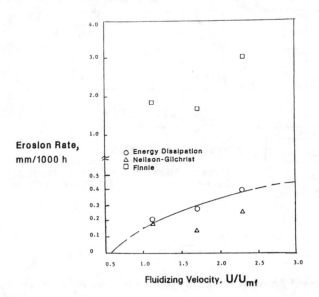

FIGURE 11 Generic Fewtube FBC Geometry Comparisons of Time-Averaged Erosion Rates for Aluminum (5 Tubes) for the Monolayer Energy Dissipation Erosion Model, Neilson-Gilchrist and Finnie Erosion Models.

velocity. The average erosion rates for the three-tube and five-tube geometries actually decrease between 1.7 and 2.3 x U_{mf} and between 1.12 and 1.7 x U_{mf}, respectively, as shown in Figures 10 and 11 for the N-G and Finnie erosion models. The N-G erosion model agrees remarkably well with the MED erosion model and their trends are similar except for the decreases at U/U_{mf} = 2.3 and U/U_{mf} = 1.7 in Figures 10 and 11, respectively. The Finnie erosion model predicts erosion rates almost an order of magnitude higher than those predicted by the N-G model and shows similar decrease trends at U/U_{mf} = 2.3 and U/U_{mf} = 1.7 depicted in Figures 10 and 11, respectively. Also shown in Figures 10 and 11, is the prediction by the MED erosion model of a monotonic increase of erosion as a function of fluidizing velocity. At higher fluidizing velocity, the rate begins to level off. The shape of the curves is remarkably similar to that determined by Parkinson et al. for 400 μm and 700 μm particles.[28] They found that their experimentally determined erosion rate curves flattened out for U/U_{mf} between 5 to 7 for those particle sizes, respectively. In this study, the prediction of this curve shape is considered to be a significant achievement. The extended Finnie erosion model was assessed at U/U_{mf} = 1.12 for the three-tube geometry and the resulting erosion rates were found to be almost identical[31] to those predicted by the original Finnie erosion model as shown in Figure 12.

The wear patterns at the lower fluidizing velocities are quite uniform, which is the same trend as in the CRE cold bed erosion experiment using PVC tubes at the lowest fluidizing velocities; approximately 2 x U_{mf}.[28] Over this range of fluidizing velocities, the erosion rates for the monolayer energy dissipation and Neilson-Gilchrist erosion models are of the same order as the experimental data (0.1-0.3 mm/1000 h)[4], while those for the Finnie erosion model are about an order of magnitude higher.

For both beds, higher tubes experience larger average erosion rates which may be explained by bubbles accelerating as they move upwards through the bed entraining with them more energetic solids. Table 1 summarizes the comparison of the erosion predictions for the three- and five-tube CRE generic FBC geometries. The results are quite close which implies that only a small tube array may be sufficient to assess FBC erosion.

7. CONCLUSIONS

Two two-dimensional idealizations of cold model fluidized beds containing three and five tubes were investigated and used to predict wear patterns. Time-averaged transient erosion rates were calculated assuming aluminum tubes and using three erosion models: the monolayer energy dissipation (MED), the extended Finnie and the Neilson-Gilchrist erosion models. The average erosion rates for all three models increase as a function of the fluidizing velocity, in agreement with experimental data. The MED and Neilson-Gilchrist erosion models are in reasonable agreement, but the Finnie erosion model predicts an order of magnitude higher erosion rates.

345

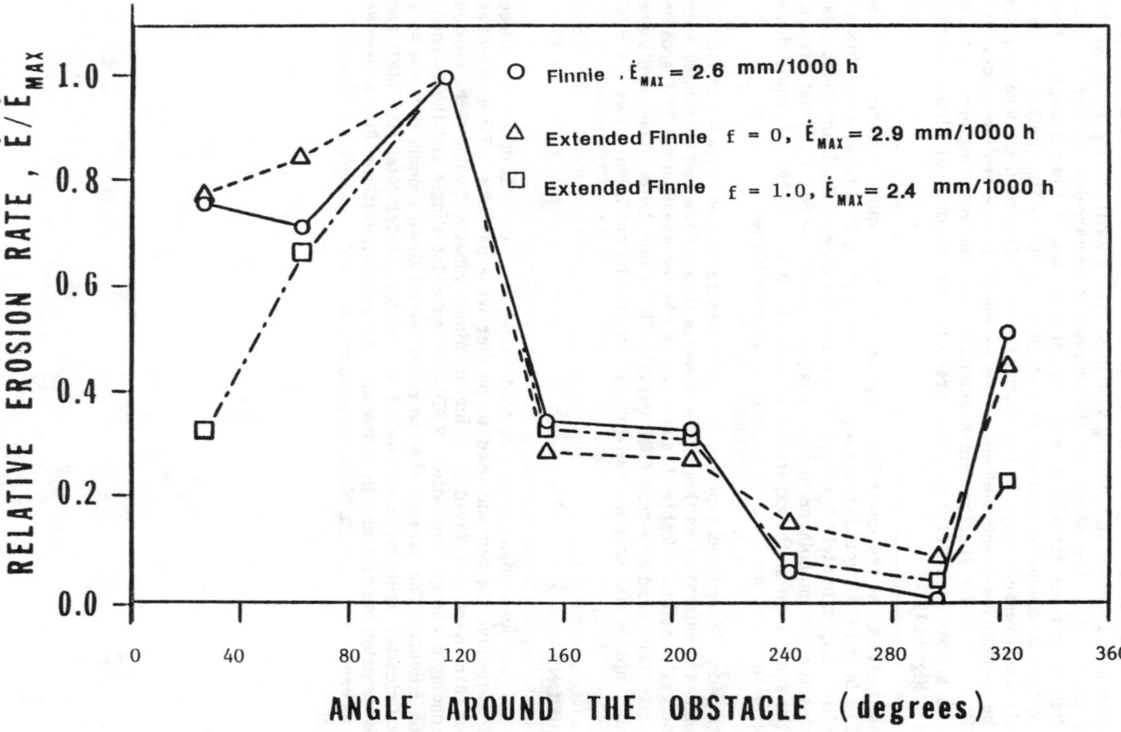

FIGURE 12 Predicted Erosion Rates for the Extended and Original Finnie Erosion Models for the Three-Tube CRE Geometry.

The legend shows:

- Finnie $\dot{E}_{MAX}$ = 2.6 mm/1000 h
- Extended Finnie f = 0, $\dot{E}_{MAX}$ = 2.9 mm/1000 h
- Extended Finnie f = 1.0, $\dot{E}_{MAX}$ = 2.4 mm/1000 h

RELATIVE EROSION RATE, $\dot{E}/\dot{E}_{MAX}$

ANGLE AROUND THE OBSTACLE (degrees)

Table 1. **Effect of Tube Number on Overall Erosion Rates**

Geometry	Erosion Model	Erosion Rate, mm/1000 h			
		Fluidizing Velocity, U/U_{mf}			
		1.12	1.7	2.3	2.7
Three Tubes	Energy Dissipation	0.18	0.35	0.44	0.47
	Neilson-Gilchrist	0.10	0.3	0.30	0.36
	Finnie	1.20	3.17	2.93	3.56
Five Tubes	Energy Dissipation	0.21	0.26	0.40	
	Neilson-Gilchrist	0.19	0.13	0.25	
	Finnie	1.04	1.71	3.01	

ACKNOWLEDGMENTS

This work was supported in part by the U.S. Department of Energy, Morgantown Energy Technology Center (METC), Contract W-31-109-ENG-38, for the Electric Power Research Institute (EPRI) under the cooperative R&D venture "Erosion of FBC Heat Transfer Tubes." The METC Project Manager for the first year (Oct. 1, 1984 through Sept. 30, 1985) was Dr. Thomas J. O'Brien. We thank him for his many suggestions and helpful comments, and particularly for his moral and financial support. The METC project was managed from Oct. 1, 1985 to Feb. 1988 by Dr. Holmes A. Webb. We thank Dr. Webb for suggesting the monolayer energy dissipation erosion model.

The EPRI project has been managed since January 1, 1987 by Dr. John Stringer. The authors would like to acknowledge Dr. Stringer's many suggestions and stimulating insights made during his frequent visits to ANL and to the Steering Committee for the Cooperative R&D Venture "Erosion of FBC Heat Transfer Tubes." Their many concerns on understanding small-scale cold model fluidized-bed hydrodynamics and their applicability to large-scale reactive FBC's helped to shape the nature of the simulations, to force us to interpret the significance and implications of the findings, and to keep the project focused.

A_t	Area of eroding target, m^2
b	Width of cutting particle face, m
c	Compaction modulus
d_p	Particle diameter, m
$\dot{E}$	Erosion rate, m/s
$\dot{E}_b, \dot{E}_d$	Brittle and ductile erosion rates in Neilson-Gilchrist erosion model, m/s
E_{sp}	Specific energy of the eroding surface, Pa
$E_{v\alpha}$	Rate of kinetic energy dissipation per unit volume, W/m^3
E_{vCF}	Rate of kinetic energy dissipation given by Equation (26), W/m^3
E_{vTOT}	Total rate of kinetic energy dissipation given by Equation (25), W/m^3
f	Sliding friction coefficient
G	Solids elastic modulus, Pa
$\vec{g}$	Acceleration due to gravity, m/s^2
I	Moment of inertia of cutting particle = $mr^2/2$, $kg \cdot m^2$
$\overline{\overline{I}}$	Unit tensor
k	Ratio of vertical to horizontal forces in Finnie erosion model
m	Cutting particle mass, kg
$\dot{m}_s$	Mass flux of solids = $\varepsilon_s \rho_s \vec{v}_s$, $kg/(m^2 \cdot s)$
N	Normal force acting on the particle, Pa
$\overline{N}$	Dimensionless normal force, $N/(m\beta^2 V)$
N_1	Defined by Equation (18b), Pa
n	Total number of phases
P	Pressure, Pa
P_t	Eroding target material pressure or hardness, Pa
p	Eroding surface flow stress or hardness, Pa

r	Particle radius, m
t	Time, s
t*	Cutting duration, s
U_{mf}	Minimum fluidizing velocity, m/s
U	Superficial gas velocity, m/s
V	Particle incident velocity, m/s
V_{el}	Threshold velocity in the Neilson-Gilchrist erosion model, m/s
V_f	Two-phase fluid-solids volume = $A_t d_p$, m^3
$\dot{V}_t$	Volumetric rate of removal of eroding surface, m^3/s
$\vec{v}$	Velocity vector, m/s
$|\vec{v}|$	Magnitude of velocity vector, m/s
W	Volume of target removed, m^3
$\bar{W}$	Dimensionless wear volume = $W\beta^2/(bV^2)$
X	Horizontal distance moved by particle, m
Z	Vertical distance moved by particle, m

Greek Letters

α	Impingement angle, degrees
α_o	Transition angle in Neilson-Gilchrist erosion model, Equations (10) and (11), degrees
β	Frequency = $(k\psi pb/m)^{1/2}$, Hz
$\bar{\beta}_{ik}$	Fluid-particle friction coefficient tensor, kg/($m^3 \cdot$ s)
δ_{ks_i}	Kronecker delta function given by Equation (4)
ε	Gas volume fraction
ε_b	Deformation wear factor, Pa
ε_s	Solids volume fraction = $1 - \varepsilon$
ε^*	Compaction gas volume fraction
θ	Angle, degrees

$\bar{\bar{\sigma}}_{ke}$	Effective stress, defined by Equation (3), Pa
σ	Yield tensile stress of target, Pa
ρ	Density, kg/m^3
ρ_s, ρ_g	Solids and gas phase densities, respectively, kg/m^3
τ	Yield shear stress of target, Pa
$\bar{\bar{\tau}}_{kc}$	Coulombic stress for solids (k = s), Pa
$\bar{\bar{\tau}}_{kv}$	Microscopic viscous phase stress, Pa
τ^*	Normalized cutting duration
μ_k	Microscopic viscosity, Pa·s (1 Pa·s = 10 poise)
ψ	Ratio of depth of contact to depth of cut in Finnie erosion model
ϕ	Cutting wear factor, Pa

Subscripts

g	Gas
k	Phase k
s	Solids
x	x-direction
y	y-direction

Superscripts

•	Denotes time rate of change
→	Denotes a vector quantity
=	Denotes a tensor quantity

Operator

| $\nabla \cdot$ | Divergence |
| ∇ | Gradient |

REFERENCES

1. Stringer, J., "Current Information on Metal Wastage in Fluidized-Bed Combustors," <u>Proc. of 1987 Intl. Conf. on Fluidized-Bed Combustion,</u> J.P. Mustonen, Ed., Vol. 2, pp. 685-696, American Society of Engineers, New York, NY (1987).

2. Lyczkowski, R.W., J.X. Bouillard, G.F. Berry and D. Gidaspow, "Erosion Calculations in a Two-Dimensional Fluidized Bed," Proc. of 1987 Intl. Conf. on Fluidized Bed Combustion, J.P. Mustonen, ed., Vol. 2, pp. 697-706, American Society of Mechanical Engineers, New York, NY (1987).

3. Anon, "Joint Program to Study Erosion in FBC Units," Chemical Engineering, Vol. 95, No. XX, p. 40 (May 1988).

4. Parkinson, M.J., J.F.G. Grainger, A.W. Jury and T.J. Kempton, "Tube Erosion at IEA Grimethorpe: Cold Model Studies at CRE, in Reports Commissioned by the Project from Outside Consultants and Others," Vol. 2, NCB (IEA Grimethorpe) Ltd., S. Barnsley, Yorkshire, U.K. (Sept. 1984).

5. Parkinson, M.J., B.A. Napier, A.W. Jury and T.J. Kempton, "Cold Model Studies of PFBC Tube Erosion," Proc. of 8th Intl. Conf. on Fluidized-Bed Combustion, Vol. II, pp. 730-738, National Technical Information Service, R.S. Department of Commerce, Springfield, VA (1985).

6. Nakamura, K., and C.E. Capes, "Vertical Pneumatic Conveying: A Theoretical Study of Uniform and Annular Particle Flow Models," Canadian J. Chemical Engineering, 51:39-46 (1973).

7. Lee, W.H., and R.W. Lyczkowski, "The Basic Character of Five Two-Phase Flow Model Equation Sets," Proc. International Topical Meeting on Advances in Mathematical Methods for the Solution of Nuclear Engineering Problems, Munich, Federal Republic of Germany, Vol. 1, pp. 489-511, American Nuclear Society, LaGrange Park, IL (1981).

8. Lyczkowski, R.W., et al., "Characteristics and Stability Analysis of Transient One-Dimensional Two-Phase Flow Equations and their Finite Difference Approximations," Nuclear Science and Engineering, 66:378-396 (1978).

9. Rudinger, G., and A. Chang, "Analysis of Non-Steady Two-Phase Flow," Physic of Fluids 7:1747-1754 (1964).

10. Lyczkowski, .R.W., "Transient Propagation Behavior of Two-Phase Flow Equations, in Heat Transfer: Research and Application," J.C. Chen, ed., AIChE Symp. Series, 75(174):165-174, American Institute of Chemical Engineers, New York, N.Y. (1978).

11. Lyczkowski, R.W., and J.X. Bouillard, "Interim User's Manual for FLUFIX/MOD1: A Computer Program for Fluid-Solids Hydrodynamics," Argonne National Laboratory, ANL/EES-TM-361, Argonne, IL (Feb. 1989).

12. Gidaspow, G., "Hydrodynamics of Fluidization and Heat Transfer: Supercomputer Modeling," Applied Mechanics Review, <u>31</u>, No. 1, pp. 1-23 (Jan. 1986).

13. Rivard, W.C., and M.D. Torrey, "K-FIX: A Computer Program for Transient, Two-Dimensional, Two-Fluid Flow," Los Alamos Scientific Laboratory Report, LA-NUREG-6623 (Apr. 1987).

14. Murray, J.D., "On the Mathematics of Fluidization. Part I: Fundamental Equation and Wave Propagation," J. Fluid Mechanics, 21(3):465-493 (1965).

15. Anderson, T.B., and R. Jackson, "A Fluid Mechanical Description of Fluidized Beds," Industrial Engineering and Chemistry Fundamentals, 6:527-534 (1967).

16. Neilson, J.H., and A. Gilchrist, "Erosion by a Stream of Solid Particles," Wear 11:111-122 (1968).

17. Finnie, I., "Erosion of Surfaces by Solid Particles," Wear, 3, pp. 87-103 (1960).

18. Engel, P.A., "Impact Wear of Materials," Elsevier Scientific Publ. Co., Amsterdam (1978).

19. Chang, S.L., "A Theory of Cutting Wears by Force Exerting Particles," submitted to the Journal of Tribology, ASME (Mar. 1986).

20. Ushimaru, K., C.T. Crowe and S. Bernstein, "Design and Applications of the Novel Slurry Jet Pump," Energy International, Inc. Report No. E-184-108 (Oct. 1984).

21. Lyczkowski, R.W., D. Gidaspow and C.W. Solbrig, "Multiphase Flow Models for Nuclear, Fossil, and Biomass Energy Production, Advanced in Transport Processes," Vol. 2, pp. 198-340, A.S. Majumdar and R.A. Mashelkar, eds., Wiley Eastern Ltd., New Delhi (1982).

22. Stewart, H.B., and B. Wendroff, "Two-Phase Flow: Models and Methods," J. Comp. Phys., 56, pp. 363-409 (1984).

23. Bird, R.B., W.E. Stewart and E.N. Lightfoot, "Transport Phenomena," (3rd Printing), John Wiley, New York (1960).

24. Bouillard, J.X., and R.W. Lyczkowski, "On the Erosion of Heat Exchanger Tube Banks in Fluidized Bed Combustors," presented at the 5th Miami Intl. Symp. on Multiphase Transport and Particulate Phenomena, Miami Beach, FL (Dec. 1988).

25. Bouillard, J.X., D. Gidaspow, R.W. Lyczkowski and G.F. Berry, "Hydrodynamics of Erosion of Heat Exchanger Tubes in Fluidized Beds," presented at the 1987 AIChE Annual Meeting, New York, NY (Nov. 15-20, 1987) and accepted by the Canadian Journal of Chemical Engineering (Dec. 1988).

26. Pourahmadi, F., and J.A.C. Humphrey, "Modeling Solid-Fluid Turbulent Flows with Application to Predicting Erosive Wear," PhysicoChemical Hydrodynamics 4, 191-219 (1983).

27. Lun, C.K.K., S.B. Savage, D.J. Jeffrey and N. Chepurniy, "Kinetic Theories for Granular Flow: Inelastic Particles in Couette Flow and Slightly Inelastic Particles in a General Flowfield," J. Fluid Mech., 140, pp. 223-256 (1984).

28. Parkinson, M.J., et al., "Cold Model Erosion Studies in Support of Pressurized Fluidized Bed Combustion," Electric Power Research Institute Draft Final Report for Project 1337-2 (Apr. 1986).

29. Grace, J.R., "Fluidized-Bed Hydrodynamics in Handbook of Multiphase Systems," G. Hetsoroni, Ed., Ch. 8.1, pp. 8-5 to 8-64, McGraw-Hill, New York, N.Y. (1982).

30. Chang, S.L., R.W. Lyczkowski and G.F. Berry, "Spectral Analysis of Dynamics in Fluidized Bed Tube Banks," presented at the 5th Miami Intl. Symp. on Multiphase Transport and Particulate Phenomena, Miami Beach, FL (Dec. 1988). To appear in the Proceedings (1989).

31. Lyczkowski, R.W., S. Folga, S.L. Chang, J.X. Bouillard, C.S. Wang, G.F. Berry and D. Gidaspow, "State-of-the-Art Computation of Dynamics and Erosion in Fluidized Bed Tube Banks," submitted to the Tenth International Conference on Fluidized Bed Combustion, San Francisco, CA, April 30 - May 3, 1989.

Investigations on the Performance of Indigenous Fluidized Bed-Heater

B. M. A. AL-ALI and D. A. FATIHY
Department of Mechanical Engineering
University of Mosul
Mosul, Iraq

Abstract

The present paper deals with the development of a fluidized bed-heater manufactured by indigenously materials and performance evaluation made on the same. The heater is tested with a gaseous fuel and also with kerosene liquid fuel. The main parameters considered include bed height and air fuel ratio. Combustion temperatures obtained are measured with the aid of a specially designed radiant heat energy unit. The test data presented include mechanical strength of the porous distributor plate, permeability of bed, pressure drop and temperature of outgoing gases.

With gaseous fuel it was found that a bed height of about 10 mm gave the highest radiant energy flux with an air fuel ratio of about 20 whereas kerosene fuel gave the highest radiant energy flux with an air fuel ratio of about 35. The fluidized bed heater developed gave a radiant energy flux three to four times higher than the conventional heater with the gaseous fuel while with kerosene fuel the heater gave two to three times higher radiant energy flux in comparison to conventional heater.

1. INTRODUCTION

One domestic heater presently used is a radiant type heater employing fire and having exposed glowing radiant surfaces with polished reflectors for directional effect and no convection element in the design. The other type of heater is a convection type heater which distributes a substantial portion of heat output by convection. Domestic fluidized bed combustor is found to successfully reduce pollution but no commercially viable type, is developed so far. The main object of this work is to investigate the feasibility of developing a domestic heater with gas or liquid fuel using the principle of fluidized bed combustion.

The phenomenon of fluidized bed combustion dates back to 1928 when Stratton developed a spouting fluidized bed boiler in which the furnace was operated near 1100°C. Ervin (1) found that the major advantage of fluidized bed combustion system is the wide range of fuels that can be used. Robert and Dorr (2) indicated that in the reactor an inert (usually silica sand) is fluidized by upflowing combustion air supplied from a blower providing an ideal environment for combustion and practically instantaneous heat transfer. Elliott (3) summarised savings of up to (20 %) in plant costs, by applying newly established fluid-bed combustion is possible. Elliott and Virr (4) studied the characteristics of gas-fired fluidized bed combustor in which gas can be burned successfully in a deep fluidized bed. They also indicated that the idea of introducing separate gas jets into an air fluidized bed was discarded because the bed depth needed for complate mixing and good combustion would be too great and require a large number of gas jets. It was suggested by them that it is better to premix gas and air before entry to the distributor plate. Their experiments were conducted with premixed gas-air mixture before entering the porous type distributor plate and they observed that the gas-air mixture must be initially within the flammability limits (2.2 to 9.5 percent by volume) for the propane fuel. Mesko and Robbin (5) studied the steam fludized-bed combustor and found that the optimum design temperature of the bed was in the range of 815 - 870°C with an excess oxygen of about 3 percent.

Elliott and Pillia (6) described the effect of particle size and found that when using Zirconia of 500 m diameter, complate combustion with stoichiometric air supply was achieved within 10 mm of bed depth. There is a necessity to have a deeper and wider understanding on fluidized bed combustion if it is to be applied to domestic heaters successfully.

Zabrodsky (7) investigated that the rise of temperature in fluidized bed reactor has an effect on the intensity of external heat transfer. Borodulya and Kovensky (8) reported that there are many techniques for computing radiation transfer in a fluidized bed over a wide range of particle properties.

2. EXPERIMENTAL PROCEDURE

Experimental work involved developing a distributor plate which can be used for combustion processes of gaseous fuel (L.P.G.) and liquid fuel (kerosene) in fluidized bed heater. The special feature of the distributor plate is that is is made of locally available materials. Different trials were made using different materials. So that a satisfactory result is obtained. The final version of the distributor plate was made of sand, molasses, linseed oil, saw dust and cement. The mixture was reinforced with mesh wire and baked in an oven to dry it at a temperature not exceeding 200°C for two hours. The compression stress of the distributor plate was tested with the help of a special compression apparatus. The permeability of the distributor plate is also evaluated using a permeability apparatus.

The dimensions of fluidized bed combustion chamber were fixed at 280 mm x 120 mm and 150 mm height. The reflector was made from a galvanised steel plate of 1.58 mm thickness. The wind box was made from the same and a slope of 1:5 was adopted for the wind box (9). According to recommendations of Baskakow et al. (10) the radiative flux is independent of the dimensions of particles. Therefore river sand 250 - 350 m similar to that chosen by Elliott (6) was used as bed material. Fig. 1 is a schematic of the combustor. Air flow measurements were made by using an orifice nozzle of 24.95 mm diameter with corner pressure tappings. The gaseous fuel (L.P.G.) flow measurement system consists

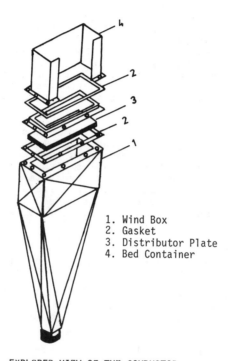

1. Wind Box
2. Gasket
3. Distributor Plate
4. Bed Container

FIG. (1) EXPLODED VIEW OF THE COMBUSTOR.

of two fuel cylinders. The fuel is passed through a flow meter
(Rotameter type) which is calibrated according to British standards by P.A. HILTON Co. Fig. 2 shows a schematic of the circuit. Liquid fuel (kerosene) flow measurement setup consisted of
a storage tank and two calibrated glass tubes. Liquid fuel was
evaporated through a specially designed preheated steel coil
through which kerosene fuel passed. The fuel vaporised by the
time it came out of the coil. Fig. 3 shows the circuit adopted
for kerosene fuel flow measurement. For evaluating the combustion
efficiency of the heaters a special radiation detector was designed and calibrated. Radiant energy emitted was evaluated in each
heater from sinusoidal signals.

3. DICUSSSION OF RESULTS

Fig. 4 shows the effect of increase of percentage of molasses with different percentages of linseed oil on the permeability and compression stress of the distributor plate. It is seen
that a mixture consisting of 5.0 percent linseed oil and 12.5
percent molasses gave good results of permeability, compression
and shear stress. This is due to good physical structure and beneficial chemical reactions that take place with these proportions.

The effect of variation of two main parameters i.e. air/fuel
ratio, and bed height were studied. Velocity of fluid flow was
neglected according to Goldobin and Makushenko's (11) recommendations. According to them radiative flux is independent of velocity of fluid. Their experimental results also indicate that the
basic characteristic that determines radiative properties is the
emissivity of particles forming the bed. For this reason river
sand was choosen as bed material. Fig. 5 shows radiant energy
as a function of air/fuel ratio at four different bed heights. It
can be seen that as air/fuel ratio increases radiant energy decreases because when air/fuel ratio exceeds the stoichiometric
limit some of the excess air does not take part in reaction and
tends to reduce bed temperature which decreases radiation energy
and increases energy carried away by convection with fuel gases.
Also it is seen that throughout air/fuel ratios are higher than
stoichiometric air/fuel ratio because of agglomeration. The best
bed height was found to be about 10 mm at an air/fuel ratio less
than 21.5.

Under the same conditions the performance of the fluidized
bed heater was compared with a commercial gas heater and it was
found that gas fluidized bed heater radiates about 3.8 times the
energy emitted by the conventional heater. However, the fuel
consumption was 2.3 times higher in the fluidized bed heater.

Fig. 6 presents radiant energy flux as a function of air/
fuel ratio at bed heights of 10 mm, 15 mm, 20 mm, and 25 mm. It
is seen that air/fuel ratio was far from stoichiometric conditions because a high percentage of fuel causes sudden combustion
melting and particles leading to severe agglomeration. It is

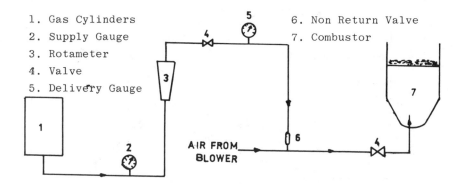

1. Gas Cylinders   6. Non Return Valve
2. Supply Gauge   7. Combustor
3. Rotameter
4. Valve
5. Delivery Gauge

AIR FROM BLOWER

FIG. (2)   SCHEMATIC DIAGRAM OF THE (L.P.G) CIRCUIT.

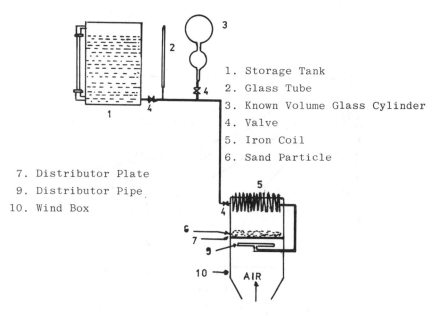

1. Storage Tank
2. Glass Tube
3. Known Volume Glass Cylinder
4. Valve
5. Iron Coil
6. Sand Particle

7. Distributor Plate
9. Distributor Pipe
10. Wind Box

AIR

FIG. (3)   SCHEMATIC DIAGRAM OF KEROSENE FUEL CIRCUIT.

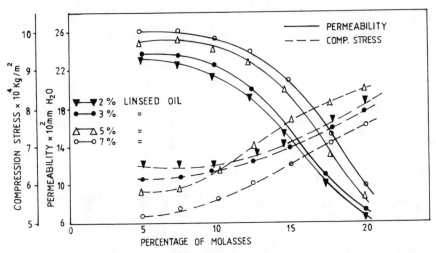

FIG. (4) COMBINATION OF PERMEABILITY AND COMPRESSION STRESS FOR
DIFFERENT PERCENTAGE OF MOLASSES AT DIFFERENT PERCENT-
AGE OF LINSEED OIL.

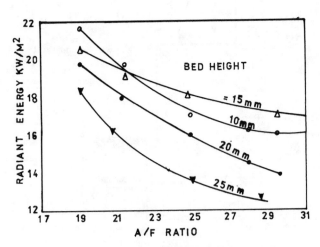

FIG. (5) VARIATION OF RADIANT ENERGY FLUX WITH A/F RATIO
AT DIFFERENT BED HEIGHTS OF GAS (F.B.H).

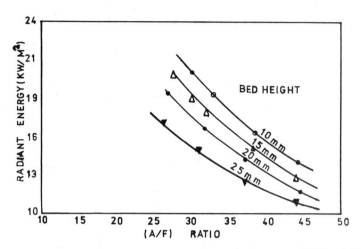

FIG. (6)   VARIATION OF RADIANT ENERGY FLUX WITH (A/F)
RATIO AT DIFFERENT BED HEIGHT OF KEROSENE
(F.B.H).

also seen that a 10 mm deep bed radiates more energy as compared with others for different air/fuel ratios. Comparison with domestic kerosene heater showed that kerosene fluidized bed heater radiates about 2.4 times the energy of a conventional type, though it consumed about 1.6 times the fuel of conventional type.

4. CONCLUSIONS

Experimental work has yielded interesting results concerning the manufacture of a distributor plate and on the radiant energy resulting from combustion of gaseous and kerosene fuels in a fluidized bed. It was found that the best percentage of mixture was 2.0 percent saw dust, 2.0 percent cement, 5.0 percent linseed oil and 12.5 percent molasses of sand weight of distributor plate. The distributor plate can be reinforced by mesh wire of size 20 mm diameter to increase the stiffness of the plate and to prevent back flow of flame in case of damage to distributor plate. It was also found that radiant energy is affected by air/fuel ratio regardless of bed height.

For the gas fluidized bed heater the best air/fuel ratio was found to be less than 21.5 at a bed height of 10 mm to get high radiant energy flux. For the kerosene fluidized bed heater the best air/fuel was found to be less than 35 at a bed height of 10 mm .

REFERENCES

1. Ervin C. Lentz, Fluidyne Engineering, Corp. Minneapolis, Minn. 55422 "Anthacite Culm Fired Fluidized-Bed Boiler" CEP January 1984.

2. Robert J. Sneyd, Dorr-Oliver, Inc., Stamford Conn. 06904 "Energy Recovery from Fluidized-Bed Combustion CEP January 1984.

3. E.E. Elliott, "Possible Use of Fluid-Bed Combustion and Heat Transfer" Paper 22, The Total Energy Conference Institute of Fuel, Nov. 1971.

4. D.E. Elliott and Virr "Small-Scale Application of Fluidized-Bed Combustion and Heat Transfer. Archer Westinghouse, Session Chairman, Report, 1968.

5. Mesko and Robbins "Coal Combustion in Limestone Bed. CEP, August Volume 74, No. 8, 1978.

6. D.E. Elliott and Pillia "Some Design Considerations for Shallow Fluid Bed Boilers, Institution of C.E.S.S. No. 38, 1972.

7. S.S. Zabrodsky, "High Temperature Fluidized Bed Plant. Izd Energiya. Moscow (1971).

8. V.A. Borodulya and V.I. Kovensky "Radiative Heat Transfer Between a Fluidized Bed and a Surface" Heat and Mass Transfer Institute, 220728, Minsk, U.S.R. July 1981.

9. British Standard, Code of Practice "Measurement of Fluid Flow in Pipes" 1042 Part (1), U.D.C. 532-54-08. 1964.

10. A.P. Baskakov, G.K. Malikov and Yu.M. Goldobin "The Effect of Radiative Heat Transfer on the Heat Transfer Coefficient in a High-Temperature Fluidized Bed, in High Temperature Endothermal Processes in a Fluidized Bed" pp. 192-196. Moscow (1968).

11. Yu. M. Goldobin and V.M. Makushenko, "An Experimental Investigation of the Effective Emissivity of a High-Temperature Fluidized Bed, in Industrial Fluidized-Bed Furnaces" pp. 23-26, Sverdlovsk (1976).

Fluidized Bed Tube Bank Erosion and Flow Visualization

H. H. J. TOSSAINT and P. P. van NORDEN
Stork Boilers
P.O. Box 20
7550 GB Hengelo, The Netherlands

P. L. F. RADEMAKERS and M. L. G. van GASSELT
TNO Apeldoorn
P.O. Box 342
7300 AH Apeldoorn, The Netherlands

Abstract

Part of the joint FBC development program of STORK BOILERS and TNO concerns the erosion resistance of heat exchanger materials in the 4 MWth AFBB test facility at the TNO laboratory. Data are obtained both from the 230 °C in-bed tube bank and from separate test tubes operated at various wall-temperatures. Previous results have indicated that tube bank erosion occurs and have shown the importance of the formation of protective oxide layers at higher temperatures. This paper describes the recent efforts in material selection, system improvements and measuring methods. Several factors affecting erosion were studied, and the behaviour of coatings is described. The competition between oxidation and erosion is discussed. Important system improvements were achieved. The temperature range of oxide layer formation and protection against erosion is extended towards lower temperatures. A method is developed to establish and investigate the flow pattern in a fluidized bed. The effect of fluidization quality, coal feed nozzle- and grit refiring nozzle-positions on the flow pattern can be distinguished now.
As a result of the investigations described a very satisfactorily tube bank lifetime is expected now for units operating at relatively high steam temperature conditions.

1. INTRODUCTION

Fluidized bed combustion of coal has the potential for integrating an attractive energy production and environmental control by simultaneous reduction of NO_x and SO_x.
Within the framework of the Dutch national Research Program on Coal Technology, NOK, various aspects of FBC are being studied at TNO together with Stork Boilers, who is developing FBC technology for industrial applications. For that purpose, a 4 MWth Atmospheric Fluidized Bed Boiler constructed by Stork Boilers is in operation since 1982 with TNO at Apeldoorn. The boiler is used as a multipurpose test facility, and is not integrated in the energy system of the building.
A schematic view of the test facility is given in Figure 1.
A description of the bubbling bed type facility and the first experimental results are given in references (1), (2) and (3).
Previous results are given in references (4), (5) and (6). The corrosion/erosion tests, which are described in this paper, are carried out in the 4 MW test facility, simultaneously with other experiments.
The results of these experiments were used for the design and operation of a 90 MWth AFBC-plant now in operation with AKZO at Hengelo. The boiler is described in reference (7).

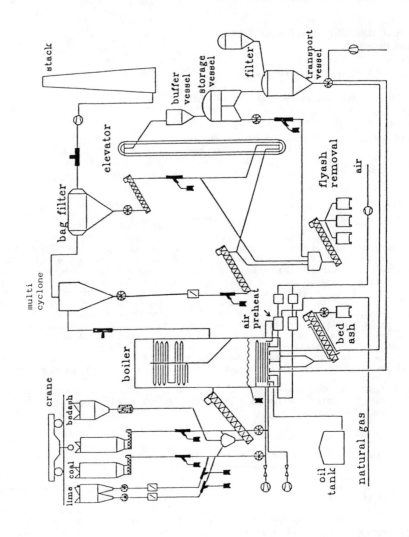

Fig. 1. - TNO - 4 MWth AFBC Boiler.

2. SCOPE OF THE CURRENT INVESTIGATIONS

The 4 MWth testfacility is operated for about 2500 hours a year with relatively low sulphur content coal (ca. 1% S), and most of the time under limestone addition. Part of the history of the erosion test program is summarized in Figure 2. During testrun (2.6) the fluid bed system was optimized and pressure fluctuations in the bed were reduced considerably. Information on corrosion/erosion rates, as well as on the mechanisms involved, is obtained from the in-bed tube bank, as well as from separate test tubes. Major operating conditions are listed in table 1. Bed temperatures range from 840 °C to 865 °C, temperatures in the freeboard are mostly 10 to 20 °C higher.

All important operating conditions, including the temperatures of the test tubes and the flue gas analyses were collected by the data acquisition system (Ref. 1). Some variations in the conditions, within specific limits, were made on purpose because this was required by different experiments, such as partial load, start-stops, bed temperature changes, operation with coal feed to one or two beds, sulphur retention studies etc. Operation periods varied from less than 20 hours to several hundreds of hours. Undesired operating conditions could be avoided by removing the test tubes. For the sulphur retention and the flue gas quality the average values can be regarded as typical values, the excess of combustion air being usually 20%.

Typical particle sizes of the bed material are 0-6 mm, with an average value of 1.1 mm. Additional information about coal, limestone and particle distribution is given in reference 3.

Tube Bank Sections

The testfacility has a closed water/steam cycle and produces 12 bar saturated steam, the steam temperature being 180 °C. The metal temperature of the in-bed tube bank will be approximately 230 °C. The tube dimensions are: 51 mm outer diameter with 4 mm wall thickness.

Various materials were used, a.o. carbon steel St.35.8II and 2¼CrlMo steel and several coatings were examined.

Intermediate inspections were carried out and ultrasonic wall thickness measurements were performed; subsequently, a micrometer was used.

In addition, metallographic examinations were carried out on cross sections.

Separate Test Tubes

Air/water cooled and uncooled bayonet-type test tubes were used to study corrosion/erosion in the bed and the freeboard. (Ref. 4 and 5).

The freeboard tube was placed 2.70 m above the bottom plate; the in-bed tube was placed 0.92 m above the bottom plate.

Intermediate wastage measurements were carried out on fixed locations by a micrometer. Final measurements were made on metallographic cross sections and results were compared with those of the original wall thickness at the same location.

Blasting tests

In order to speed up the erosion investigation blasting tests were performed.

The arrangement of the blasting test equipment is represented in Figure 3, the measuring instrument in Figure 4 and the test conditions are mentioned in

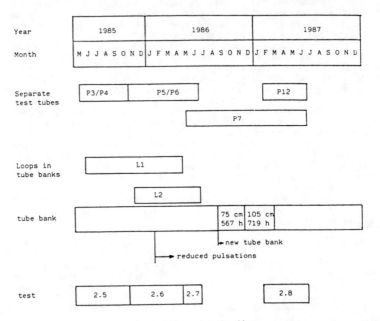

Fig. 2. - Review of Material Studies.

TABLE I: REVIEW OF TEST CONDITIONS FOR TEST TUBES

Test Nr.		2.5	2.6	2.7	2.8
Exposure time	(h)	600	900	300	1400
Limestone		+	+	+	+
Temp. Bed/Freeb.	(°C)	845/855	845/855	845/845	840/865
Fluid. velocity	(m/s)	2.3	2.0	2.0	2.1
Ca/S ratio		2.4	2.8	2.7	2.5
S capture	(%)	75	70	60	80
Coal *		P	P	P	C/P

* P = Polish   C = Columbian
  During test 2.6 pressure fluctuations are reduced.

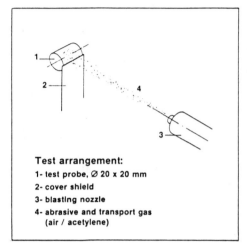

Test arrangement:
1- test probe, ⌀ 20 x 20 mm
2- cover shield
3- blasting nozzle
4- abrasive and transport gas
   (air / acetylene)

Fig. 3. – Sand-blasting Erosion Test.

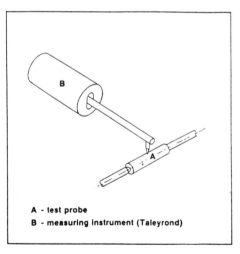

A - test probe
B - measuring instrument (Taleyrond)

Fig. 4. – Measuring Arrangement.

Figure 5. An example of a measuring result is shown in Figure 6. Some erosion rates established during blasting tests were compared with those of identical materials in the bed. The ratios between the erosion rates established are given in Figure 7. These ratios considerably differ mutually. This indicates that the materials cannot be selected on the basis of blasting tests.

Development Of An Accurate Erosion Measuring Device for the in-bed tube bank

First, measurements were performed with an ultrasonic wall thickness gauge. However, this procedure proved to be insufficiently accurate. In the case of surface-treated steels, this even led to a completely incorrect interpretation of the results. Next, two instruments were used, which measure the tube diameters, the first being a micrometer equipped with a Mahr sensor, which has an accuracy of 1 micron. The micrometer has a template, which can be fitted on the in-bed tubes. The sensor is connected with read-out equipment having an accuracy of 1 micron. The second measuring instrument is a measuring ring (Figure 8) of our own design, which accommodates eight Mahr sensors arranged in such a way that any two sensors are at an angle of 45° relatively to one another. The measuring ring is of the divisible type and is connected with the read-out equipment. This enables the tube diameters to be measured into four directions in a simple manner.
Both measuring instruments are calibrated prior to the measurements. In actual practice the accuracy of the measurements proves to be approximately 10 microns due to surface irregularities and positioning problems.

Surface treated and coated materials

An important part of the material selection was focused on surface-treated and coated steels.
It was found that under relatively severe conditions surface treatments, such as nitriding and boriding, could extend the lifetime considerably. Before further optimizations were adopted, the applicability of coatings was studied.
A separate loop (L2) being part of the in bed bundle, consisted of coated samples as shown in Figure 9.
Intermediate and final measurements were carried out, and some of the results are shown in Figure 10. The hardness of the coatings seems to be important up to values of 600 to 700 HV. Similar values were shown previously by the erosion-resistant oxide scales at about 450 °C. A further hardness increase has only a minor effect.
It is important to notice that solid coatings without pores behaved best. Under the conditions encountered metal losses of 0.03 mm/1000 hours are observed for the best coatings.

Flow patterns within the bed

Many wastage failures can be attributed to local effects: inadequately diffused jets from various feed and recycle ports; local leaks; local excessive particle velocities.
The material research has shown that the erosion rate of one and the same material may differ considerably from place to place in the bed bundle. On this account, a new bed bundle was placed in September 1986, which entirely consisted of poorly erosion-resistant mild steel. Figure 11 schematically represents the tube bank and bottom plate. The local presence of the erosion is shown by Figure 12. Under these conditions, it is not very useful to give an average diameter decrease.

Test specification

Test probe temperature : 350 ° C
Test period   : 1/4 - 5 hrs.

Abrasives:
- milled slags of PCF-combustors
- bed ash of AFBC

    size: 0.1 - 1.0 mm
    flow: 200 g / min
    velocity: 40 m / s

Transport gas: air / acetylene

Fig. 5. - Test Specification.

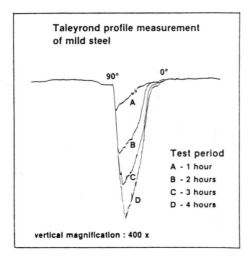

Fig. 6 - Typical Test Results.

371

Relationship between erosion velocity of materials tested in the TNO 4 MWth rig and materials tested during shotblasting tests at elevated temperature with bedmaterial

Material	ratio	erosion velocity bed test / erosion velocity shotblasting test
Mild steel		1,0
Stainless steel		1,8
Coatings - NiCr 80/20		2,9
- Stellite		4,5
- WC		0,6
Surface treated - 2 1/4 % Cr steel		2,0
- mild steel		0,1

Fig. 7. - Comparison Between Test Results.

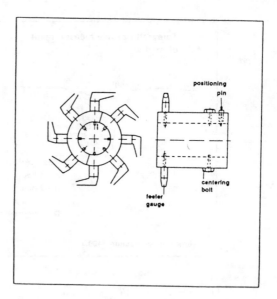

Fig. 8. - Measuring-instrument for Determining Tube Diameter.

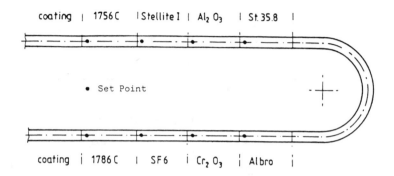

Fig. 9. - Test Loop No. 2.

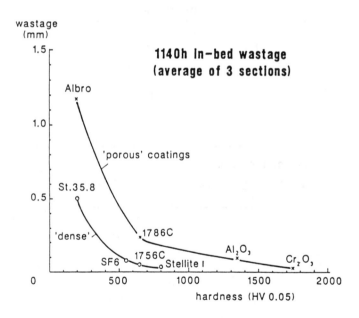

Fig. 10. - Results Test Loop No. 2.

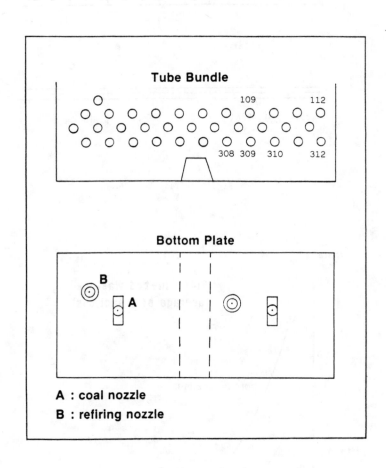

Fig. 11 - Tube and Nozzle Arrangement.

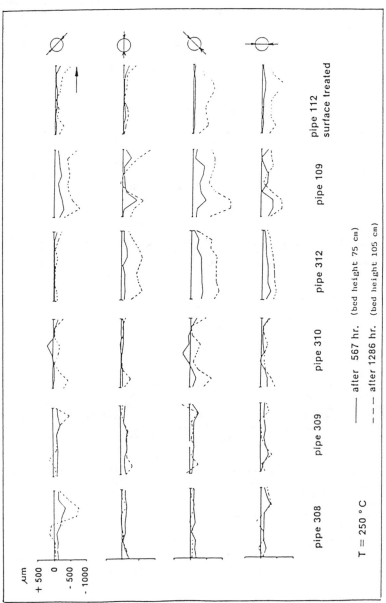

Fig. 12 - Erosion Data Tube Bank.

375

As from September 1986 the research focuses on factors that may have an impact on the fluidization behaviour of the bed material. In order to be able to establish target values for the various parameters, the AFBC of Babcock Power in Renfrew was taken as a reference, since no significant bed tube erosion occurred over there. All bundles are horizontally mounted. The investigation includes the following points:

- Reduction of the amplitude of the 1Hz pulsation in the combustion air duct. The amplitude of this pulsation at TNO turned out to be much higher than that of the Renfrew boiler. Adaptations have led to an important decrease of the amplitude. This has led to a ca. 50% reduction of the erosion rate of the bed bundle (Ref. 8).

- Impact of the composition of the bed material. In the fourth quarter of 1987 the impact of non-dosage of limestone is examined, however no difference in erosivity was noticed. A further examination of plant experience has not revealed a trend between erosivity of bed material and tube wear.

- British Coal has performed a study of the rising velocity of bubbles on the TNO boiler. The rising velocity of the bubbles appears to be a parameter which has hardly any significance and both above and under the bundle this velocity is at a comparable level. The bed is in the slow-bubble regime and the rising velocity varies from 1/3 x the superficial velocity in the bed centre to 1 x the superficial velocity at the boiler walls.

- The impact of the bed height has been studied at a dynamic bed height of 75 cm and 105 cm. The results of some measurements are represented in Figure 12. This shows that bed height reduction does not have a large impact on the erosion rate under testing conditions.

- Impact of the bundle geometry. A staggered bundle configuration has been adopted, in which the vertical and horizontal pitch, as well as the tube diameter, has been varied. Both measures showed no impact on erosion. In the original configuration of the tube banks the lacking of tubes above the coal nozzles has turned out to be very unfavourable, however.

- Impact of the ratio between the pressure drop across the bed and the pressure drop across the distributor plate. In the current research the pressure drop across the bottom plate is strongly increased. Temperature measurements show that the temperature in the bed is homogeneous and that there are no places where the material is stationary. The relatively low pressure drop across the bottom plate may have provoked partial instability. Hereby, the bed material tends to get stationary, but this is prevented by the radial mixing in the bed. However, this phenomenon has an impact on the flow behaviour of the bed material.

- Figure 13 shows a wastage reduction of almost 90%, half of it contributed by a reduction of fluidizing velocity from 2.6 to 1.9 m/s and half of it contributed to the reduction of the amplitudes of the 1 Hz pulsation.

From the above it emerges that a large number of factors may have an impact on the erosion. In order to understand the differences in erosion rate between the different boilers, it is important to have sufficient data available, both on the boiler and the erosion phenomena.
So, a list containing information on combustion air, bed geometry, tube bank design, nozzles and bed material is available on request.

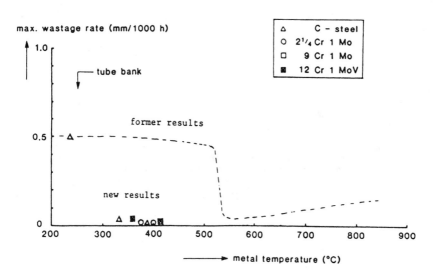

Fig. 13 - Results Of AFBB In-bed Tests.

3. EROSION-RESULTS

Tube bank erosion

The erosion of the tube bank, which operates at about 230 °C, was published in the past (Ref. 5). After the optimization of the tube bank configuration and the position of the tube bank, much emphasis was put on material selection, which was mostly carried out with separately cooled or uncooled tube specimen-holders.

Separately Cooled Tubes Erosion

Program number P5 and P6 (see fig. 2).

Bayonet type air + water cooled tubes were exposed in the freeboard and the bed (Ref. 5). The tube samples consisted of the following materials: 1Cr½Mo, 2¼CrlMo, 9CrlMo and 12CrlMoV.
The temperatures were 550 °C (first 250 hours) and 425 °C (additional 850 hours). After 1100 hours the maximum metal loss measured at the underside reached the following values:

	Metal losses (mm/1000 hours)	
Material	in-bed	freeboard
1Cr½Mo	0.12	0.19
2¼CrlMo	0.25	0.25
9CrlMo	0.035	0.035
12CrlMoV	0.045	0.027

The observations have indicated/confirmed the following:
. the presence of hard oxide layers (hardness: 600-700 HV);
. oxide layers can offer protection against in-bed erosion;
. pre-oxidation occurred during the first period, and this explains the scaling and relatively high losses of the low alloyed materials;
. the promising behaviour of the 9%/12% Cr steels;
. the reduction of the pressure fluctuations carried out during test 2.6 is beneficial.

Program Number P7

In view of the erosion, which was observed previously at low metal temperatures, the temperature of the bayonet type test tubes has been lowered as much as possible (Figure 14). The effect of the metal temperature was studied on a 12CrlMoV test tube exposed in the bed for 1670 hours. The results for the underside of the tube are given in the following table.

Metal temp. (°C)	Metal loss (mm/1000 hours)	Oxide
420	< 0.005	+
410	0.020	±
360	0.048	−

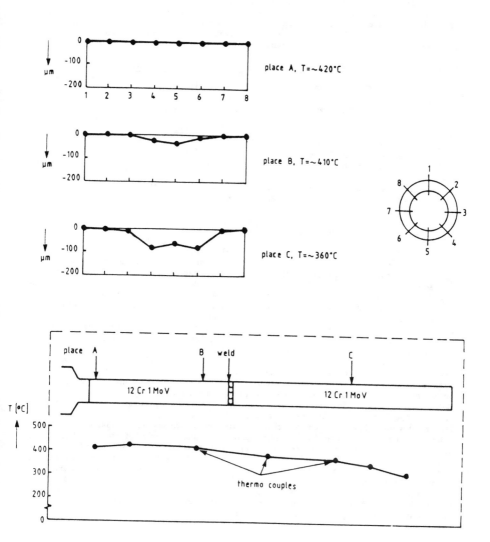

Fig. 14. - Tube 7, Exposure Time 1670 hrs.

379

In spite of the detrimental effect of the ash refiring nozzle, resulting in the removal of parts of the oxide layer, the metal losses are very promising.

Program number P12

This tube made from low-alloyed ferritic steels was exposed in the bed for 1350 hours, to verify the previous results at the lowest possible temperatures. The tube arrangement, the temperature profile and the results (metal losses) are given in Figure 15. The highest metal losses observed at the underside of the tube are given in the following table:

Material	Temperature (°C)	Metal loss (mm/1000 h)	Oxide
C-steel (35.8)	430	0.007	+
2¼CrlMo	400	< 0.005	+
C-steel (35.8)	390	< 0.005	+
2¼CrlMo	380	0.018	+
C-steel (35.8)	330	0.044	+

So far, the results confirmed:
- that oxide layer formation is possible under optimized fluidization conditions;
- that although oxide layers are relatively thin and subject to some wear, they obviously can offer protection against erosion down to temperatures of about 350 °C.

4.   FLOW VISUALISATION

As to the mechanism of erosion the following:
Particles in the bed vary in size, composition and hardness. The hardness varies from 160 HV for limestones and 700 HV for silicates to 1300 HV for some sulfate-surfaces. The moving particles can cause an abrasive erosion on the oxide layers or on the base material of the tube banks. Bad fluidization quality such as slugs, high amplitude pulsations, non uniform flow distributions (due e.g. to low pressure drop across distributior plate) can cause local velocities of the particles which exceed the average and can cause significant local erosion. So it is believed now that next to controlling the wall temperature, it is essential to know the local flow pattern and local flow velocities at positions in the bed which are difficult to forecast. Cold model studies give only restricted information on this matter. However when we improved the erosion-rate measuring technique by putting centre dots on certain positions on the tubes to be measured, we noticed that these centre dots showed very specific wear patterns due to the local particles-flow direction. So in this way the flow pattern of the bed material has been examined.
The resultant flow pattern of the bed material across the upper-most row of tubes is shown in Figure 16. It seems that the bed material flows at this level to the place where the bubbles rise up from the coal nozzle. In view of the relatively large air flow being blown into the bed together with the coal (15%), a large number of bubbles may occur at this place, which may provoke flow patterns above the bundle, as they also can be found in a spouted bed.
As a consequence of these observations, the tubes in the bundle have been provided with many centre dots in order to obtain more information about flow patterns especially on problematic area's like grit refiring nozzles and around the coal feed nozzles.

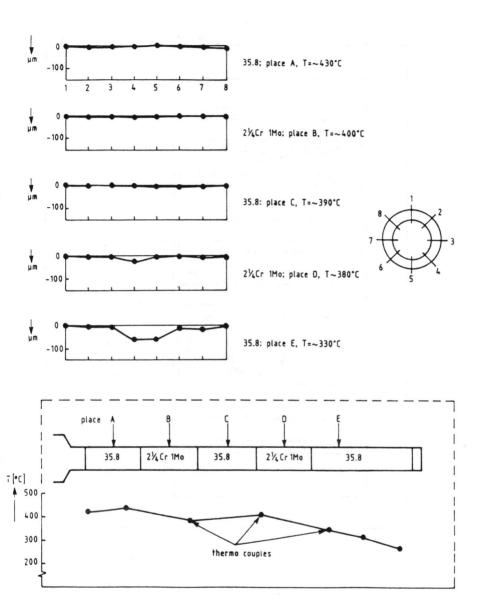

Fig. 15. - Tube 12, Exposure Time 1350 hrs.

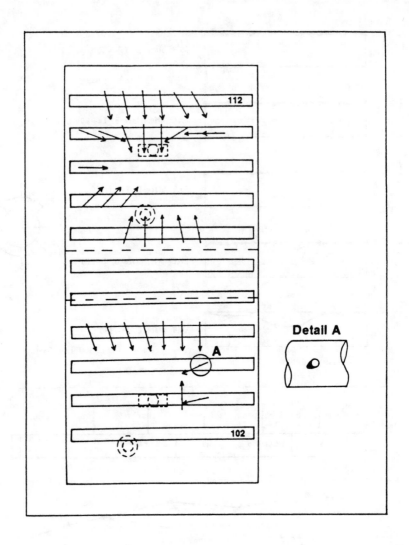

Fig. 16. - Wear pattern of Centre-dots Of Upper Tube Row.

5. DISCUSSION AND CONCLUSIONS

Previous results have indicated a kind of transition temperature. Below this temperature, metal loss rates are high due to erosion of the tube wall. Above this temperature, protection can be obtained by the formation of relatively hard oxide scales.
On the tube wall competition takes place between oxide formation and the erosion process. The rate at which the erosion process takes place is dependent on factors that have an impact on the fluidization behaviour and on the properties of the tube material. The initial rate at which an oxide is formed (Vox) mainly depends on the tube wall temperature. (see Figure 17). This figure shows that at a given fluidization behaviour of the bed material the erosion rate depends on whether or not oxide scale is present and, thus, on the tube wall temperature. Suppose that the erosion rate for a non-oxidized material has the value E_4 mentioned in the figure. The impact of the temperature on this rate will yet be subject to further examination. Here, it is assumed that the erosion rate is independent of the temperature.
If the tube wall temperature exceeds T_4 ($Vox > E_4$), oxide formation takes place and the erosion rate will be determined by the hard oxide scale and drop from E_4 to E_4'.

If the tube wall temperature is lower, no oxide will be formed and the erosion rate will retain the value E_4.
If for instance, at a tube wall temperature T3 and an erosion rate E_4 the erosion rate drops to a value E_2 owing to a changed fluidization behaviour of the bed material, oxide scale will be formed and the erosion rate will drop further to value E_2'. Consequently, the large drop of the erosion rate from E_4 to E_2' must not entirely be attributed to the changed fluidization behaviour of the bed material, but also to the formation of oxide scale.
In Figure 13 recent results are compared with previous ones. In the case of the materials now tested at approx. 400 °C the fluidization behaviour in the bed has changed because the fluidization velocity has been reduced from 2.6 to 2.1 m/s and the amplitude of the 1Hz frequency has been stepped down considerably. At those earlier tests the test pieces were bright and now they are covered with oxide scale.
This has led to a reduction of the erosion rate from 0.5 mm/1000 hours down to less than 0.05 mm/1000 hours.

Figure 17 further shows that if the erosion rate has a value E_4' and the tube wall temperature a value between T_1 and T_4, it may be useful to apply pre-oxidation of the tubes. Thanks to the oxide scale provided beforehand, the erosion rate is reduced such (E_4') that it is smaller than the oxide formation rate and, thus, a stable lower erosion rate is obtained.
The above may be the explanation why no large erosion rate reductions were established at the bed bundle of the TNO AFBB in case of variations affecting the fluidization behaviour. The tube wall temperature is 230 °C, at which oxide scale is only formed slowly. Consequently, research on the bed bundle will focus on the 350 °C loop incorporated and temperatures in between will be included in the research program.
Low metal temperatures, as they mostly prevail at small units raising hot water or low-pressure steam, may cause erosion of the in-bed tubing. Several options are available to achieve an economic operation of these units, such as:
1. application of coatings or surface treatments;
2. pre-oxidation;
3. reduction of fluidization velocity.
In order to be able to assess the effect of local flow patterns in the fluidized bed, more experimental flow pattern information is needed on the fluid bed with its tube banks as a whole and on problematic area around feed and recycle ports.

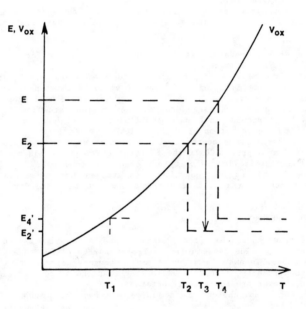

Fig. 17 - Relationship between Initial Oxide Film Growth (V_{ox}) and Erosion Velocity (E).

By applying centre-dots on the various places in the fluid bed we are now able to obtain realistic flow patterns-information which are highly necessary for design purposes.

Our results have indicated a kind of transition temperature. Below this temperature, metal loss rates are high due to erosion of the tube wall. Above the temperature, protection can be obtained by the formation of relatively hard oxide scales. For large steam raising units operating at medium or high pressure and the wall temperature exceeding 350 °C, our results are very promising. Based on our observations, of controlled wall-temperatures and good fluidization quality, it is expected that acceptable life times can be achieved with common ferritic boiler steels for in-bed components, such as evaporators and super-heaters.

ACKNOWLEDGEMENT

This research is carried out under contract with the Project Office for Energy Research, PEO, NOVEM the framework of the Dutch National Research Program on Coal Technology, NOK, financed by the Ministry of Economic Affairs, co-sponsored by the EC.
Part of the TNO program is carried out under the European COST 501 collaboration, project NL-9.
The authors would like to acknowledge the support of the Project Office for Energy Research, NOVEM, and the support of the European Community (EC).
The authors would also like to thank their colleages at STORK BOILERS and TNO-APELDOORN for all their contributions, without which it would have been impossible to reach these results.

REFERENCES

1.   F. Verhoeff, M. van Gasselt,
     "TNO/Stork Fluidized Bed Combustion Development", paper presented at the 7th Int. Conference on Fluidized Bed Combustion, Philadelphia, Pa., USA, October 25-27, 1982.

2.   H.M.G. Temmink, J. Meulink,
     "Operation experiences with the TNO 2mx1m Atmospheric Fluid Bed Boiler Facility", paper presented at Coal Technology Europe, 3rd European Coal Utilization Exhibition and Conference, Amsterdam, the Netherlands, October 11-13, 1983.

3.   J.P. Meulink, A.W.M.B. van Haasteren, H.M.G. Temmink,
     "Operating Experience with a 4 MWth AFBC Research Facility", 8th International Conference on fluidized Bed Combustion, Houston, USA, March 18-21, 1985.

4.   P.L.F. Rademakers et al,
     "Corrosion/erosion tests in a 4MWth Atmospheric Fluidized Bed Boiler", paper presented at the 3rd International FBC Conference, London, October 16 and 17, 1984.

5.   P.L.F. Rademakers, J. Meulink,
     "Erosion/Corrosion under AFBC Conditions, Experience from a 4 MWth Test Facility", paper presented at the EPRI Workshop on FBC Materials Issues, Port Hawkesbury, Nova Scotia, Canada, July 29 – August 1, 1985.

6.   P.L.F. Rademakers, P.O. Kettunen,
     "Materials Requirements and Selection for Fluidized Bed Combustors", paper presented at the conference High-Temperature Alloys for Gas Turbines and Other Applications 1986, Liege, Belgium, October 6-9, 1986.

7. F. Verhoeff,
    "Design and Operation of the 115 t/h FBC Boiler for AKZO-Holland",
    paper presented at the 1987 International Concerence on Fluidized Bed
    Combustion, Boston, USA, May 3-7, 1987.

8. W.G.J. Little,
    "Pulsation Phenomena in Fluidized Bed Boilers", paper presented at the
    1987 International Conference on Fluidized Bed Combustion, Boston, USA,
    May 3-7, 1987.

An Optical Probe for the Detection of Bubbles in High Temperature Fluidized Beds

A. H. GEORGE
Department of Mechanical Engineering
Montana State University
Bozeman, Montana 59717, USA

Abstract

An optical probe consisting of a light source and a photodetector separated by a gap has been developed for use in fluidized beds which operate at combustion level temperatures. Assuming opaque particles, essentially no transmission of light occurs across the gap when the emulsion phase fills the gap. For common gas fluidized systems, the bubble phase is essentially optically clear, and light will be transmitted across the gap to the photodetector when a bubble contacts the probe. The restrictions are that the particles be opaque to the light used and that the gas be optically clear at least for distances of a few millimeters. Optical probes are, therefore, suitable to a wide range of fluidized systems and operating conditions.

Infrared light (880 nm spectral peak) produced by a high intensity LED is conducted down a Vycor glass rod, across a gap and through another Vycor glass rod to a silicon infrared phototransistor. The Vycor glass rods can operate at combustion level temperatures without softening or undergoing a significant loss of transmittance at infrared wavelengths. A water-cooled mounting is provided to avoid excessively high temperatures at the LED and phototransistor. After a few hours of operation in a fluidized bed, during which the glass tips of the probe are subject to abrasion by the bed particles, the calibration of the probe remains essentially constant.

Radiant emission from the high temperature bed to the probe contains a significant component in the infrared spectrum. This radiant emission from the bed to the probe is a disturbance input to the probe since it is sensed by the phototransistor. A signal conditioning technique is described which essentially eliminates this disturbance input from the probe's output.

Actual data obtained in several fluidized systems or high temperature environments with operating temperatures up to approximately 930°C are presented. The experimental results obtained indicate that the probe's output is insensitive to bed temperature or particle type (for opaque particles).

1. INTRODUCTION

Probes to detect the local presence of bubbles in fluidized beds have been used for over 30 years. With appropriate statistical techniques for data reduction, quantities such as bubble frequency, bubble size distribution, bubble rise speed and bubble volume fraction can be determined [1, 2]. The probes presence in the bed does disturb the characteristics of the bubbles. Recent work has, however, concluded that accurate bubble properties can be established using probes but that corrections to the measured values are generally required [3].

Probes based on detecting the change in electrical capacitance--due to the probe tip being surrounded by a bubble--have been utilized by many investigators [2, 3, 4, 5, 6, 7, 8]. Capacitance probes must be designed for the particular solid-gas system of interest and may need to be calibrated for each operating temperature and pressure considered [9]. Probes based on sensing local capacitance would be cumbersome to use in beds that operate over a range of temperatures and particle compositions such as is the case in fluidized bed combustion.

Electroresistivity probes have been used in fluidized beds at elevated temperature but are only useful if the bed particles are electrically conducting [1]. Electric discharge probes measure the dielectric breakdown voltage between two electrodes immersed in the bed [10]. Since this breakdown voltage--at least for the sand-air system studied in the reference cited--is different for the bubble phase than for the emulsion phase, the device functions as a bubble detector. Both the inductance probe [11] and the impedance probe [12] also depend on local electrical characteristics of the fluidized bed to determine bubble contact.

In summary, all of the probes described above depend on local electrical properties of the fluidized bed in order to detect bubble presence. These properties depend on the type of particles used, the fluidizing gas and the operating temperature and pressure. The discrimination voltage used to distinguish emulsion phase contact from bubble phase contact is a function of particle type, the fluidizing gas used, and operating conditions.

The optical probe described in reference [13] is composed of a bundle of quartz fibers one of which is used to project light into the fluidized bed while the other fibers transmit light reflected from the fluidized bed to phototransistors. This type of probe senses reflected light and is, therefore, sensitive to the reflectivity of the particles. Probes of this design have, however, operated in fluidized beds at elevated temperature (up to 500^0 C in the above reference). Application of the probes described above is normally limited to the fluidized system for which their operation has been validated.

Optical probes which detect bubbles by distinguishing between the light transmission characteristics of the bubble phase and the emulsion phase consist of a light source and a photodetector separated by a gap. The word "light" as used in this discussion refers not only to the visible spectrum but to the infrared and ultraviolet spectrum as well. Clearly, the photodetector must be matched to the light source used. Assuming opaque particles, when emulsion phase fills the gap, essentially no transmission of light occurs across the gap. For common gas fluidized systems, the bubble phase is essentially optically clear, and light will be transmitted across the gap to the photodetector when a bubble contacts the probe. It is important to note that the calibration of this type of probe is not a function of particle material or what fluidizing gas is used. The restrictions are that the particles be opaque to the light used and that the gas be optically clear at least for distances of a few millimeters. Optical probes of this type are, therefore, suitable to a wide range of fluidized systems and operating conditions.

Application of this type of probe has been limited to low temperatures since, in previous designs, either the light source or photodetector was immersed directly in the fluidized bed [9, 14]. Modification of the existing optical probes of this type is required to extend their use to high temperature fluidized beds. The technique reported here utilizes a glass light guide to transmit the light from the source into the fluidized bed and another glass light guide to transmit the light from the bed to a photodetector.

Yasui and Johanson [14] immersed a small light bulb directly in a low temperature fluidized bed and utilized a 4 mm diameter 1.1 m long quartz rod wrapped with aluminum foil as a light guide to transmit the signal out of the bed to phototubes. Vacuum tube amplifiers and an oscillograph were used to amplify and record the output signal. Bubble rise speed, bubble frequency and bubble volume flow rate were determined in a low temperature fluidized bed for a range of operating conditions by using two such probes.

With proper design, the light guides can be made to operate at combustion level bed temperatures and the temperature sensitive electrical components isolated outside the bed or placed in a cooled enclosure within the bed.

2. DESIGN

Figure 1 shows the design of an optical probe for detecting the local presence of bubbles in a high temperature fluidized bed. The probe has been designed to mount vertically, using a compression fitting, in a small high temperature fluidized bed. This fluidized bed is electrically heated and has a cross-section of 104 x 104 mm. Probes based on the same operating principles could be constructed in a range of sizes and configurations.

Infrared light (880 nm spectral peak) produced by a high intensity LED is conducted down a Vycor glass rod, across a gap and through another Vycor glass rod to a silicon infrared phototransistor. The phototransistor is very fast responding with a rise time of approximately 8 µs. The Vycor glass rods can operate with temperatures up to approximately 1670^{0}C without softening or undergoing a significant loss of transmittance at infrared wavelengths. A water-cooled mounting is provided to avoid excessively high temperatures at the LED and phototransistor.

After a few hours of operation in a fluidized bed, during which the glass tips of the probe will be subject to abrasion by the bed particles, the calibration of the probe remains essentially constant. Radiant emission from the high temperature bed to the probe contains a significant component in the infrared spectrum. This radiant emission from the bed to the probe is a disturbance input to the probe since it will be sensed by the phototransistor. The signal conditioning technique described below essentially eliminates this disturbance input from the probe's output.

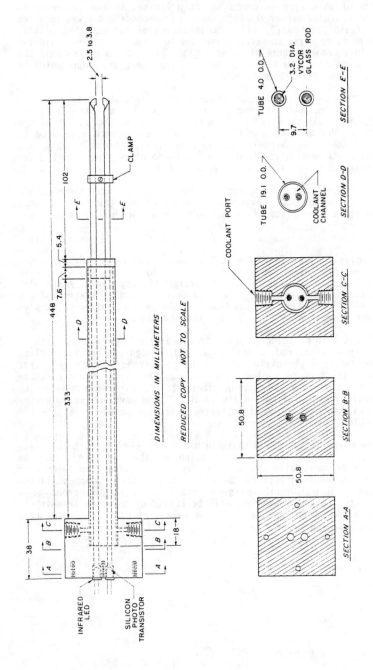

DIMENSIONS IN MILLIMETERS

REDUCED COPY NOT TO SCALE

Figure 1. Optical probe for detecting bubbles in a high temperature fluidized bed.

Analog Signal Conditioning and Data Acquisition

The excitation current to the LED is provided as a square wave at approximately 10 kHz with a constant nominal current large enough that the current supplied is never negative. With light transmitted across the gap, the corresponding output of the phototransistor is approximately a square wave but offset by some constant. This phototransistor output is processed by a high pass filter which will remove all low frequency components but allow the frequency components at 10 kHz or higher to pass through without attenuation. This filtering will remove the constant or slowly fluctuating component of the phototransistor output due to radiant emission from the high temperature fluidized bed. Almost all noise pick-up due to AC power lines operating at 50 Hz or 60 Hz will also be removed by this filtering.

The remaining square wave signal is amplified, full-wave rectified and filtered to yield a voltage which can be recorded on an oscillograph or by a digital data acquisition system. This voltage is offset to be nearly zero when no transmission of light occurs across the gap and have a positive value when light transmission occurs across the gap.

In summary, the signal conditioning method utilizes the amplitude modulation technique. The useful information is transmitted by a square wave carrier with frequency components of approximately 10 kHz and higher. Frequency components much lower than this are removed by filtering. Essentially all of the useful signal retained after the filtering has its source at the LED and is not due to radiant emission from the bed as transmitted to the phototransistor through the glass light guide.

The settling time (time interval required for the voltage output to reach 98% of its steady value after the gap is opened) of the combined probe and associated signal conditioning circuit was determined experimentally to be approximately 2 ms.

3. DATA

The voltage output of the probe as a function of time is shown in Figure 2. The low voltage level (0 v) corresponds to a closed gap condition at the probe tip while the high voltage level (5.4 v) corresponds to an open gap condition at the probe tip. A digital storage oscilloscope was used to record the data. The signal was subsequently plotted on a x-y recorder. Glass beds of approximately 150 μm diameter were used as particles. The data shown are for a bed temperature of 389^0 C and a superficial gas velocity just sufficient to produce isolated bubbles.

It should be mentioned that the maximum voltage output (5.4 v) provided by the probe is at least a factor of 10 higher than is needed for recording purposes. A probe of similar design but with smaller diameter light guides would be less intrusive but still provide an adequate voltage output. Longer light guides could also be used.

391

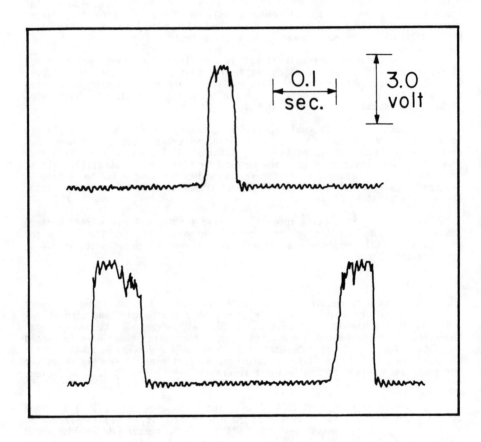

Figure 2. Voltage output of the optical probe in a fluidized bed of 150 μm glass beads at 389°C.

In actual use, a discrimination voltage must be specified and used as a criteria to distinguish emulsion phase contact from bubble phase contact at the probe tip. If the voltage output of the probe was strictly stepwise, any discrimination voltage between the minimum and maximum voltage output could be used without influencing the bubble detecting capability of the probe. Figure 3 shows the time fraction of bubble contact indicated by the probe for two operating conditions and discrimination voltages between 0.6 v and 1.6 v. The indicated fraction of bubble contact is reduced by approximately 23% as the discrimination voltage is increased from 0.6 v to 1.6 v. The fact that the probe tip disturbs the local flow of bubbles within the fluidized bed causes some particles to fall through the interior of the bubble and attenuate the passage of light between the light guides. This attenuation causes the voltage output of the probe to be somewhat less than the open gap value and also causes the voltage vs. time trace to be different from the ideal stepwise change between the open gap voltage and closed gap voltage. The small size of the fluidized bed apparatus used (104 x 104 mm cross-section) allows only corresponding small bubbles to be produced. It is probable that the time fraction of bubble contact would show less variation with changing discrimination voltage if the mean bubble size was larger than in the test facility.

No fluidized bed capable of operation at combustion level temperatures was available to test how well the signal conditioning circuit described above kept the voltage output of the probe constant over a wide range of bed temperatures. Therefore, the probe tip and 200 mm of the light guides nearest the tip were placed in a tubular furnace and allowed to come to thermal equilibrium with the furnace interior. Cooling water was supplied to the probe during this high temperature testing. Figure 4 shows the ratio of voltage output to voltage output at 20^0 C for open gap conditions and a range of furnace temperatures from 20^0 C to 930^0 C. The highest temperature considered is somewhat beyond that normally utilized in fluidized bed combustion of coal. As shown in Figure 4, the open gap voltage varies from 108% to 89% of the open gap voltage at 20^0 C. This relatively small variation of open gap voltage with temperature would allow a fixed discrimination voltage to be used for the entire range of temperatures shown in Figure 4.

4. CONCLUSIONS

An optical probe utilizing Vycor glass light guides has been developed for detecting bubbles in high temperature fluidized beds. Probes based on the same operating principle could be constructed in a variety of sizes and configurations. In particular, probes with smaller diameter glass rods than those used in the probe described in this work could certainly be used since the current design provides an open gap voltage output at least 10 times that required for recording purposes.

The use of the amplitude modulation technique in the signal conditioning circuit provides a relatively constant value of the open gap voltage--within approximately ±10% of the room temperature value--for the range of probe tip temperatures 20^0 C to 930^0 C. This should allow a fixed discrimination voltage to be used for the entire temperature range indicated.

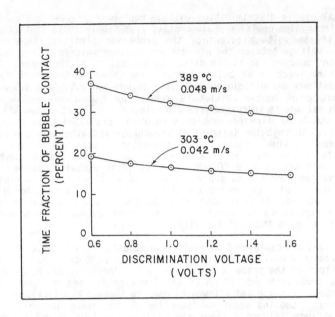

Figure 3.   Time fraction of bubble contact indicated by the optical probe as a function of discrimination voltage for two operating conditions.   Fluidized bed composed of 150 μm glass beads with the temperature and superficial gas velocity as shown above.

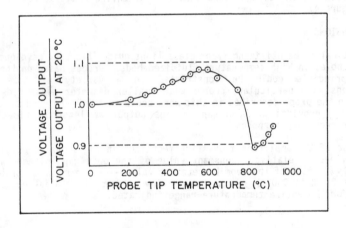

Figure 4.   Ratio of voltage output to voltage output at 20°C for open gap conditions and a range of probe tip temperatures.

For fluidized systems which operate over a range of temperatures or bed compositions, the technique reported here may prove superior to the inductance or capacitance probes currently in use.

Acknowledgements

The author wishes to thank Messrs. Pat Vowell and Gordon Williamson for fabricating the probe described above and Mr. John Rompel for designing and building the associated signal conditioning circuit. The data shown in Figures 2 and 3 were taken by Mr. John Roy, undergraduate assistant. This work was supported, in part, by the National Science Foundation (grant number CBT-8801618).

REFERENCES

1. Park, W.H., Kang, W.K., Capes, C.E. and Osberg, G.L., "The Properties of Bubbles in Fluidized Beds of Conducting Particles as Measured by an Electroresistivity Probe," Chem. Engr. Sci., 24, 851-865, (1969).

2. Wittmann, K., Helmrich, H. and Schugerl, K., "Measurements of Bubble Properties in Continuously Operated Fluidized Bed Reactors at Elevated Temperatures," Chem. Engr. Sci., 36, 1672-1677, (1981).

3. Gunn, D.J. and Al-Doori, H.H., "Measurement of Bubble Flows in Fluidized Beds by Electrical Probe," Int. J. Multiphase Flow, 11, 535-551, (1985).

4. Morse, R.D. and Ballou, C.D., "The Uniformity of Fluidization--Its Measurement and Use," Chem. Engr. Prog., 47, 199-204, (1951).

5. Tomita, M. and Adachi, T., "The Effect of Bed Diameter on the Behavior of Bubbles in Gas-Solid Fluidized Beds," J. Chem. Engr. Japan, 6, 196-201, (1973).

6. Tone, S., Seko, H., Maruyama, H. and Otake, T., "Catalytic Cracking of Methylcyclohexane over Silica Alumina Catalyst in Gas Fluidized Bed," J. Chem. Engr. Japan, 7, 44-51, (1974).

7. Werther, J., "Bubbles in Gas Fluidized Beds--Part I," Trans. Instn. Chem. Engr., 52, 149-159, (1974).

8. Otake, T., Tone, S., Kawashima, M. and Shibata, T., "Behavior of Rising Bubbles in a Gas-Fluidized Bed at Elevated Temperature," J. Chem. Engr. Japan, 8, 388-392, (1975).

9. Dutta, S. and Wen, C.Y., "A Simple Probe for Fluidized Bed Measurements," Canadian J. Chem. Engr., 57, 115-119, (1979).

10. Yoshida, K., Sakane, J. and Shimizu, F., "A New Probe for Measuring Fluidized Bed Characteristics at High Temperature," Ind. Engr. Chem. Fund., 21, 83-85, (1982).

11. Cranfield, R.R., "A Probe for Bubble Detection and Measurement in Large Particle Fluidized Beds," Chem. Engr. Sci., 27, 239-245, (1972).

12. Boelens, G., Liefhebber, F. and Van Koppen, C.W.J., "A High Temperature Impedance Probe for Local Porosity Measurements in a Fluidized Bed Combustor," Chem. Engr. Sci., 40, 365-373, (1985).

13. Ishida, M., Nishiwaki, A. and Shirai, T., "Movement of Solid Particles Around Bubbles in a Three-Dimensional Fluidized Bed at High Temperatures," in Fluidization Proceedings of the 1980 Fluidization Conference, Grace, J.R. and Matsen, J.M., eds., Plenum Press, New York (1980).

A Shallow Fluidized Bed Heat Exchanger for Domestic Heating

B. M. MUSTAFA and B. M. A. AL-ALI
Department of Mechanical Engineering
University of Mosul
Mosul, Iraq

Abstract

This work is concerned with the heat transfer performance of finned tubes immersed in shallow beds of sand particles fluidized with air. An experimental test rig is constructed to test the factors, which are expected to influence its heat transfer performance such as particle size, element position from distributor plate, fluidizing velocity and the angle of element inclination in the bed. Experimental results showed that the element position in the bed, influences its performance. The best performance is obtained when the element is placed at a distance of (40 - 50 mm) from the distributor plate. It was also observed that the angles ranging between 0° and 10° with the horizontal enhance the heat transfer coefficient, when the element is placed at 90 mm from the distributor plate.

Also the results indicated that larger heat transfer rates can be obtained with the fluid-bed technique than that obtained without the solid particles at the same operating conditions. However, in this case the auxiliary fan power was found to be four times that of the conventional type exchanger.

1. INTRODUCTION

Due to the advantageous heat transfer properties of gas-fluidized beds, some of which were summarized by Botterill [1], Davidson and Harrison [2] and Kunii and Levenspiel [3], they are widely used for chemical reactors and dryers. Hence, this work can be considered as an attempt to make use of this mechanism in heating or cooling units used for air conditioning purposes, particularly for fan-coil units and cooling towers, where the water sources are very limited, also where high quality water is not required, according to the work done by Dickey et al. [4].

In fluidized beds, it could be very convenient to use finned tubes to increase the effective surface area of the tubes by an amount such that unfinned area of the same effective surface would have a diameter in the range (4 - 12) times the diameter of finned tubes employed as suggested by Elliott [5]. Al-Ali [6] recommended that, in case of horizontal tubes, it would be a good practice to use extended surfaces, if the arrangement helps to eliminate or even to minimize the defluidized cap on the top of the tubes which is normally present with unfinned tubes.

397

In recent years, a number of workers such as Petrie et al. [7], Bartel et al. [8], Vreedenberg [9] and Bartel and Genetti [10] have studied the effect of particle size, fin thickness, fin height and fluidizing velocity on the performance of finned tubes in deep beds. They found that increasing the number of fins decreases the overall effectiveness [7], and that the fin effectiveness factor decreases as fin height increases, and it is increased when fluidizing velocity is increased. It is also found that, tubes with longer fins are less sensitive to tube spacing than those with shorter fins.

Al-Ali [6] studied the performance of different types of finned tubes in shallow beds taking into consideration the main geometric variables which may influence the heat transfer coefficient such as, basic tube diameter, fin height and fin efficiency. The best heat transfer coefficient is obtained when the tube was resting on the distributor plate, which is the same result of that obtained by Atkinson 6 , due to the restrictions of bubble coalesceness entry between adjacent fins.

Chen [11] observed for verticle tube with spiral fins in glass spheres fluidizing beds, that for a given geometry the gain in performance reached a maximum over some ratios of fin spacing to particle size diameter and when the number of fins per unit length of tube increased the maximum performance also increased. Genetti et al. [12] studied the effect of tube orientation on the heat transfer coefficient. They observed that tube positions of 45° and 60° from the horizontal produce minimum values of heat transfer coefficients for bare and finned tubes respectively.

Several aspects of fluidized beds and applications are reviewed by Botterill [13, 14]. In spite of the larger order of improvement in heat transfer, upto now no example of gas to liquid heat exchanger, such as economizers based on the process, as reported by Elliott et al. [15, 16, 17 and 5]. The method of testing these types of heat exchangers is either to compare the reduction in surface area with the increase in power requirements for pumping the air through the bed or to compare the heat transfer rates with that of a conventional type without the solid particles at the same flow rates, as suggested by Noe" et al. [18].

2. EXPERIMENTAL APPARATUS

The heating element is of the type commercially used in central heating systems It is constructed from seamless double copper tubes mechanically expanded into 108 smooth rectangular aluminium fins. Its dimensions are shown in figure 1.

The experiments are carried out in a plexiglass bed column of rectangular cross-section (0.1 x 0.587 m) and of overall height of 0.505 m. The two sides of the container are made so that the heating element could be fixed in any position between the distributor plate and a distance about 270 mm from it. The air is supplied from a 2.2 kW fan unit, and the flow rate is measured by calibrated orifice nozzle of 24.95 mm in diameter with corner pressure tappings and inclined manometer. The closed loop ABCD shown in figure 2 is the main water circuit supplying the hot water to the element. The water is pumped round by 0.125 kW water pump passing through 1 kJ/hr electric heater. The flow rate is controlled by the valve V_m and is measured by the calibrated rotameter R_m.

Two thermometers are inserted at points B and C to measure the water temperature before and after the test element. Four calibrated copper constantan thermo couples are suspended in the bed such that they are free to move freely in the bed and are used to indicate the temperature gradient along the bed.

The particles used are sand of mean diameter of 197 μm and 283 μm. Superficial air velocity ranged from minimum fluidizing velocity of 0.059 m/s and of 0.0944 m/s for the two sizes fractions to 0.65 m/s. Two static bed heights 40 and 65 mm for the smaller size and three bed heights 40, 65 and 90 mm for the larger one are used. For each size and each bed height the effect of element position vertically from distributor plate is investigated together with the fluidizing velocity.

All heat transfer coefficients are evaluated from the measured temperatures, hot water flow rates, heat capacity and tube dimensions using an iterative process of the following equation:

$$\frac{1}{H_{ov} \cdot A_T} = \frac{1}{H_i A_i} + \frac{1}{H_{BM} \cdot A_{eff}} \tag{1}$$

which is obtained from the rate of heat transfer to the bed, that is:

$$Q = H_{ov} A_T (T_w - T_b) \tag{2}$$

$$Q = H_i A_i (T_w - T_m) \tag{3}$$

$$Q = H_{BM} A_{eff} (T_m - T_b) \tag{4}$$

At certain positions of the heating element and at the optimum fluidizing velocity, for each size, the effect of element angle of inclination at different bed heights is also investigated.

3. RESULTS AND DISCUSSION

The effect of tube position is examined for each particle size and each bed height beginning from the distributor plate. The results of maximum heat transfer coefficients are obtained at different bed heights for the two sizes and are shown in figures 3 and 4. It can be observed from figure 3, that the heat transfer

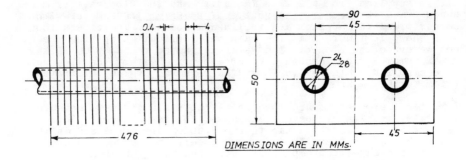

Fig. 1. - The Heating Element and Fin Dimensions.

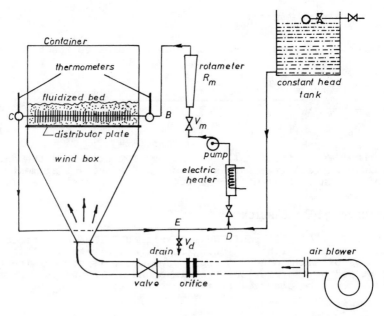

Fig. 2. - Schematic Diagram of Water and Air Circuits.

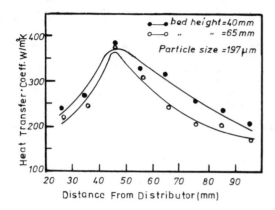

Fig. 3. - Variation of Heat Transfer Coefficient
with Tube Position for Different Bed
Heights.

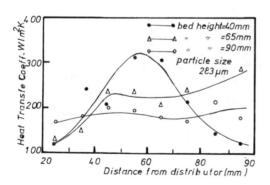

Fig. 4. - Variation of Heat Transfer Coefficient
with Tube Position for Different Bed
Heights.

401

coefficient at primary positions for the two bed heights are approximately the same, which means that the motion of the bubbles and the particles circulation follow the same trend between the fins. The best results are obtained when the element is placed at 45 mm from the distributor plate due to the good circulation of the particles resulting from bursting bubbles between the fins and the larger number of particles coming in contact with the element surface. After that as the distance from the distributor plate increases, the element becomes at the place of the bed where less numbers of particles can reach the heating element. But the curve of 65 mm bed height is lower than that of 40 mm, this is because that the element is still in bed which tends to restrict the particles motion and may be there is a defluidized cap on the top of tubes.

Finally, figure 4 indicates that as the bed height increases, the heat transfer coefficient tend to become more stable because the change in tube position may be considered without effect since it is so small compared to the bed height itself.

To investigate the effect of element inclination on heat transfer rate, the heating element is placed at 90 mm from the distributor plate in order to test the larger range of angles possible at a given position since the tilting of the element is restricted by both sides of the container. Keeping the mid-point of the element always at 90 mm from the distributor plate, however, four angles are tested when the bed is 65 mm in height and at the optimum air velocity for the larger size. The results of this test are shown in figure 5. They indicate that the heat transfer coefficient increases within the angles ranging between 0° and 10° with the horizontal, while a reduction is clearly noticeable at larger angles. Since the heating element is placed in the lean phase of the bed (i.e., above the bed surface), therefore, at the small angles, the recirculating particles may slide quickly on the fin surface. Consequently, the probability of contacts between the particles and the surfaces will be increased, then better heat transfer coefficient could be expected. As the angle increases, it is expected that the heat transfer coefficient would be decreased, since the horizontal gap between the fins becomes smaller than before and most of the particles can't penetrate freely between the fins. This is in a good agreement with the results obtained by Genetti et al. [12] within the angles ranging between 0° and 30° in deep beds.

As a reference test, the heating element was tested without the sand particles. Its results, then are compared with that which are obtained at best position and best bed height for both particle sizes. From figure 6, it is observed that the increase in the output heat transfer rate is about 79 % with the larger size and about 81 % with the smaller one.

4.   CONCLUSION

1. At distances 40 - 60 mm from the distributor plate, maxi-

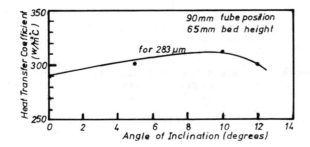

Fig. 5. - Variation of Heat Transfer Coefficient
with the Element Angle of Inclination.

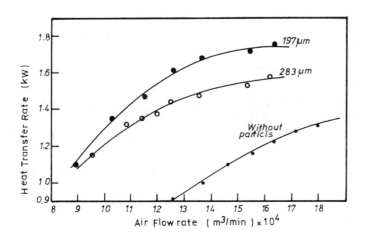

Fig. 6. - Comparison of Heat Transfer Performance
between the Fluid-Bed Heat Exchanger
and the Conventional Exchanger.

403

mum heat transfer coefficients are obtained for all of the three bed heights tested.

2. Larger heat transfer coefficients are obtained with the 40 mm bed height than those obtained with deeper beds, when the element is placed at the bed surface.

3. An increase of about 8 % in heat transfer coefficient is obtained with the 65 mm bed height when the element inclination angle is between 0° and 10° with the horizontal.

4. It is found that the heat transfer rates are about 80 % larger than those obtained with the conventional type heat exchanger with the same unit using shallow bed of 40 mm height of sand fluidized with air.

5. NOMENCLATURE

A_{eff}:   Effective surface area of finned tubes, m^2.

A_i: Inside surface area of the tubes, m^2.

A_T: Total surface area of finned tubes, m^2.

H_{BM}:   Metal-to-bed heat transfer coefficient, W/m^2K.

H_i: Tube inside heat transfer coefficient, W/m^2K.

H_{ov}:   Overall heat transfer coefficient, W/m^2K.

Q: Heat transfer rate, W.

T_b: Average bed temperature, K.

T_m: The mean of the tube surface temperature, K.

T_w: Average water temperature, K.

REFERENCES

1. Botterill, J.S.M. "FLUID BED HEAT TRANSFER", Chapter 1, Academic Press, 1975.

2. Davidson, J.F. and Harrison, D., "Fluidization", Academic Press, 1971.

3. Kunii, D. and Levenspiel, O. "Fluidization Engineering", John Wiley and Sons Inc., 1969.

4. Dickey, B.R., Grimmett, E.S. and Kilian, D.C., Chem. Engg. Progress, Vol. 170, No. 1, pp. 61-65, 1974.

5. Elliott, D.E., UNITED STATES PATENT OFFICE, No. 3, 912,002, Oct. 1975.

6. Al-Ali, B.M.A., Ph.D. Thesis, The University of Aston in Birmingham, 1976.

7. Petrie, J.C., Freeby, W.A. and Buckham, J.A. "In-Bed Heat

Exchangers". Chem. Engg. Progress, Vol. 64, No. 7, pp. 45-52, 1958.

8. Bartel, W.J., Genetti, W.E. and Grimmet, E.S., A.t.ch.E. Symp. series, Vol. 67, No. 116, 1971, pp. 90-96.

9. Vreedenberg, H.A., Chem. Engg. Science, Vol. 9, pp. 52-60, 1958.

10. Bartel, W.J. and Genetti, W.E., A.I.ch. Symp. Series, Vol. 69, No. 128, pp. 85-93, 1973.

11. Chen, J.C., ASME Publication, Paper 76, H.T.-75, pp. 1-7, 1976.

12. Genetti, W.E., Schmall, R.A. and Grimett, E.S., A.I.ch.E. Symp. Series, Vol. 67, No. 116, pp. 90-93, 1971.

13. Botterill, J.S.M., Br. Chem. Engg., Vol. 13, No. 8, pp. 1121-1126, 1968.

14. Botterill, J.S.M., "FLUID BED HEAT TRANSFER", Chapter 5, Academic Press, 1975.

15. Elliott, D.E., Paper 22, The Total Energy Conference, Institute of Fuel, pp. 361-375, Nov. 2971.

16. Elliott, D.E., Healey, E.M. and Roberts, A.G., Paper 17, Presented at the Heat Exchanger Conference, Inst. of Fuel, Paris, 12 pages, 1971.

17. Elliott, D.E. and Hume, B.G., Symposium on Heat Transfer - Institution of Chem. Engineers, 15 pages, March, 1976.

18. Noe", A.R. and Knudsen, J.G., Chem. Engg. Progress. Symp. Series, Vol. 64, No. 82, pp. 202-211, 1968.

Spectral Analysis of Dynamics in Fluidized Bed Tube Banks

S. L. CHANG, R. W. LYCZKOWSKI, and G. F. BERRY
Argonne National Laboratory
Energy and Environmental Systems Division
9700 South Cass Avenue
Argonne, Illinois 60439, USA

ABSTRACT

The advancement of fluidized bed combustors, which has attracted intense commercial interest for burning high-sulfur coal economically and in an environmentally acceptable manner, has been impeded by severe erosion on the tubes. To understand the erosion problems through detailed knowledge of the complex phenomena of solids circulation and bubble motion, Argonne National Laboratory (ANL) has developed a methodology to predict local erosion rates in a two-dimensional fluidized bed by coupling hydrodynamic and erosion models. The ANL state-of-the-art hydrodynamic computer model (FLUFIX) is capable of computing the temporal and spatial flow property distributions in the bed. By using the computed results from the FLUFIX code, several erosion models have been used to predict local erosion rates in fluidized beds. Power spectral analysis can be used to correlate the flow dynamics and erosion. In this paper, the spectral dynamics of two-dimensional fluidized bed tube banks will be presented. The spectral analysis includes autocorrelation, cross-correlation and power spectral density of computed hydrodynamic properties, i.e., pressure, porosity, and gas and solids velocities.

NOMENCLATURE

A	Power spectral density
G	Solids elastic modulus, Pa
g	Acceleration due to gravity, m/s^2
I	Unit tensor
n	Total number of phases
P	Pressure, Pa
t	Time, s
U_{mf}	Minimum fluidizing velocity, m/s
V	Velocity, m/s
$\vec{v}$	Velocity vector, m/s

Greek Letters

δ	Kronecker delta function
ε	Gas volume fraction or porosity
ε_s	Solids volume fraction = $1 - \varepsilon$
ε^*	Compaction gas volume fraction

$\bar{\bar{\sigma}}_{ke}$	Effective stress, Pa
ρ	Density, kg/m^3, autocorrelation, or cross-correlation
$\bar{\bar{\tau}}_{kc}$	Coulombic stress for solids, Pa
τ	Time lag, s
σ	Variance
ω	Angular frequency, rad/s
$\bar{\bar{\beta}}_{ik}$	Fluid-particle friction coefficient tensor, kg/(m$^3 \cdot$ s)

INTRODUCTION

The fluidized bed combustion technology attracts intense commercial interest for its capability of burning high-sulfur coal in a more economic and environmentally acceptable manner, despite the unsolved erosion problems.[1] Detailed knowledge of the complex phenomena of solids circulation and bubble motion in the bed would help to understand the problems and to find proper solutions. Argonne National Laboratory has been developing a methodology to investigate local erosion phenomena in bubbling fluidized bed systems:[2] a state-of-the-art two-phase two-dimensional hydrodynamic computer model, FLUFIX, is capable of computing the temporal and spatial distributions of the flow properties and erosion model, EROS, consisting of several erosion submodels using the computed hydrodynamic results to predict local erosion patterns and rates. In addition to the erosion models based on various single particle wear theories, the technique of power spectral analysis has been used in the erosion calculation because it can not only correlate the flow dynamics with the erosion patterns and rates but also determine the bubble speed. This paper will present the spectral analysis of the flow dynamics in a generic fewtube fluidized bed combustor (FBC) system.

The statistical description of flow dynamics, which has been widely used in the analyses of turbulence, chaotic dynamics and combustion instability[3,4], is also important to our understanding of erosion in fluidized bed combustors. This approach examines how fluctuations of hydrodynamic properties are distributed around an average value and how adjacent fluctuations (next to each other in time or space) are related.   The autocorrelation and its Fourier transform, the energy spectrum, show the relationship between neighboring fluctuations, from which the major oscillation modes of the bed are determined, and the cross-correlation between adjacent fluctuations in space can be used to compute the fluctuation propagation speed.

FORMULATION

FLUFIX code uses the principles of conservation of mass, momentum and energy. The general mass conservation equations and the separated phase momentum equations for transient and isothermal fluid-solids non-reactive multiphase flow (in vector notation) are written in conservation-law form for the hydrodynamic model as follows:

Continuity

$$\frac{\partial}{\partial t} \left(\epsilon_k \rho_k \right) + \nabla \cdot \left(\epsilon_k \rho_k \vec{v}_k \right) = 0 \tag{1}$$

Momentum

$$\frac{\partial}{\partial t}\left(\varepsilon_k \rho_k \vec{v}_k\right) + \nabla \cdot \left(\varepsilon_k \rho_k \vec{v}_k \vec{v}_k\right) = \nabla \cdot \overline{\overline{\sigma}}_{ke} + \varepsilon_k \rho_k \vec{g}$$
$$\qquad\qquad \text{Acceleration} \qquad\qquad\qquad \text{Stress} \qquad \text{Gravity}$$

$$+ \sum_{i=1}^{n} \overline{\overline{\beta}}_{ik} \cdot \left(\vec{v}_i - \vec{v}_k\right) \tag{2}$$
$$\text{Interphase Drag}$$

Each phase (gas or solids) is denoted by the subscript k and the total number of phases in n.

The effective stress tensor, $\overline{\overline{\sigma}}_{ke}$, contains pressure, viscous and coulombic components according to the convention

$$\nabla \cdot \overline{\overline{\sigma}}_{ke} = -\left(1 - \delta_{ks_i}\right)\nabla \cdot \left(P\overline{\overline{I}}\right) + \nabla \cdot \left(\varepsilon_k \overline{\overline{\tau}}_{kv}\right) - \delta_{ks_i}\nabla \cdot \overline{\overline{\tau}}_{kc} \tag{3}$$

The Kronecker delta function is given by:

$$\delta_{ks_i} = \begin{cases} 1 \text{ if } k = s_i \\[8pt] 0 \text{ if } k \neq s_i \end{cases} \tag{4}$$

where $k = s_i$ is a solids phase. The set of momentum equations given by Eqs. (2) and (3) were first given by Rudinger and Chang[5], and was extended by Lyczkowski.[6]

The solids elastic modulus, $G(\varepsilon_k)$, is defined as the normal component of the solids coulombic stress through the following relationship:

$$\nabla \cdot \overline{\overline{\tau}}_{kc} = G(\varepsilon_k)\overline{\overline{I}} \cdot \nabla \varepsilon_k \tag{5}$$

where $\overline{\overline{I}}$ is the unit tensor. $G(\varepsilon_k)$ is the only component of the solids coulombic stress presently used. Simple semi-empirical formula of the solids elastic modulus may be written as:

$$G(\varepsilon_s)/G_o = \exp\left[c(\varepsilon_s - \varepsilon_s^*)\right] = \exp\left[-c(\varepsilon - \varepsilon^*)\right] \tag{6}$$

In this study, we use c = 600, ε^* = 0.367, and G_o = 1 Pa.

For a continuous function of time, f(t), the mean and variance of the function and its autocorrelation and power spectral density are commonly defined[7] as follows:

Mean

$$\bar{f} = \frac{1}{T} \int_{t_o}^{t_o + T} f(t) \, dt \tag{7}$$

Variance

$$\sigma_f = \left[\overline{f(t) \, f(t)} \right]^{1/2} \tag{8}$$

Autocorrelation

$$\rho_f(\tau) = \overline{f(t) \, f(t+\tau)}/\sigma_f^2 \tag{9}$$

Power Spectral Density

$$A(\omega) = \frac{2}{T} \int_o^T \rho(\tau) \cos (\omega\tau) \, d\tau \tag{10}$$

As for the cross-correlation between two fluctuations f(t) and g(t), it is computed using the following equation:

Cross-Correlation

$$\rho_{f,g}(\tau) = \overline{f(t) \, g(t)}/\left(\sigma_f \sigma_g\right) \tag{11}$$

Since the fluctuations of pressure, porosity, and velocities computed from the FLUFIX code are digital data in a finite time interval, the statistical results computed from Eqs. (7)-(11) are integrated numerically using trapezoidal approximation.

RESULTS AND DISCUSSIONS

Figure 1 shows the schematic of a two-dimensional fewtube fluidized bed combustor system with a three-tube arrangement under investigation. The bed, which is 0.816 m in height and 0.306 m in width, is simulated using the computational grid of 48 x 18 nodes. At each node, the FLUFIX code computed porosity, pressure, solids velocity and gas velocity fluctuations consisting of 1400 time steps in a 7 s time period or 5 ms interval between consecutive steps. There are three shorter periods in this seven second time period: an initial 2 s at a jet velocity of 1.12 minimum fluidizing velocity (U_{mf}), the following 2.5 s at 1.7 U_{mf}, and the last 2.5 s at 2.3 U_{mf}. In Fig. 2, time-averaged porosity and solids velocity results of these three periods are plotted. These results have been spectrally analyzed and are to be discussed in the following.

First, at each node, the statistical mean, variance, autocorrelation, and power spectral density of fluctuation of each hydrodynamic property, i.e., pressure, porosity, velocities (gas and solids), or axial solids momentum flux, were computed. Figure 3

410

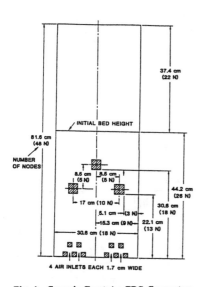

Fig. 1. Generic Fewtube FBC Geometry.

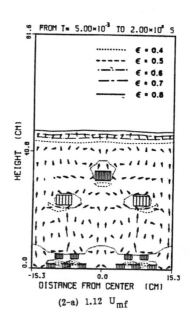

(2-a) 1.12 U_{mf}

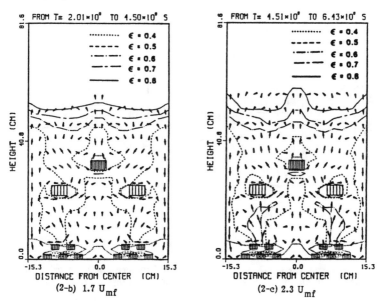

(2-b) 1.7 U_{mf}

(2-c) 2.3 U_{mf}

Fig. 2. Computed Time-Averaged Porosity and Solids Velocity

411

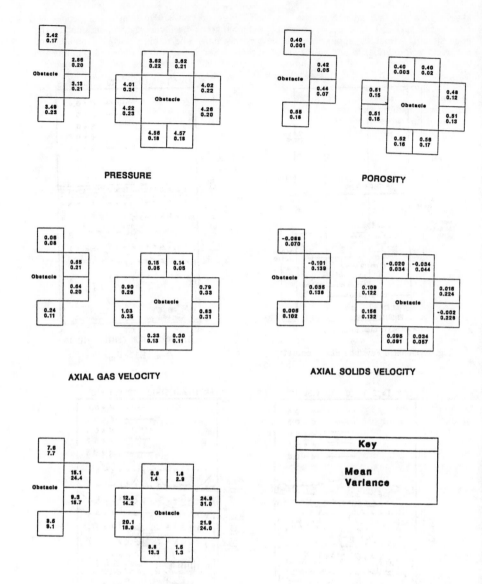

PRESSURE

POROSITY

AXIAL GAS VELOCITY

AXIAL SOLIDS VELOCITY

AXIAL SOLIDS MOMENTUM FLUX

Key

Mean
Variance

Fig. 3. Means and Variances of Fluctuations of Pressure (kpa), Porosity, Axial Gas and Solids Velocity (m/s), and Axial Solids Momentum Flux (J/g).

412

summarizes the statistical means and variances of the above-mentioned properties around the tubes at 1.12 U_{mf}. Pressure mean seems to be inversely proportional to the node height. Bubbles appear active below the tube where high porosity means and variances are calculated and solids are accumulated on top of tubes where low means and near zero variances are the computed results. Around the tubes, axial gas velocity means are generally higher than their variances, indicating little chance of back flow of gas, but most axial solids velocity variances are larger than their means, indicating the existence of strong circulations and back flows of solids. Large axial solids momentum fluxes are found on the sides of the lower tube. Since some experimental data show that high erosion rates occur near these locations, these large momentum values may be associated with the high erosion rates.

These hydrodynamic fluctuations are not random but possess some major oscillatory modes. In Fig. 4, the porosity-time data of the node at the lower left corner of the lower tube shows several high porosity waves (bubbles) passing the node in 0-2 s period. By taking spectral calculations of this porosity fluctuation, a good autocorrelation at 0.33 s time lag and a corresponding 3.1 Hz oscillation mode on power spectral density curve are found. In other words, one bubble oscillatory mode at this location is 3.1 Hz in 0-2 s period. Figure 5 summarizes the major oscillatory modes of porosity, axial solids and gas velocities, and axial solids momentum in a time period from 0 to 2 s at 1.12 U_{mf}. These modes range from 0.9 to 5 Hz. Possible causes of these oscillations may include bubble formation, breakup and coalescence.

Spectral dynamics of an FBC is not only strongly influenced by the combustor geometry and tube arrangement, but also by jet velocity. In Fig. 6, the jet effect of the porosity means around the tubes are compared. In general, the porosity mean increases with the jet velocity. For example, below the center tube, the porosity increases from 0.55 to 0.66 as the jet velocity increases from 1.12 to 2.3 U_{mf}. In short, higher jet velocity may cause larger bubbles to form.

In an FBC, fluctuations of hydrodynamic properties at neighboring locations have similar wave form and such similarity can be statistically correlated. In Fig. 7, porosity fluctuations from 0 to 2 s at nodes F and G are compared and excellent cross-correlation between these two fluctuations are found. Node G's fluctuation lags node F's by 0.16 s. In addition, excellent cross-correlations of porosity, pressure and solids velocity fluctuations between neighboring nodes are found, as shown in Fig. 8. The propagation velocity of these fluctuations ranges from 0.4 to 1.06 m/s.

Besides the cross-correlations in space, cross-correlations between porosity and pressure at the same node are also performed. Similar good cross-correlations are found. For example, the porosity at node G lags the pressure by 0.07 s during the time period 2-4.5 s, as shown in Fig. 9. Similar calculations which were carried out at various locations and subject to various jet velocities indicate that pressure and porosity are very well correlated and porosity lags pressure. The lag times summarized in Fig. 10 range from 0.06 to 0.375 s. These important findings indicate that simple diagnostic measurements of pressure can be used to correlate bubble dynamics and erosion patterns.

CONCLUSIONS

Power spectral analyses including statistical means, variances, autocorrelations and cross-correlations have been performed to study the results of the hydrodynamic computations. The means and variances of hydrodynamic properties are strongly dependent on the bed geometry, tube arrangement and jet velocity. Major oscillatory modes in the bed are identified. Bubble propagation velocities are computed. Using

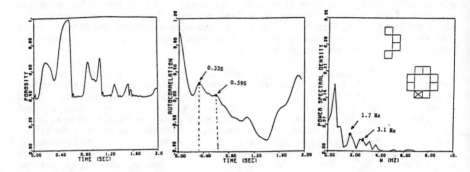

Fig. 4. Porosity Fluctuation and its Autocorrelation and Power Spectral Density, 0-2 s (1.12 U$_{mf}$).

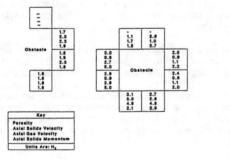

Fig. 5. Major Oscillation Modes Around Obstacles.

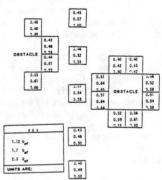

Fig. 6. Jet Velocity Effects on Porosity Means.

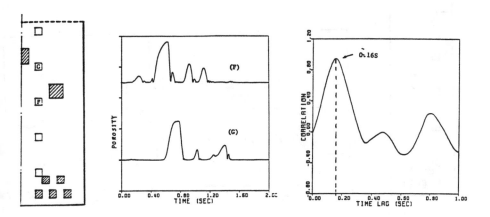

Fig. 7. Cross-Correlation of Porosity Fluctuation in Space, 0-2 s (1.12 U_{mf}).

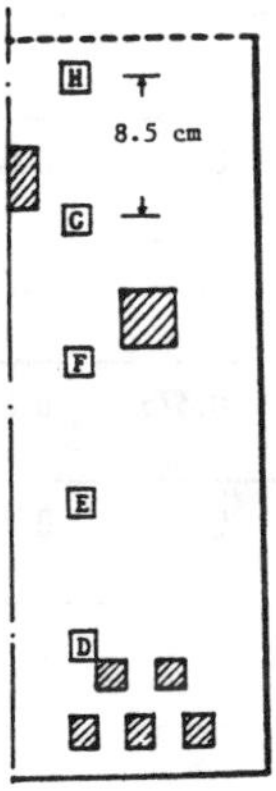

NODES	CORRELATED TIME LAG (s)			PROPAGATION VELOCITY (W/s)		
	ϵ	V_S	P	ϵ	V_S	P
D-E	0.165	0.160	0.08	0.515	0.531	1.06
E-F	0.170	0.090	0.15	0.500	0.944	0.57
F-G	0.160	0.200	0.21	0.531	0.425	0.40
G-H	0.190	0.125	0.18	0.447	0.680	0.47

Fig 8. Propagation of Fluctuations, 0-2 s (1.12 U_{mf}).

415

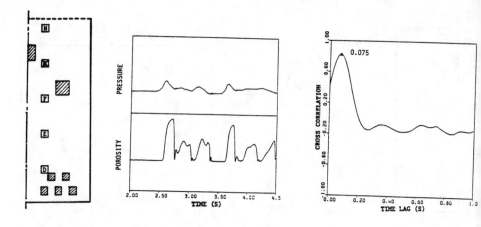

Fig. 9. Pressure and Porosity Cross-Correlation at Node G, 2-4.5 s (1.7 U$_{mf}$).

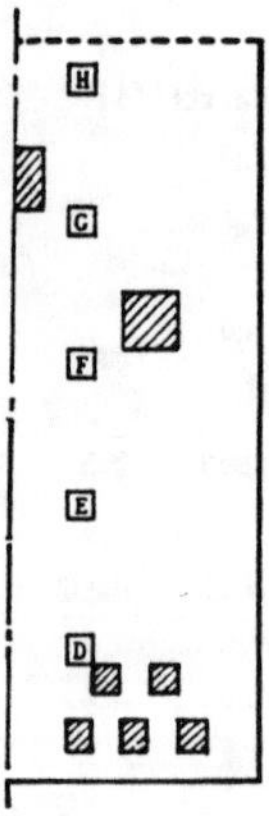

U/U$_{MF}$	1.12		1.7		2.3	
NODE	τ	ρ	τ	ρ	τ	ρ
D	0.175	0.64	0.095	0.70	0.1	0.66
E	0.31	0.72	0.06	0.59	0.31	0.72
F	0.125	0.69	0.075	0.80	0.375	0.84
G	0.115	0.76	0.075	0.83	0.285	0.66
H	0.11	0.78	0.11	0.87	0.10	0.71

NOTE: τ IS TIME LAG IN SECONDS
ρ IS CROSS-CORRELATION COEFFICIENT

Fig. 10. Computed Time Lags between Pressure and Porosity Fluctuations.

simple diagnostic measurements of pressure to correlate bubble dynamics and erosion patterns becomes more reasonable because of good pressure and porosity correlations. These spectral results are to be used to develop correlations of the hydrodynamics and erosion in a fewtube FBC.

ACKNOWLEDGMENTS

This work was performed for the U.S. Department of Energy, Morgantown Energy Technology Center (METC), Contract W-31-109-ENG-38, and for the Electric Power Research Institute (EPRI) under the cooperative R&D venture "Erosion of FBC Heat Transfer Tubes."

REFERENCES

1. Stringer, J., *Current Information on Metal Wastage in Fluidized-Bed Combustors*, Proc. of 1987 Intl. Conf. on Fluidized-Bed Combustion, J.P. Mustonen, ed., 2:685-696, American Society of Mechanical Engineers, New York, NY (1987).

2. Lyczkowski, R.W., J.X. Bouillard, G.F. Berry and D. Gidaspow, *Erosion Calculations in a Two-Dimensional Fluidized Bed Combustion*, Proc. of 1987 Intl. Conf. on Fluidized-Bed Combustion, J.P. Mustonen, ed., 2:697-706, American Society of Mechanical Engineers, New York, NY (1987).

3. *Turbulence and Chaotic Phenomena in Fluids*, T. Tatsumi, ed., Proc. of the Intl. Symp. on Turbulence and Chaotic Phenomena in Fluids, Kyoto, Japan, September 5-10, 1983.

4. Fan, L.T., T.C. Ho, S. Hiraoka and W.P. Walawender, *Pressure Fluctuations in a Fluidized Bed*, 27(3):388-396, AIChE Journal (1981).

5. Rudinger, G. and A. Chang, *Analysis of Non-Steady Two-Phase Flow*, Physics of Fluids, 7:1747-1754 (1964).

6. Lyczkowski, R.W., *Transient Propagation Behavior of Two-Phase Flow Equations*, in Heat Transfer: Research and Application, J.C. Chen, ed., AIChE Symp. Series 75(174):165-174, American Institute of Chemical Engineers, New York (1978).

7. Tennekes, H., and J.L. Lumley, *A First Course in Turbulence*, pp. 197-222, The MIT Press (1980).

On the Erosion of Heat Exchanger Tube Banks in Fluidized Bed Combustors

JACQUES X. BOUILLARD and ROBERT W. LYCZKOWSKI
Argonne National Laboratory
9700 South Cass Avenue
Argonne, Illinois 60439, USA

ABSTRACT

A simplified version of the monolayer energy dissipation erosion models developed at Argonne National Laboratory (ANL) is presented and used to explain the erosion rate dependency upon the particle diameter, the fluidizing velocity and the bed porosity. Good erosion trends are obtained and agree qualitatively and quantitatively with maximum experimental erosion data obtained at the Grimethorpe cold model facility.

1.   INTRODUCTION

Excellent heat transfer from a gas fluidized bed to the surfaces of immersed heat exchanger tubes is recognized as being one of the most attractive features of atmospheric and pressurized fluidized bed combustors (AFBC and PFBC). Unfortunately, erosion of such tube banks by vigorous solids scouring sometimes plague their operation and must be overcome to guarantee reasonable lifetimes of commercial FBC units (~ 100,000 h).

In previous erosion studies, it was shown that one of the determining factors affecting the erosion of ductile and brittle materials is the particle velocity impinging on the target. The erosion rate was shown to be proportional to the particle velocity raised to the power 2 to 2.5.[1,2] In recent erosion studies of immersed heat exchanger tubes, simple models have been developed which attempt to relate the solids velocity to more easily measurable quantities such as the bubble diameter, the bubble velocity, the bed porosity and the fluidizing velocity.[3-4]

In this study, a simplified version of the general Monolayer Energy Dissipation (MED) Erosion Model is presented. This model evaluates the kinetic energy dissipated by one layer of particles in contact with the tube surface, which is then converted into erosion rates by means of a thermodynamic energy balance. This model is shown to predict reasonable dependency of erosion rates upon particle diameter, fluidizing velocity and bed porosity. Predicted erosion rates compare favorably to available experimental data and several useful guidelines concerning the design and operating conditions of FBC are discussed in view of this newly developed erosion model.

2.   THEORY: THE MONOLAYER ENERGY DISSIPATION (MED) EROSION MODEL

The MED erosion model is an extension of the so-called power dissipation erosion model for low angle erosion caused by slurries, proposed by Ushimaru et al.[5] A review of the recently developed general MED erosion models was given by Bouillard et al.[6] These models are based on the premise that the solids mechanical energy is irreversibly dissipated in the neighborhood of stationary surfaces by three competitive mechanisms:

a.   Heat transfer between the gas and solids particles, between the gas and tube surfaces, and between the solids and the tube surfaces,

b. Erosion of obstacles, and

c. Attrition of solid particles.

Thus, the rate of energy dissipated in erosion represents only a fraction of the total power dissipation (which is related to the total entropy production). In the following analysis, a brief derivation of one MED erosion model is given and the model dependency upon the particle diameter, the fluidizing velocity and the bed porosity is studied.

In the absence of solids viscous stresses, one may write the momentum equation for the solids phase in nonconservation law form as follows:[6-9]

$$\epsilon_s \rho_s \frac{d\vec{V}_s}{dt} = -\epsilon_s \nabla P + \beta_a \left(\vec{V}_g - \vec{V}_s\right) + \epsilon_s \rho_s \vec{g} - G \nabla \epsilon_s \tag{1}$$

Equation (1) is called hydrodynamic model A by Bouillard et al.[8] By dropping the pressure gradient in Eq. (1), assuming the entire pressure gradient is in the gas phase, and modifying the interphase friction coefficient, β_a, one obtains hydrodynamic model B.[8] For more information on these two hydrodynamic models, the reader should refer to Refs. 6, 7, 8 and 9.

Such hydrodynamic models are used to calculate the solids velocity, $\vec{V}_s$, in the fluidized bed and especially in the neighborhood of immersed surfaces. Equation (1) describes the momentum equation of a dissipative system whose dissipative function D is given by[10,11] (after dropping superscripted arrows for clarity):

$$D = 1/2 \, \beta_a \left(v_g - v_s\right)^2 \tag{2}$$

To estimate the power dissipation per unit of volume in the vicinity of a tube surface in the monolayer control volume shown in Fig. 1, we assume that $V_g \sim 0$ in the monolayer control volume and thus, Eq. (2) becomes

$$D = 1/2 \, \beta_a \, v_s^2 \tag{3}$$

We imagine that the solids collide inelastically with the tube surface so that the particle velocities before (V_{s1}) and after (V_{s2}) collision with the tube surface are such that $V_{s2} = e \, V_{s1}$, or $\delta V_s^2 = V_{s1}^2 - V_{s2}^2 = (1-e^2) \, V_{s1}^2$. The variation of power dissipation per unit volume, δD, associated with the collision of the particles onto the tube surface may be expressed as

$$\delta D = 1/2 \, \beta_a \, \delta V_s^2 = 1/2 \, \beta_a \left(1-e^2\right) V_{s1}^2 \tag{4}$$

where β_a was assumed to be constant during the collision, and e is the coefficient of restitution of the particle collision onto the tube surface. Typically $e^2 \sim 0.90$.[12] After dropping the subscript 1 ($V_{s1} = V_s$) in Eq. (4), the change in power dissipation δD is

420

integrated over the monolayer control volume $V = A_t \cdot d_p$ where A_t is the local external tube surface (see Fig. 1). The particle diameter, d_p was chosen as one dimension of this control volume since d_p is on the order of the particle mean free path.[13]. The resulting power dissipated is assumed to represent the power available for erosion, $\dot{P}_e$, which may be expressed as

$$\dot{P}_e = \int_V \delta D \ dV = \delta D \cdot V = \frac{\beta_a}{2} \left(1 - e^2\right) V_s^2 \ d_p \cdot A_t \tag{5}$$

This dissipated power is related to the volume of material removed from the target, V_t, by the following thermodynamic energy balance:

$$\dot{P}_e = P_t \frac{dV_t}{dt} \tag{6}$$

The right-hand side of Eq. (6) represents the plastic compressive work per unit of time expressed in terms of the target erosive flow stress P_t which is related to the target hardness. The term $\dfrac{dV_t}{dt}$ is the rate of volume removed during plastic deformation and is expressed as

$$\frac{dV_t}{dt} = A_t \dot{E} \tag{7}$$

where A_t is the external surface of the target as shown in Fig. 1 and $\dot{E}$ is the erosion rate (m/s). Combining Eqs. (5), (6) and (7) results in

$$\dot{E} = \frac{1}{2} \frac{\left(1 - e^2\right)}{P_t} \beta_a V_s^2 \ d_p \tag{8}$$

To express the solids velocity V_s in terms of the solids volume fraction, ε_s, we use Gidaspow's et al. approach.[14] By assuming that the solids momentum is much larger than that of the gas phase and that the solids Coulombic stress term, $G\nabla\varepsilon_s$, is negligible near and above minimum fluidization. The total momentum equation for the mixture (gas and solids phases) reads:

$$\rho_s \varepsilon_s \frac{dV_s}{dt} = - \nabla P + \rho_s \varepsilon_s g \tag{9}$$

Gidaspow derived a simple expression for the solids circulation rate in a fluidized bed.[14] In his model, the bed was divided into two distinct regions: one region influenced by a jet and the other region in the vicinity of the side walls. These two regions are called the "jet" region and the "downcomer" region having the solids volume fraction ε_{sj} and ε_{sd}, respectively. By neglecting the solids acceleration in the "downcomer" region and setting the pressure drops in the two regions to be equal, we may obtain the square of the solids velocity in the "jet" region from the circulation rate obtained by Gidaspow as:[14]

$$v_{sj}^2 = g\, x_d \left(\varepsilon_{sd} - \varepsilon_{sj}\right)/\varepsilon_{sj} \tag{10}$$

where x_d is the height of the downcomer region. In extending Gidaspow's simple model, we will replace the "jet" region with a region characteristically lean in solids, and the "downcomer" region with a region packed with solids in a nearly defluidized state. The height of the downcomer becomes a distance characteristic of the problem under consideration. It could be the tube spacing or the distance of the tube bank to the distributor of the expanded bed height. In this study x_d will be taken to be the distance between the tubes, which is believed to represent the averaged acceleration distance of the solids phase moving through the tube bundle. Setting $V_s^2 = V_{sj}^2$ and combining Eqs. (8) and (10), one obtains:

$$\dot{E} = \frac{1}{2}\, \frac{(1-e^2)}{P_t}\, \beta_a g\, x_d \left(\varepsilon_{sd} - \varepsilon_{sj}\right) d_p/\varepsilon_{sj} \tag{11}$$

Substituting for β_a (Gidaspow[9]) into Eq. (11), the general expression for the erosion rate, $\dot{E}$, becomes

$$\dot{E} = \frac{1}{2}\, \frac{(1-e^2)}{P_t}\, g\, x_d\, d_p\, \frac{\left(\varepsilon_{sd} - \varepsilon_{sj}\right)}{\varepsilon_{sj}} \left[\frac{150\, \mu_g\, \varepsilon_{sj}}{(1-\varepsilon_{sj})\,(\phi_s d_p)^2} + 1.75\, \frac{\rho_g \left(U_f - \varepsilon_{sj}V_s\right)\varepsilon_{sj}}{(1-\varepsilon_{sj})\,(\phi_s d_p)} \right] \tag{12}$$

Equation (12) represents an upper bound of the erosion rates because the expression used for the solids velocity (Eq. (10)) neglects all solids viscous effects. Hence, Eq. (12) is a simplified expression which relates the maximum erosion rates to the system variables such as the lean and dense region solids volume fractions, ε_{sj} and ε_{sd}, the particle diameter, d_p, the fluidizing velocity, U_f, and the characteristic distance, x_d. The dependence of the erosion rates upon these system variables will be studied in the following sections.

3. EROSION RATE DEPENDENCY UPON FLUIDIZING VELOCITY, PARTICLE DIAMETER AND BED POROSITY

The solids acceleration length, x_d, may be thought as the vertical tube spacing or the distance between the distributor and the tube bank. Equation (12) suggests that a reduction of erosion may be obtained by reducing the distance between the distributor and the tube bank and/or by reducing the tube spacing in the tube bank. Such conclusions were proven effective at the Grimethorpe facility.[15-16]

a. Small Particles (Geldart's Class A and C)[17]

For small particles, the second term in brackets of Eq. (12) may be neglected. The erosion rate $\dot{E}$ is then expressed as

$$\dot{E} = \frac{1}{2}\, \frac{(1-e^2)}{P_t}\, g\, x_d\, d_p \left[\frac{150\, \mu_g\, \varepsilon_{sj}\left(\varepsilon_{sd} - \varepsilon_{sj}\right)}{(1-\varepsilon_{sj})\,(d_p \phi_s)^2} \right] \tag{13}$$

The erosion rate $\dot{E}$ has a maximum at $\varepsilon = 1 - \varepsilon_{sj} = \left[\varepsilon_{gd}\right]^{1/2}$ where $\varepsilon_{gd} = 1 - \varepsilon_{sd}$. The porosity at which this maximum is reached is $\varepsilon = 0.62$ for $\varepsilon_{gd} = 0.38$. Figure 2 shows the

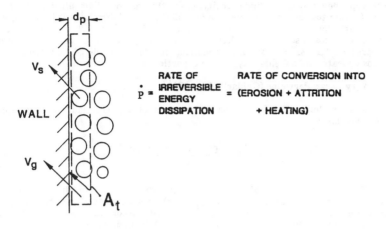

FIGURE 1 Conceptual Picture of Monolayer Energy Dissipation Erosion Model.

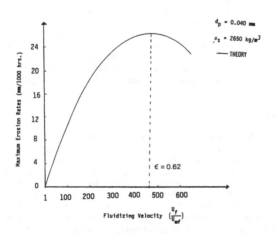

FIGURE 2 Predicted Dependency of Erosion Rates with the Particle Diameter, the Fluidizing Velocity and Bed Porosity for Small Particle Size (Geldart's Class A and C Type).

erosion rate dependency with the bed porosity, ϵ, for the case of 40 μm silica sand hitting a PVC tube of Shore test hardness of about A80-A90 ($P_t \sim$ 40 MN/m^2).[18] This hardness is about 7 times smaller than that of aluminum ($P_t \sim$ 294 MN/m^2). The coefficient of restitution, e, was chosen so that e^2 = 0.90.[12] At ϵ = 0.62, the erosion rate is maximum and may represent typical erosion rates occurring in plastic bends of fast circulating fluidized beds operated with Geldart type A or C particles.

b. Very Large Particles (Geldart's Class D)[17]

For large particles, the first term in brackets of Eq. (12) may be neglected. Equation (12) becomes:

$$\dot{E} = \frac{1}{2} \frac{(1-e^2)}{P_t} g \, x_d \, 1.75 \left[\frac{\rho_g \, (U_f - \epsilon_{sj} \, V_s) \, (\epsilon_{sd} - \epsilon_{sj})}{(1 - \epsilon_{sj}) \, \phi_s} \right] \tag{14}$$

As shown by Eq. (14), the erosion rate is no longer directly dependent on the particle size but mainly depends on the fluidizing velocity U_f and the bed porosity ϵ. As the fluidizing velocity U_f increases, the erosion rate increases. A similar trend was reported by Yates[3] who found a linear dependency of the erosion rate upon the fluidizing velocity.

c. Medium Size Particles (Geldart's Class B)[17]

For Geldart class B particles, both terms in brackets of Eq. (12) are important. In this case, the erosion rate dependency upon the particle diameter, fluidizing velocity and bed porosity becomes more difficult to express explicitly. Equations (10) and (12) can be quickly evaluated using a desktop calculator.

Such calculations were performed for the cold fluidized bed (Task 1) of Parkinson et al.[15-16] The vertical spacing of the staggered tube array was 0.16 m, and three main particle sizes were used (700, 1000 and 1300 μm). Erosion rates calculated by the simplified MED erosion model given by Eq. (12) are presented in Figures 3 and 4 as a function of the fluidizing velocity and the bed porosity. Experimental erosion rates presented in Figs. 3 and 4 are the maximum erosion rates experimentally determined in the tube bank.[19] A Shore test hardness of ($P_t \sim$ 40 MN/m^2) for the PVC tubes is used in the predicted results. As can be seen, the predicted erosion rate trends as a function of the particle diameter, fluidizing velocity and bed porosity agree reasonably well with experimental data. In view of the assumptions made in this simplified MED erosion model, and although some uncertainties still lie in the material properties used for the PVC tubes, the predicted trends and rates are remarkably similar to that of the experimental data.

4. CONCLUSIONS

A simplified monolayer energy dissipation erosion model was developed and tested against some limited heat exchanger tube erosion data in cold model fluidized beds. This simple erosion model predicts that erosion rates increase whenever the tube spacing or the distance between distributor and tube bank increases. The model shows that for Geldart's class A or C particles (fine particles), the erosion rate presents a maximum for a bed porosity, ϵ, of about 0.62. A bubbling or circulating fluidized bed operating at this porosity might suffer serious erosion. To reduce tube erosion in beds operated with Type A or C particles, it is recommended to either decrease the fluidizing velocity (make the bed less energetic) or to increase the fluidizing velocity so that the bed porosity is above 0.7.

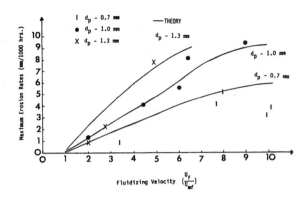

FIGURE 3   Predicted Erosion Rates Versus the Fluidizing Velocity for Several Particle Sizes (Geldart's Class B). (Experimental Data is from Parkinson et al. [19]).

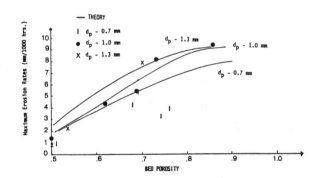

FIGURE 4   Predicted Erosion Rates Versus the Bed Porosity for Several Particle Sizes (Geldart's Class B). (Experimental Data is from Parkinson et al.[19]).

It was also shown that, for a given particle distribution at a given fluidizing velocity, smaller particles are usually more energetic than larger particles because of their larger drag. This situation is common in fluidized beds with large particle size distributions, for which the fluidizing velocity necessary to fluidize the largest particles can be well above the minimum fluidization of the smallest particles. In this condition, smaller particles may become more energetic and, thus, may greatly contribute to tube erosion. The control of the small size particle content in the bed would, thus, be a key factor to controlling the overall erosion rate. Hence, tube erosion would appear as a small/large size particle synergism.

Finally, this study revealed that the erosion rate is generally an increasing function of the mean particle size, the fluidizing velocity and the bed porosity. In view of the lack of good material properties, and of the simplifications of the MED erosion model, predicted erosion rate trends using this simple erosion model are considered adequate and are in good agreement with available experimental data.

5. NOMENCLATURE

A_t Target surface, m^2

d_p Particle diameter, m

$\dot{E}$ Erosion rate, m/s

e Coefficient of restitution of particle-target collision

D Onsager's dissipation function, W/m^3

G Solids elastic modulus, Pa

g Acceleration due to gravity, m/s^2

M Momentum sources and sink, N/m^3

P Pressure, Pa

$\dot{P}$ Total dissipated power, W

$\dot{P}_e$ Total dissipated power converted into erosion

P_t Eroding flow stress or hardness of target, Pa

U_f Fluidizing velocity, m/s

V_s Solids velocity, m/s

V_{sj} Solids velocity in the jetting region, m/s

V_t Target volume removed, m^3

x_d Downcomer height, m

Greek Letters

β	Fluid-particle friction coefficient, $kg/(m^3 \cdot s)$
ε	Gas phase volume fraction
ε_{gd}	Downcomer gas volume fraction = $1 - \varepsilon_{sd}$
ε_s	Solids volume fraction = $(1 - \varepsilon)$
ε_{sd}	Downcomer solids volume fraction
ε_{sj}	Jet phase solids volume fraction
ρ	Density kg/m^3
ϕ_s	Sphericity

Operator

$\nabla \cdot$	Divergence
∇	Gradient
$\dfrac{d\vec{V}_s}{dt}$	Total derivative following the solids = $\dfrac{\partial \vec{V}_s}{\partial t} + \vec{V}_s \cdot \nabla \vec{V}_s$, cm/sec^2

ACKNOWLEDGMENTS

This work was performed for the U.S. Department of Energy, Morgantown Energy Technology Center, Contract W-31-109-ENG-38, and for the Electric Power Research Institute under the cooperative R&D venture "Erosion of FBC Heat Transfer Tubes."

REFERENCES

1.   Finnie, I., *Erosion of Surfaces by Solid Particles*, Wear, 3, pp. 87-103 (1960).

2.   Neilson, J.H., and A. Gilchrist, *Erosion by a Stream of Solid Particles*, Wear 11:111-122 (1968).

3.   Yates, J.G., *On Erosion of Metal Tubes in Fluidized Beds*, Chem Eng. Sci., 42, No. 2, pp. 379-380 (1987).

4.   Wood, R.T., and D.A. Woodford, *Tube Erosion in Fluidized-Beds*, ERDA Report 81-12 911/ET-FUC/79) prepared for New York State Energy Research and Development Authority by General Electric Co., Schenectady, NY (Dec. 1980).

5.   Ushimaru, K., C.T. Crowe and S. Bernstein, *Design and Applications of the Novel Slurry Jet Pump*, Energy International, Inc., Report No. E184-108 (Oct. 1984).

6. Bouillard, J.X., R.W. Lyczkowski, S. Folga, D. Gidaspow and G.F. Berry, *Hydrodynamics of Heat Exchanger Tubes in Fluidized Bed Combustors*, accepted for publication in the Canadian Journal of Chemical Engineering (Jan. 1989).

7. Lyczkowski, R.W., J.X. Bouillard and D. Gidaspow, *Computed and Experimental Motion Picture Determination of Bubble and Solids Motion in a Two-Dimensional Fluidized Bed with a Jet and Immersed Obstacle*, Proc. of the Eighth Intl. Heat Transfer Conf., C.L. Tien, V.P. Carey and J.K. Ferrell, eds., Vol. 5, pp. 2593-2598, Washington, D.C. (1986).

8. Bouillard, J.X., R.W. Lyczkowski and D. Gidaspow, *Hydrodynamics of Fluidization: Time-Averaged and Instantaneous Porosity Distributions in a Fluidized-Bed with an Immersed Obstacle*, paper submitted to AIChE Journal (July 1987), revised and resubmitted (July 1988).

9. Gidaspow, D., *Hydrodynamics of Fluidization and Heat Transfer: Supercomputer Modeling*, Applied Mechanics Review, 31, No. 1, pp. 1-23 (Jan. 1986).

10. Onsager, L., *Theories and Problems of Liquid Diffusion*, Ann. New York Acad. Sci., 46, pp. 241-265 (1945).

11. Goldstein, H., *Classical Mechanics*, Addison-Wesley, Cambridge, MA (1956).

12. Savage, S.B., "Granular Flow at High Shear Rates," Theory of Dispersed Multiphase Flow, Academic Press, New York, pp. 339-357 (1982).

13. Lun, C.K.K., S.B. Savage, D.J. Jeffrey and N. Chepurniy, *Kinetic Theories for Granular Flow: Inelastic Particles in Couette Flow and Slightly Inelastic Particles in a General Flowfield*, J. Fluid Mech., 140, 223-256 (1984).

14. Gidaspow, D., and B. Ettehadieh, *Fluidization in Two-Dimensional Beds with a Jet 2. Hydrodynamic Modeling*, I&EC Fund, 22, pp. 193-201 (1983).

15. Parkinson, M.J., B.A. Napier, A.W. Jury and T.J. Kempton, *Cold Model Studies of PFBC Tube Erosion*, Proc. of 8th Intl. Conf. on Fluidized-Bed Combustion, Vol. II, pp. 730-738, National Technical Information Service, U.S. Department of Commerce, Springfield, VA (1985).

16. Parkinson, M.J., J.F.G. Grainger, A.W. Jury and T.J. Kempton, *Tube Erosion at IEA Grimethorpe: Cold Model Studies at CRE*, in Report Commissioned by the Project from Outside Consultants and Others, Vol. 2, NCB (IEA Grimethorpe) Ltd., Barnsley, S. Yorkshire, U.K. (Sept. 1984).

17. D. Geldart, *Types of Gas Fluidization*, Powder Technology 7, 285-292 (1973).

18. Plastics, Edition 8, *Thermoplastics and Thermosets*, Desk-Top Data Bank produced by DATA, Inc. (1986).

19. Parkinson, M.J., B.A. Napier, A.W. Jury and T.J. Kempton, *Cold Model Studies of PFBC Tube Erosion*, in "Proc. of 8th Intl. Conf. on Fluidized-Bed Combustion," Vol. II, pp. 730-738, National Technical Information Service, U.S. Department of Commerce, Springfield, VA (1985).

COMBUSTION

A Numerical Study of the Mixing Chamber in Y-Type Burners

SONG-SHOU ZHANG, JUN AI, and JIE ZHOU
The Shanghai Institute of Mechanical Engineering
516 Jun Gong Road
Shanghai 200093, PRC

Abstract

   Based on experimental results, a three-dimensional two-phase flow physical model of the mixing chamber in y-type burner is set up. A mathematical model related to the physical model is also established. The numerical computation method is used to investigate and predict the flow fields. In order to consider the effect of the interface wave between gas and liquid phase, an "Interface function" is proposed, the effect on effective viscosity caused by the existing liquid droplets in the gas phase is also considered in the numerical computation. A $K-\varepsilon$ turbulent model is used in the computation of the gas phase. The numerical computation method is also employed to solve the equations of three-dimensional continuity, momentum, turblent energy and dissipation. It shows that the computational program can be used to handle the problems of pressure, velocity, turbulent fields and liquid film thickness of the mixing chamber in y-type burner. A comparision is made between the prediction of the pressure fields and the experimental data. Finally, some ideas are proposed to improve the mathematical model.

1.   INTRODUCTION

   To improve the performances and design method the flow in mixing chamber of y-type burner must be taken further steps to study. To obtain it's numerical calculation results an "interface function" is set up in paper[1] . Althought it's convenient to handle the interface problem between gas and liquid, but it exists the problems as follow:1).the function is set up by turbulent model; 2). the wavement and entrainment in interface between gas and liquid is considered by constant E. These make the "interface function" is simple for the use of numerical calculation. To improve the "interface function" a new "interface function" is set up. The flow fields in mixing chamber of y-type burner is sloved by numerical calculation.

2.   THE PHYSICAL MODEL

   The geometry and the flow situation of the mixing chamber in y-type burner is show in fig. 1. According to the paper[1], the three dimensional two-phase flow model of the mixing chanber is as follow:

431

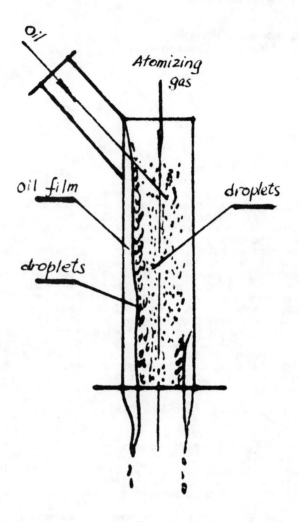

Fig. 1 The geometry and the flow situation of mixing chamber in y-type burner

1).The flow in gas phase or liquid phase is steady and these fluids are incompressible.
2).The atomizing gas flows through the core of the mixing chamber and oil flows along the wall to form the oil film.
3).There are wave in the interface between gas and liquid and droplets entrained by the atomizing gas concentrate on the interface.
4).The axial Reynolds number is greater than other directions.
5).The temperatures are uniform in whole flow fields.
6).The flow in gas phase is turbulent and the flow in liquid phase is laminar.

3.   MATHEMATICAL STATEMENT OF THE PROBLEM

Governing Equations

Based on the paper[1], the governing equations of the flow in the mixing chamber are as follow:

$$\frac{1}{r}\frac{\partial}{\partial r}(ru_r\,\rho\phi) + \frac{1}{r}\frac{\partial}{\partial\vartheta}(u_\vartheta\,\rho\phi) + \frac{\partial}{\partial z}(\rho\phi u_z)$$

$$= \frac{1}{r}\frac{\partial}{\partial r}\left(\Gamma_\phi r\frac{\partial\phi}{\partial r}\right) + \frac{\partial}{r\partial\vartheta}\left(\Gamma_\phi\frac{1}{r}\frac{\partial\phi}{\partial\vartheta}\right) + S_\phi \qquad (1)$$

Conservation of		ϕ	Γ_ϕ	S_ϕ
Governing eg. of liquid phase	mass	1	0	0
	axial (Z) momentum	u_z	μ_{eff}	$\rho g - \dfrac{\partial P}{\partial Z}$
	radius (r) momentum	u_r	μ_{eff}	$-\mu_{eff}\dfrac{2}{r^2}\dfrac{\partial u_\vartheta}{\partial\vartheta} - \mu_{eff}\dfrac{u_r}{r^2} + \rho\dfrac{u_\vartheta^2}{r}$
	angular (θ) momentum	u_ϑ	μ_{eff}	$-\mu_{eff}\dfrac{u_\vartheta}{r^2} + \mu_{eff}\dfrac{2}{r^2}\dfrac{\partial u_r}{\partial\vartheta} - \rho\dfrac{u_r\,u_\vartheta}{r}$

In liq. film $K = 0$, $\varepsilon = 0$, $\mu_{eff} = \mu_l$

Governing eg. of gas phase	mass	1	0	0
	axial (Z) momentum	u_z	μ_{eff}	$-\dfrac{\partial P}{\partial Z}$
	radius (r) momentum	u_r	μ_{eff}	$\dfrac{1}{r}\dfrac{\partial}{\partial r}\left(r\mu_{eff}\dfrac{\partial u_r}{\partial r}\right) + \dfrac{1}{r}\dfrac{\partial}{\partial\vartheta}\left(r\mu_{eff}\dfrac{\partial}{\partial r}\left(\dfrac{u_\vartheta}{r}\right)\right)$ $+ \dfrac{2\mu_{eff}}{r^2}\dfrac{\partial u_\vartheta}{\partial\vartheta} - \dfrac{2\mu_{eff}}{r^2}u_r + \dfrac{u_\vartheta^2}{r}\rho$

433

angular (θ) momentum	u_θ	μ_{eff}	$\dfrac{1}{r}\dfrac{\partial}{\partial r}\left(\mu_{eff}\,r\left(\dfrac{1}{r}\dfrac{\partial u_r}{\partial\vartheta}-\dfrac{u_\theta}{r}\right)\right)-\rho\,\dfrac{u_\theta u_r}{r}$ $+\dfrac{1}{r}\dfrac{\partial}{\partial r}\left(\dfrac{\mu_{eff}}{r}\left(\dfrac{\partial u_\theta}{\partial\vartheta}+2u_r\right)\right)$ $+\dfrac{1}{r}\mu_{eff}\left(\dfrac{\partial u_\theta}{\partial r}+\dfrac{1}{r}\dfrac{\partial u_r}{\partial\vartheta}-\dfrac{u_\theta}{r}\right)$
kinetic energy	K	μ_{eff}/σ_k	$G_k-\rho\varepsilon$
dissipation rate	ε	$\mu_{eff}/\sigma_\varepsilon$	$\dfrac{\varepsilon}{K}\left(c_1 G_k-c_2\,\rho\varepsilon\right)$

$$G_k=\mu_{eff}\left\{2\left[\left(\frac{\partial u_r}{\partial r}\right)^2+\left(\frac{1}{r}\frac{\partial u_\theta}{\partial\vartheta}+\frac{u_r}{r}\right)^2\right]+\left(\frac{\partial u_z}{\partial r}\right)^2\right.$$
$$\left.+\left(\frac{1}{r}\frac{\partial u_z}{\partial\vartheta}\right)^2+\left(\frac{1}{r}\frac{\partial u_r}{\partial\vartheta}+\frac{\partial u_\theta}{\partial r}-\frac{u_\theta}{r}\right)^2\right\}$$

$$\mu_{eff}=\mu_\varepsilon+c_\mu\,\rho K^2/\varepsilon$$

$$c_1=1.43,\ c_2=1.92,\ c_\mu=0.09,\ \sigma_k=0.9,\ \sigma_\varepsilon=1.1$$

The Interface Conditions

According to the paper[1] , without the surface tension, the interface condition are as follow:

$$\vec{V}_\varepsilon=\vec{V}_I$$
$$\mu_\varepsilon\,S_{ij}{}^{(\varepsilon)}\,t_i\,n_j=\mu_I\,S_{ij}{}^{(I)}\,t_i\,n_j \tag{2}$$
$$P^{(\varepsilon)}=P^{(I)}$$

The Interface Equqtion

Fig.2. shows the cylindrical polar cordinate $(r\theta z 0)$ of the mixing chamber in y-type burners. According to paper[1] :

$$\frac{\partial h}{\partial z}\Big|_{z\to 0}=-\frac{u_r}{u_z}\Big|_h \tag{3}$$

$$\frac{1}{r}\frac{\partial h}{\partial\vartheta}\Big|_{z\to 0}=\frac{u_r}{u_\theta}\Big|_h \tag{4}$$

The Interface Function

There are two reasons for the use of the "interface function".
1). There are waves in the interface and it must be considered in calculation of the flow fields. It is very expensive and di-

fricult to slove the wave problem, if the influence of waves is directly represented in the interface equation. 2). These are entrainments at the interface and in the calculation it must be considered too, all these can be expressed easily by the interface function.

Assume that these are several points along Y direction, as shown in fig.3. Here: S_a represents the interface. S_0 represents the outline of the viscosity layer.

a). The Former Interface Function

Based on paper[1], the former interface function is as fol-low:

$$u_w - u_p = \frac{\rho c_\mu^{\frac{1}{2}} K_w (y_s - y_p)}{u_l} + \frac{c_\mu^{\frac{1}{4}} K_w^{\frac{1}{2}}}{k} \ln \Big(E (y_w - y_s) -$$

$$- \frac{\rho (c_\mu^{\frac{1}{2}} K_w)^{\frac{1}{2}}}{\rho v_{eff}} \Big) \tag{5}$$

The former interface function is set up on the basis of the "wall function" of single phase turbulent flow and assumptions as follow : 1) the shearing stress τ_s is equal to $\rho c_\mu^{\frac{1}{2}} K_w$. 2). E is a constant. 3). the shearing stress between P and W is constant. These make the former interface function too simple and it doesn't represent the characteristics of the interface flow. It must be improved.

b). The Improved Interface Function

The main reason that the former interface function can't be used is that the shearing stress $\tau_s = \rho c_\mu^{\frac{1}{2}} K_w$ is incorrect. Thus, for improving the interface function, the expression of τ_s must be corrected.

i). The Shearing Stresses At The Interface

Fig.4. shows the sketch map of the one-dimenssional gas-liquid flow.

In the paper[2]. It assumes that $a/b \approx 1$, $P_q/P_L \leq 1$, $(-\frac{dp}{du}) \gg g l_L$, $u_L \gg u_L$ then the expression of shearing stresses τ_s is

$$\tau_s = (-\frac{dp}{dx})(\frac{b - looh}{2}) \times 10 \tag{6}$$

$$(-\frac{dp}{dx}) = \Big[g \big(\frac{G_L}{G_L + G_q} \cdot \frac{1}{P_L} + \frac{G_q}{G_L + G_q} \cdot \frac{1}{P_q} \big) \Big]$$

$$+ \frac{0.43}{(2b)^{1.2}} (\frac{\sigma}{73})^{0.4} \Big[(G_L + G_q)^2 (\frac{G_L}{G_L + G_q} \cdot \frac{1}{P_L} + \frac{G_q}{G_L + G_q} \cdot \frac{1}{P_q}) \Big]^{0.75} \tag{7}$$

Here P, $dyne/cm^2$; G_L, $g/cm^2 \cdot s$; σ, $dyne/cm$; $g = 980 \ cm/s^2$

ii). The Derivation Of The Interface Function

It is assumed that a logarithmic velocity profile prevails in

435

the region between the interface and the node W. the expression for the velocity is

$$\frac{u_w - u_s}{\sqrt{\tau_s/\rho}} = \frac{1}{k} \ln \left(E \cdot (y_w - y_s) \sqrt{\tau_s} \cdot \rho / \mu_{eff} \right) \tag{8}$$

It is then assumed that the velocity distribution in the region between the interface and the node P. is linear. Then:

$$\tau_s = \mu_{\ell} \frac{u_s - u_p}{y_s - y_p} \tag{9}$$

By the use of this expression in conjunction with eq(8) .The improved interface function is obtained:

$$u_w - u_p = \frac{\tau_s (y_s - y_p)}{\mu_{\ell}} + \frac{\sqrt{\tau_s/\rho}}{k} \ln \left(E \frac{y_w - y_s}{\rho \nu_{eff}} \sqrt{\tau_s} \cdot \rho \right) \tag{10}$$

Here [1]

$$\left. \begin{array}{l} \rho = \frac{1}{2} (\rho_L + \rho_G) \\[2mm] \nu_{eff} = \frac{\mu_g + \mu_\ell}{\rho_L + \rho_G} \end{array} \right\} \tag{11}$$

C). The Wall Function

If it is assumed that the velocity distribution between the wall and node W. (seeing fig.5) is as follow:

$$u_w = \frac{C_\mu^{1/2} K_w^{1/2}}{k} \ln \left(E y_w \cdot \frac{\rho (C_\mu^{1/2} K_w)^{1/2}}{\rho \nu_{eff}} \right) \tag{12}$$

d). Kinetic Energy And Dissipation Rate In The Region Near The Wall Or The Interface

$y^* = \frac{y v^x}{\nu_{eff}} > 30$ K and ε are determined from the transport equations for K and ε.

$y^x = \frac{y v^*}{\nu_{eff}} \leq 30$ K distribution is linear. ε distribution is linear too.

e). Boundary Conditions

Fig.6. shows the geometry of the mixing chamber and its cylindrical polar coordinate $(r \theta Z O)$.

$$\theta_1 = \cos^{-1} \left(\frac{\cos \alpha \cdot (Z - Z_0) + ((Z - Z_0)^2 + (R^2 - R_1^2))^{1/2}}{R \sin \alpha} \right) \tag{13}$$

i). $r = R$, $Z = B$, $-\pi + \theta_1 \leq \theta \leq \pi - \theta_1$

$u_z = 0$, $u_r = 0$, $u_\theta = 0$, $K = 0$, $\varepsilon = 0$

ii). $r = k$, $Z \geq B$

$u_z = 0$, $u_r = 0$, $u_\theta = 0$, $K = 0$, $\varepsilon = 0$

436

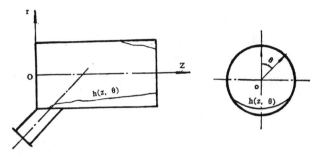

Fig. 2.　The cylindrical polar coordinate
of mixing chamber in y-type burners

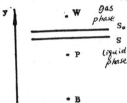

Fig. 3.　The near-interface nodes

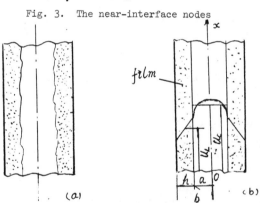

Fig. 4.　The sketch map of one-dimensional
gas-liquid film flow

Fig. 5.　The near-wall nodes

iii). $\vartheta = 0$, $\theta = \pi$, $U_\theta = 0$, $\frac{\partial \phi}{\partial \theta} = 0$ Here $\phi = U_z, U_r, U_\theta, K, \varepsilon$

iv). $r = 0$, $\frac{\partial \phi}{\partial r} = 0$, Here $\phi = U_z, U_r, U_\theta, K, \varepsilon$.

v). Entrance Bondary Condition Of Gas Flow

It is assumed that the flow of gas at the entrance is one-dimensional developed turbulent.

$$U_z = \frac{60 \cdot G_q}{49 \cdot \pi} \cdot R^{-15/7} (R-r)^{1/7}$$

$$U_r = 0, \qquad U_\theta = 0$$

$$K = a + br + dr^2 + er^3$$

Here: $a = 0.0005 U_{max}^2 + \frac{3R r_c^2 - R^2}{(R-r_c)^3} (0.0005 U_{max}^2 - C_\mu^{1/2} (\tau/\rho)_w)$

$b = -\frac{6R}{(R-r_c)^3} (0.0005 U_{max}^2 - C_\mu^{1/2} (\tau/\rho)_w)$

$d = 3(R+r_c)(0.0005 U_{max}^2 - C_\mu^{1/2} (\tau/\rho)_w / (R-r_w)^3)$

$e = -2(0.0005 U_{max}^2 - C_\mu^{1/2} (\tau/\rho)_w)/(R-r_w)^3$

$\varepsilon = C_\mu^{3/4} K^{1/2} / (\hat{k}(R-r))$

vi). Entrance Bondary Condition Of Oil Flow

It is assumed that the flow of oil at the entrance is one-dimensional developed laminar.

$$r = R, \quad Z \leq B, \quad \pi - \theta_1 \leq \theta \leq \pi$$

$$U_r = \frac{2G_L}{\pi R_1^4} (R_1^2 - (2Sin\alpha + RCos\alpha \cdot Cos\theta_1 - Z_0 Sin\alpha)^2 - R^2 Sin\theta_1) Sin\alpha$$

$$U_2 = \frac{2G_L}{\pi R_1^4} (R_1^2 - (2Sin\alpha + RCos\alpha \cdot Cos\theta_1 - Z_0 Sin\alpha)^2 - R^2 Sin\theta_1) Cos\alpha$$

$$U_\theta = 0, \quad K = 0, \quad \varepsilon = 0$$

4). RESULTS AND DISCUSS

The authors improve the program which can be used to calculate the flow field in the mixing chamber of y-type burner. A new "interface function" is adopted in the improved program. Tested the reliability, the program is used to calculate the flow field in the mixing chamber whose geometric sizes are: radius R=6.75 mm. , Length L=31.52 mm. under gas flow rate $G_G = 0.3503 \times 10^{2}$ m³/s, oil flow rate $G_L = 0.9197 \times 10^{-4}$ m³/s, and gas pressure at entrance is P =98661 Pa.

Fig.7. presents the liquid film variations along the axial under two kinds of interface functions.

Fig.8. presents the film distribution at three cross sections 1) Z/L =0.25, 2) Z/L =o.5, 3) Z/L =0.75

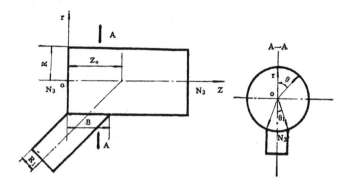

Fig. 6. The geometry of the mixing
chamber and its cylindrical
polar coordinate (r θ zo)

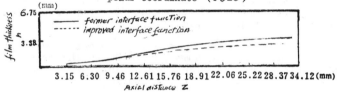

Fig. 7. The liquid film variation along the
axial under two kinds of interface
function

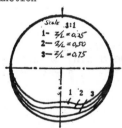

Fig. 8. The film distribution at three cross
section 1) z/L=0.25, 2) z/L=0.5,
z/L=0.75

439

Fig.9. presents the axial velocity distribution located on a vertical diametral plane.
Fig.10. presents the sketch map of the radical velocity at several different angle in a cross section
Fig.11. presents the sketch map of angular velocity at several different angle in a cross section.
Fig.12. presents the kinetic energy distribution in radical direction at certain angle in a cross section.
Fig.13. presents the sketch map of the dissipation rate distribution in radical direction at several different angle in a cross section.
Fig.14. presents the comparison between the experimental results and the numerical results about the pressure in axial direction.

In Fig.7. It can be seen that the liquid films are different in the axial direction and they are also different at the same location under different interface function. The liquid film increses in axial direction. It also can be seen that the liquid film thickness increases slower at axial direction in new interface function than that in former interface function and the liquid film thickness in new interface function is slighter than that in former interface function. These phenomenons occur because the new interface function is derivated by the semi-emprical formula.

It can be seen that the program can be used to calculate qualitatively the flow field turbulent field, pressure field and the liquid film in Fig.7, Fig.8, Fig.9, Fig.10, Fig.11, Fig.12, and Fig.13.

The Fig.14 shows that these are differences between the experimental results and the numerical results. The reason is mainly because that the actual flow in the entrance of the mixing chamber is elliptical other than parabolic.

5   CONCLUSIONS

The three-dimensional two phase flow field in the mixing chamber of y-type burner has been presented. The discussion of the predictions indicates that:
1) New interface function can be used to consider the entrance and wave in the interface between gas and liquid.
2) The improved program can be used to calculate qualitatively the flow field in the mixing chamber of y-type burner.

Further works need to be done at least in the following aspects.

1) To further improve the program for calculating the three dimensional two-phase elliptical flow field in mixing chamber of y-type burner.
2) To consider the droplets inside the gas phase.

6   NOMENCLATURE

u, velocity [m/s]

k, kinetic energy [J]

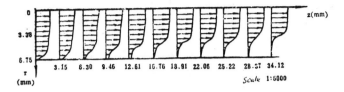

Fig. 9. The axial velocity distribution
located on a vertical diametral
plane.

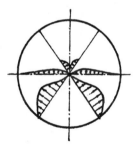

Fig. 10. The sketch map of the radial
velocity at several different angle
in a cross section.

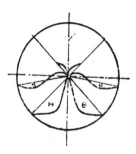

Fig. 11. The sketch map of the angular
velocity at several different
angle in a cross section.

441

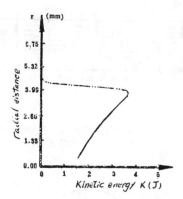

Fig. 12. The kinetic energy distribution in
radical direction at certain
angle in a cross section

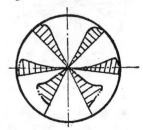

Fig. 13. The sketch map of the dissipation
rate distribution in radial direction at several different angles
in a cross section

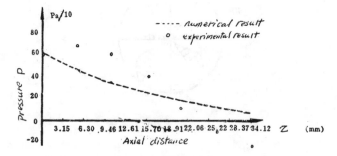

Fig. 14. The comparison between the experimental results and the numerical
results about the pressure in
axial direction.

$\rho,$ local density $[kg/m^3]$

$\tau,$ shearing stresses $[N/m^2]$

V_f^* friction velocity $[m/s]$

$\sigma,$ surface tension $[N/m]$

$G,$ volumetric flow rate $[m^3/s]$

$R,$ radius of the mixing chamber $[m]$

$R,$ radius of the oil pipe $[m]$

$a,$ radius of core $[cm]$

$b,$ pipe radius $[cm]$

$\varepsilon,$ dissipetion rate

$\mu,$ viscosity $[Pa.s]$

$\nu,$ kinematic viscosity $[m^2/s]$

$P,$ pressure $[Pa]$

$h,$ film thinckness $[m]$

$\vec{V},$ velocity vector $[m/s]$

$k,$ constant $k = 0.4$

$Y,$ distance from the interface

Subscripts

$G,$ gas

$L,$ liquid

$Z,$ axial direction

$r,$ radial direction

$\theta,$ angular direction

$e,$ effective

$l,$ liquid

$S,$ the interface

Superscripts

g,   gas

l,   liquid

Acknowledgements

This work was carried out with the financial support from the scientific research fund of Shanghai China.

REFERNCES

1). Ai Jun, Zhang Songshou, Huang Weimin
     A Numerical Study Of The Mixing Chamber In Y-jet (I).
     Journal of Shanghai Institute Of Mechanical Engineering
     Vol. 9. No.9. Sep. 1987 pp 19-30.
2). S. Levy,
     Prediction of Two-phase Annular Flow with Liquid Entrainment
Int.J. Heat Mass Transfer, Vol.9. pp 171-187, 1966

A New Gas Burner with a Swirling Central Flame

A. TAMIR and I. ELPERIN
Department of Chemical Engineering
Ben Gurion University of the Negev
Beer Sheva, Israel

Abstract

A new gas burner generating a swirling central flame has been described. The application of a swirling flame in the burner enhances considerably the heat transfer between the combustion products and a heated vessel. The thermal efficiency of the new burner is higher by 10 to 30% than that of the conventional one. The effect of design parameters on the performance characteristics of the burner under various operating conditions is discussed.

1. INTRODUCTION

Domestic gas hotplates equipped with conventional burners are widely used in most households due to their well-known merits such as: simplicity, low cost of manufacture and operation, wide range of available operating heating speeds, reliability, etc. However, along with these advantages, gas hotplates have a relatively low thermal efficiency of burning, which is of the order of 55-60%. The thermal efficiency is defined as the ratio between the heat absorbed by a vessel being heated and the heat of combustion of the burned gaseous fuel. In order to improve the thermal efficiency of burners, a unique construction of the burner was developed [1] as shown in Fig.1. The thermal efficiency of the new burner was found higher by 10-30% as compared to that of a conventional burner, due to the implementation of a swirling central flame. The swirl flame enhances the heat transfer between the combustion products and a heated vessel and reduces the heat losses.

2. DESCRIPTION OF THE NEW BURNER.

The essence of the new burner lies in the implementation of a rotating vertical flame located in the center of the burner rather than a band-form flame generated by conventional gas burners. The rotating central flame is generated by feeding of gas-primary air mixture through ports in the burner's cap. The axes of the ports are oriented (Fig.1) towards the center of the burner at an angle β with respect to the cap's horizontal plane. Projections of these axes on a horizontal plane are tangential to the surface of a central vertical flame with the angle α (Fig.1) to the radius of the burner's cap. The jets of the ignited gas-primary air mixture flowing out of the

445

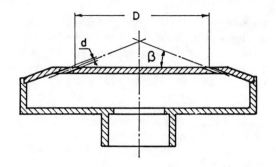

A-A' CROSS- SECTION SIDE VIEW

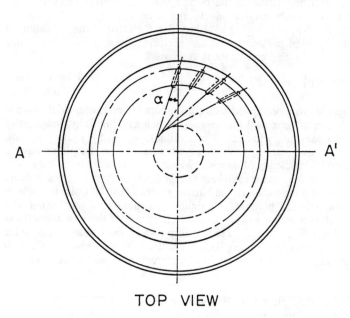

TOP VIEW

Fig. I Scheme of the new gas burner

cap's ports generate the vertical rotating flame over the burner's cap. This swirling flame, which is impinging on the bottom of the heated vessel, forms the rotating radially diverging jet of hot gases.

The new burner is characterized by the following design parameters shown in Fig.1: α - the swirl angle between the projection of the ports axes and the radius of the burner's cap plane; β - the angle between the axes of the burner's ports and the cap's horizontal plane; D - the distance between two opposite ports along the burner's diameter; d - the diameter of a port; n - the number of ports. The distance H between the burner's cap and the heated vessel's basis also affects the burner's performance.

The experiments resulted in the optimum combination of these parameters which yield the highest thermal efficiency for combustion of synthetic and natural gases. The performance characteristics of the new burner were compared with those of conventional burners.

3. EXPERIMENTAL PROCEDURE

Hot plates for synthetic gas (a mixture of propane and butane) and natural gas (mainly methane) were employed. Thermal efficiencies were determined by procedures outlined in Israeli (for synthetic gas) and British (for natural gas) Standards for domestic cooking appliances burning gas fuel. The process of the heat exchange between the burned gaseous fuel and the vessel being heated is associated with the following heat losses: Q_2 - heat losses with flue gases; Q_3 - heat losses caused by chemically incomplete combustion; Q_4 - heat losses caused by convective and radiative heat exchange between the hot combustion products and the surroundings. Thus, if Q_1 is the amount of heat absorbed by a heated vessel, the heat balance for the gas burner will read:

$$Q_1 + Q_2 + Q_3 + Q_4 = Q \qquad (1)$$

where Q is the heat of combustion of the gaseous fuel. Thus the burner's thermal efficiency η is defined by:

$$\eta = Q_1 / Q \qquad (2)$$

For the test presented in this work, heat losses in combustion and heat exchange processes were determined by the procedure described in ref. [2].

4. RESULTS AND DISCUSSION

Some results of experiments [2-4] performed during the investigation of the characteristics of the new burner's performance are presented below.

Effect of Gas Flow-Rate on Thermal Efficiency.

Fig.2 demonstrates the variation in the thermal efficiency against the input flow rate of natural gas for a small vessel (finjan) with a basis diameter of 11.5 cm. Since it was noticed that the construction of the pan support affects the thermal efficiency, the original pan support was modified (Fig.2, top right) in order to attain the maximal possible thermal efficiency. The major observations in Fig.2 are that the new burner increases the efficiency by approx. 25% and that the pan support improves the efficiency by about 12%. In addition, a characteristic behaviour is revealed, namely when the gas flow-rate is increased, the burner's efficiency increases until reaching the optimum range. An additional increase in the gas input rate will cause the thermal efficiency to decrease. An explanation for this variation in thermal efficiency is as follows. At relatively low gas input rates the heat transfer coefficient between the combustion gases and the bottom of the vessel is relatively small because of the low velocity of the gas flow. The process of heat transfer is enhanced by raising the gas input rate until reaching optimal conditions, i.e., when the thermal efficiency reaches its maximal values. From this point on, the surface area of the vessel's bottom which is available for heat transfer becomes the controlling factor for some definite conditions (temperature, heated mass). In other words, the rate of heat losses with the flue gases grows faster than the heat absorption rate by the vessel. Thus, the efficiency decreases by increasing gas flow rate.

Effect of the Swirl Angle α

The variation of the thermal efficiency as a function of the swirl angle α (Fig.1) for two different input rates of natural gas is shown in Fig.3. The rotation of the flame is caused by the tangential component of the gas-air jets velocity which is increased by increasing α. Consequently, the residence time of the hot combustion products in the vicinity of the vessel's bottom is increased, which improves the thermal efficiency of the burner. The pressure drop caused by the flame vortex due to rotation, promotes suction of secondary air into the combustion zone, where a larger amount of secondary air improves combustion by reducing heat losses because of chemically incomplete combustion. However, secondary air at ambient temperature reduces the temperature of the hot gases in the flame. The existence of an optimal value of α, where a maximal thermal efficiency is achieved (Fig.3), takes into account the above mentioned counteracting effects.

Effect of the Angle β.

Fig.4 demonstrates the effect of the angle β (Fig.1) on the thermal efficiency of the new burner for synthetic gas. As observed, two optimal ranges exist where the thermal efficiency reaches a maximum value. This can be explained as follows. The angle β defines the residence time of the gas-air mixture in the reaction zone.

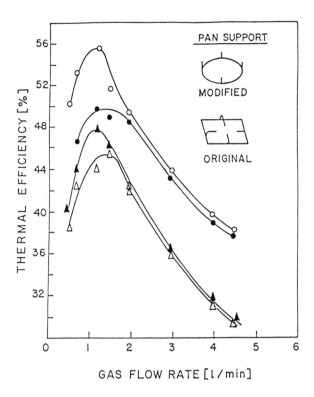

FIG.2 EFFECT OF GAS FLOW RATE ON THERMAL EFFICIENCY OF BURNERS TESTED
ON FINJAN

o - NEW BURNER WITH MODIFIED PAN SUPPORT

● - NEW BURNER WITH ORIGINAL PAN SUPPORT

▲ - CONVENTIONAL BURNER WITH MODIFIED PAN SUPPORT

△ - CONVENTIONAL BURNER WITH ORIGINAL PAN SUPPORT

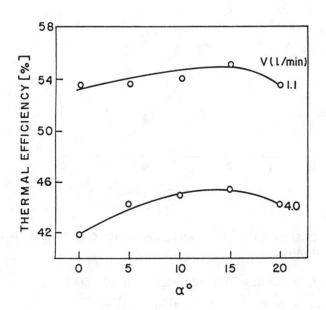

Fig.3 Effect of angle α on thermal efficiency

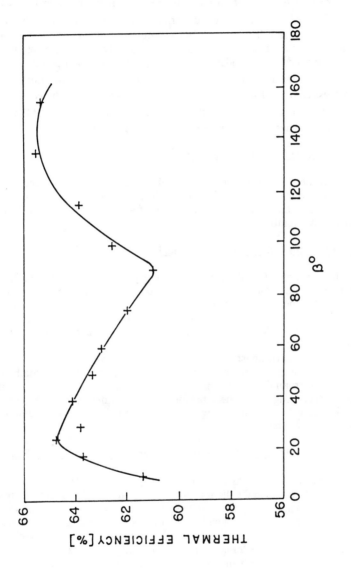

Fig.4 Effect of angle β on thermal efficiency

451

for small values of β ($\beta < 90°$), the residence time is longer since the mixture flows first towards the burner's center and then spreads out of the reaction zone over the basis of the heated vessel. It may then be expected that heat losses caused by chemically incomplete combustion will be smaller than for $\beta > 90°$. When $\beta > 90°$, the flame resembles the band-form flame of the conventional burner but swirls. However, the larger is the angle β, the easier is for the secondary air to be sucked and diffused into the reaction zone of the swirling flame. Thus, on the one hand, more air results in minor heat losses because of chemically incomplete combustion but, on the other hand, this secondary air cools the hot combustion products. Thus, for $\beta < 90°$, longer residence time for the gas mixture in the combustion zone is achieved but less amount of secondary air reaches the reaction zone; however, for $\beta > 90°$, the opposite is true. Consequently, two optimal ranges of values of β may be anticipated (Fig.4) where the thermal efficiency is the highest.

Effect of the Distance D Between Two Opposite Ports.

The experimental results presented in Fig.5 demonstrate the effect of the distance D (Fig.1) on the thermal efficiency of the burner for synthetic gas. As can be seen, a change in the distance D in the range of 40-80 mm does not affect the thermal efficiency significantly.

Effect of the Burner's Number of Ports, n and Their Diameter, d.

The effect of n and d on the thermal efficiency for two flow rates of natural gas shown in Fig.6 is similar, namely, increasing both n and d results in an increase in the thermal efficiency . An increase in n or d increases the cross-sectional area of ports available for flow of gas-primary air mixture and therefore decelerates the velocity of the mixture. Consequently, the mean residence time of mixture in the reaction zone is increased, thereby producing a decrease in heat losses because of chemically incomplete combustion. The reduction in the velocity of the mixture also diminishes heat losses caused by convective heat exchange between combustion products and the surroundings. Heat losses with flew gases are not affected by n or d [2]. Hence, according to the heat balance of gas burners (Eq.1) the thermal efficiency increases with increasing burner's number of ports,n, and their diameter,d.

Effect of the Distance H Between the Burner's Cap and the Heated Vessel's Basis.

The above effect is demonstrated in Fig.5. It was concluded that there is an optimum distance H at which the thermal efficiency attains its maximum because of the following reasons. For very small values of H, the residence time of gas in the combustion zone is relatively small and hence, the chemical reaction of combustion is not completed. The flame is also quenching as it approaches the relatively cold bottom of the vessel and the temperature of flame does not reach its maximum value. On the other hand, large values of H cause the combustion production to cool before reaching the pan's bottom.

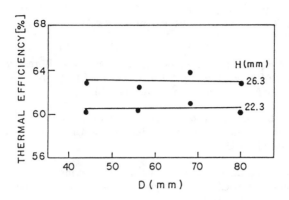

Fig.5 Effect of distance D betwen two opposite ports on
thermal efficiency

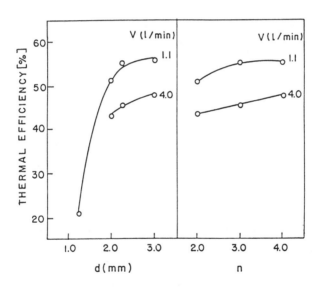

Fig.6 Effect of ports diameter d and number
of ports n on thermal efficiency

REFERENCES

1   Elperin, I.T., and Tamir,A., "Burner for Gaseous Fuel", Israeli Patent No.66538 (1985), United States Patent No.4,583,941 (1986), and other important countries.

2   Tamir, A and Elperin, I., "Development of a Gas Range Burner for Natural Gas". Research Report, Department of Chemical Engineering, Ben Gurion University of the Negev, Beer Sheva, Israel (1985).

3   Yotzer, S., "Improvement of Household Gas Stoves", M.Sc. Thesis, Department of Chemical Engineering, Ben Gurion University of the Negev, Beer Sheva, Israel (1983).

4   Tamir, A and Elperin, I., "Semi-Industrial Gas Saving Burner", Research Report, Department of Chemical Engineering, Ben Gurion University of the Negev, Beer Sheva, Israel (1986).

Combustion Characteristics of Pulverized Coal in a Free Jet

H. A. MONEIB
Department of Mechanical Power
Engineering
Helwan University
Cairo, Egypt

Y. H. EL-BANHAWY
Department of Mechanical Power
Engineering
Ain Shams University
Cairo, Egypt

H. A. KHATER
Department of Mechanical Power Engineering
Cairo University
Cairo, Egypt

Abstract

The present paper describes temperature measurements in a pulverized coal/gas free reacting jet. The experimental set up comprises a coal feeding system, air and gas supply systems and a gas/coal burner. Temperature measurements are carried out for two load ratios, 24% and 34% and the results are compared against pure gas flame measurements. The results showed that the coal/gas flame could be divided into three distinct zones (i) preheating zone where coal particles are heated up by the gas flame, (ii) devolatilization zone where coal devolatilization and combustion take place and (iii) flame tail zone where radiative heat loss dominates the flame characteristics. The results also indicate that temperature levels are generally lower in the coal/gas flames than in the pure gas flame.

1. INTRODUCTION

The ever growing energy demand , and the desire, from an economical point of view, to use national energy resources, have prompted the government of Egypt to consider the use of coal-fired power stations. Several plans are presently being considered to increase the number of power stations and boilers which utilize coal as the main fuel source. The present work has been motivated by this growing interest in the use of coal in such application and, in particular, the use of El-Maghara coal; a mine in the Sinai peninsula. Table 1 shows the properties of El-Maghara coal together with other Egyptian coal.

Coal-fired flames have several features which differ largely from the extensively used gaseous and liquid fuel flames. These include, for example, the basic flame structure, energy release pattern, ash formation and pollutants emission. Understanding these characteristics of coal-fired flames is therefore highly required prior to the use of coal flames in furnaces and boilers since this would provide the necessary data to control their application.

455

Table 1: Properties of Egyptian Coal.[*]

	BEDAA& THAWRA	O'YOUN MOUSSA	EL-MAGHARA
Calorific Value (kj/kg)	16534-19067	22311-26414	30432
Chemical Composition %			
-Fixed carbon	30.0-40.0	34.6-36.1	37.9
-Volatile matter	17.0-27.0	35.7-46.5	50.7
-Moisture content	2.0-3.8	7.2-12.4	4.9
-Ash	39.0-49.0	8.9-23.4	6.5
-Sulfur	0.4-1.0	1.3-4.9	2.97
	-low rank -high ash -high volatiles	-bituminous -high volatiles -low cocking properties	-homogenous -Sub-bitomi-nous -high volatiles -has cocking properties

* Egyptian Organization of Mining.

456

The present work investigates the dual fuel (coal/gas) free jet diffusion flame, which simulates the primary jet stream in a coal-fired furnace. The idea of using coal as a dual fuel stems from two considerations: (1) it is a way of optimizing the use of the available resources of natural gas in Egypt. The injection of fine coal particles in a gaseous flame zone increases the radiative heat transfer and thus increases the furnace capacity at reduced running cost, and (2) the burning of a gas-coal mixture in furnaces partially eliminates many problems in coal burning such as complexities of furnace design, coal flame stabilization, fouling and slagging, and ash removal problems.

2. LITERATURE REVIEW

Considerable effort has been expended in studying the mixing process in cold flow particle-laden jets, however basic investigations in reacting particle-laden free jets are lacking.

For cold flow jets, the surveyed experimental studies have investigated the mixing characteristics of the gaseous and particulate phases of a central primary jet, where particles are injected, as it mixes with a clear gaseous secondary jet. Many of the experimental studies, e.g. Refs. [1-4] have presented measurements for the radial profiles of gas composition, gas velocity and particle mass flux across the mixing zone at various axial positions downstream from the jet exit. Three classifications of test sections have been used extensively, namely the parallel injection system, the angular secondary injection system and the angular secondary injection system with swirl.

Many parameters affect the mixing phenomena, the most significant are: particle type, particle size, primary jet temperature, primary jet velocity, secondary jet temperature, secondary jet velocity, secondary jet injection angle, secondary to primary gas flow ratio, secondary to primary velocity ratio, secondary to primary density ratio, mixing chamber diameter and primary jet solids loading level. Table 2 lists the range of the significant parameters in the cold flow tests.

Comparing the data for silicon powder and pulverized coal having the same particle size , see Ref. [1], shows that the initial gas mixing rate is slightly higher for coal particles (about 25% higher) in both parallel and angular secondary injection systems. However the increase is more pronounced (66%) when an expanded mixing chamber is used with angular secondary injection. The effect of particle type on initial particle mixing rate is less significant being less than 20% for the former case. The changes in the overall mixing length for complete mixing of both the gas and particles are generally very small being within the uncertainity of the data.

Table 2: Cold flow particle-laden jets test parameters.

| | PARTICLES | | | PRIMARY JET | | | SECONDARY JET | | | SECONDARY/PRIMARY RATIO | | | NOMINAL PRIMARY SOLIDS LOADING LEVEL (%wt) | MASS FLOW RATE (kg/s) | | | | DIMENSIONS | | |
REF	TYPE	SIZE (μm)	GAS	TEMP (K)	VEL (m/s)	TEMP (K)	VEL (m/s)	ANGLE (deg)	GAS	VELOCITY	DENSITY		PRIMARY AIR	PRIMARY HELIUM OR ARGON	PRIMARY PARTICLES	SECONDARY AIR	PRIMARY JET DIAM (mm)	SECONDARY JET DIAM (mm)	MIXING CHAMBER DIAM (mm)
[1]	Aluminum	6 & 30	Air & Helium	284 & 487	300	288	30 & 122	0	2.65 to 42.8	0.103 to 0.436	1.12 to 4.66	20	.0316 & .154	.0024 & .01	0.01 & 0.04	.425 & 1.80	25.4	130	130
[2]	Silicon & Pulverized Coal	19 to 54.1	Air & Argon	283	30.5	283 & 370	38.1 & 61.0	30	15.5 to 32.0	1.3 to 2.0	0.6 to 0.8	0, 40 & 60	.0054	.0174	0.0152 & 0.0342	.353 to .730	25.4	130	130
[3]	Silicon	38.6 to 54.1	Air & Argon	283 & 360	30.5	283	25 to 61.0	0	24.6 & 39.3	1.25 & 2.0	0.79	0, 40 & 60	.0052	.0169	0, 0.0147 & 0.0331	.54 & .870	25.4	130	206 & 343
[4]	Silicon & Pulverized Coal	24, 46 & 48	Air & Argon	280	30.5	280	38.1 & 61.0	30	26.5 & 42.6	1.25 & 2.0	0.79	40	.0053	.0143	0.0146	.520 & .835	25.4	130	260
[10]	Sand	180 to 1200	Air	280	6 to 9.6	—	—	—	—	—	—	114 to 350	.00362 to .00419		.00413 to .015	—	15	—	—

The use of smaller particles , as shown in Ref. [2], reduces the drag effect and tends to make the two-phase mixture more homogeneous thus increasing the initial mixing rate of the particles, while having a small effect on the initial gas mixing rate and overall length required for complete mixing of both the gas and particles.

In practice the secondary jet temperature can be varied somewhat for the purpose of improving the combustion efficiency; the maximum temperature is limited by material problems. Varying the secondary gas density (by changing its temperature) in non-parallel systems does not significantly affect the mixing rate of either the particles or the gas [2]. On the other hand, lowering the gas density by 25% in parallel systems decreases the gas and particle mixing rates by 25 to 35%.

Increasing the secondary velocity with angular injection [3] generally increases the completely mixed core lengthes for both the gas and particles, indicating a reduction in overall gas and particles mixing rates. In both non-expanded mixing chambers, the initial gas mixing rate is incraesed by increasing the secondary velocity. The initial particle mixing rate is similarly affected but to a lesser degree. Comparing the mixing rate data for parallel systems and systems with 30 secondary jet injection angle [2], shows enhancements in both the gas and particle mixing rates using the latter system. However, little effect is shown on the overall length required for complete mixing.

In addition, it has been observed [2] that for parallel injection systems particle mixing rate is practically independent of gas mixing rate. However, a strong dependence has been observed for non-parallel injection systems.

Incraesing the particle solids loading level of the primary jet (within the range 0-60%) decreases the gas mixing rate [2], however this decrease is small and can be considered negligible. In fact the effect is also negligible on the particle mixing rate. In practice, typical pulverized coal gasifiers and furnaces use solids loading levels between 40 to 90% by weight. The available test data do not cover the upper limit where the effect may be significant and may lead to quite different results.

The use of expanded mixing chamber [4] causes flow recirculation downstream the exit plane of the primary jet and hence it improves the gas mixing rate and more dramatically it increases the particle mixing rates. However, increasing the mixing chamber diameter beyond a certain size, when recirculation flow is established, does not cause practically any further increase in the mixing rates.

The structure of turbulent particle-laden jets has been studied by many investigators. Refs. [5] through [10] provide a good review of such jets.

When the mixing is accompanied by combustion, see Ref.[11], the flow temperature rises rapidly in the mixing zone of the jet which in turn causes the density to decrease practically to less than one-third its cold value. In addition, the pressure is reduced, while the velocity gradients increase causing an increase in the turbulence level in comparison with cold flow mixing, and hence it improves the mixing rates.

A number of simple analytical approaches appeared in the literature [11, 12]. These approaches applied the concept of eddy viscosity to study the effect of temperature rise. Generally, these methods agree that an increase in temperature will increase the eddy viscosity which in turn will increase both the jet angle and the rate of entrainment. However, it appears that there is no conclusive experimental evidence that the jet angles of spread in cold flow jet and in burning jet, under the same conditions, are different. Thus many researchers assume that the angles are practically the same. this is also true for the axial decay constants for velocity and concentration. In addition to the change in density due to temperture rise other changes are possible due to chemical reaction in the flame, for example the presence of volatile matter and the delayed combustion of carbon monoxide.

Non-swirled burners have been used in the past by many investigators [13- 16] for studies on the combustion of pulverized coal. The majority of the pulverized coal burning combustors constructed in recent years have used the swirled burners [17-20]. With the application of swirl, high combustion intensities and wide operating limits can be acheived. Syred and Beer [21] have discussed the various salient features of swirl burners. Refs. [22-24] showed that the process of flame stabilization varies widely with the design of the swirl buner, which naturally affects the stability. Other interesting material in pulverized coal combustion can be found in Refs. [25] through [28].

In the following sections a description of the present reacting free-jet experimental facility, the measurement techniques and the results obtained are presented.

2. EXPERIMENTAL

The experimental set-up comprises a gas-coal burner , a screw feeder, mixing sections, an air line and a gas line, as shown in Fig. 1. Pulverized coal is filled continuously from a storage hopper into the main hopper. A screw feeder having a variable speed motor feeds the coal into a vibrating duct. The coal flowing out of the vibrating duct drops into a mixing section where it is entrained by the primary combustion air; supplied from a laboratory compressor. The coal-air mixture flows then into another mixing section where it mixes with the secondary gaseous fuel (butane) which is introduced through a small tube of 12 mm diameter which has a closed end and a number of small

Figure (1) Experimental Set-Up

461

diameter holes, 1.5 mm each, along its periphery. The fuel jet issuing from the small diameter holes allow a better fuel-air mixing pattern. The mixture then flows to the gas-coal burner.

The gas-coal burner, shown in Fig. 2, consists mainly of two concentric tubes of 12.6 and 22 mm internal diameters. The coal-gas mixture flows through the central tube while a stabilizing fuel (hydrogen) flows through the annulus between the two tubes. The latter provides a pilot flame to stabilize the main coal-gas diffusion flame at very high Reynolds number. It is not , however, being used herein to eliminate the effect of hydrogen burning at the early stage of flame development. Small lift-off distance is however being anticipated.

The above design offers enormous flexibility in varying controllable parameters such as premixed air flow rate, gas flow rate and coal flow rate.

The flow rates of both the gas and the air are measured using calibrated orifice plates, whereas the pulverized coal flow rate is measured by measuring the weight of coal collected during a given period of time. Platinum/Platinum 13%-Rhodium fine wire thermocouples (40 um) are used to measure the flame temperature. A special traversing mechanism which enables movement in three directions has been designed and used for temperature measure-ments. Three experimental runs were made, one with pure gas and the others using pulverized coal/gas mixture at different load ratios. The coal had a median size of 160um. Figure 3 shows a representative size distribution curve. Table 3 shows the relevant test parameters.

3. RESULTS AND DISCUSSION

Figure 4 shows the axial temperature distribution for the gas and coal/gas flames with different load ratios. The temperature distribution along the axis of the pure gas flame agrees broadly with previous lifted flame measurements (see Refs. [29] and [30]). The lift-off region was estimated at about 5 buner diameters and no measurements were made within this distance. Along the centre line between X/D=5 and X/D=20 (X is the axial distance from burner outlet and D is its diameter) the jet is characterized by partial premixing and relatively slow reaction rate as revealed by the low temperature gradient shown in this region (Ref [30]). This region is followed by the main reaction zone where rapid increase in temperature is seen to occur and the flame is behaving in a diffusion-like manner.

Axial temperature distribution of the coal/gas flames may be divided into three different zones according to the rate of temperature increase as well as temperature levels as shown in Figure 4.

Zone (1) between the burner exit and nearly X/D =15. The temperature distribution in this region is markedly lower than

462

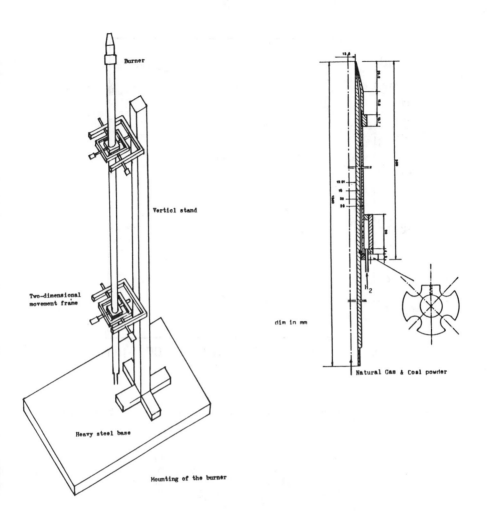

Figure (2) Details of Gas/Coal Burner

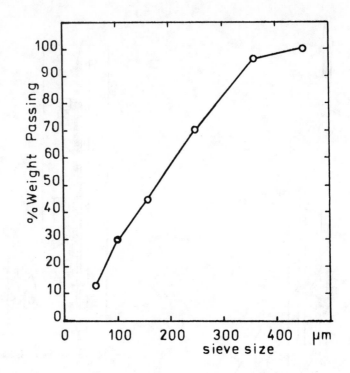

Figure (3)   A Representative Size Distribution of the Present
Pulverized Coal

Table 3: Summary of Test Parameters

Run No.	Gas Flow Rate (kg/hr)	Coal Flow Rate (kg/hr)	Loading Ratio %	Air Flow Rate (kg/hr)	Particle Size (um)
1	3.683	--	--	2.09	--
2	2.824	0.9	24.16	2.09	160
3	2.446	1.29	34.5	2.09	160

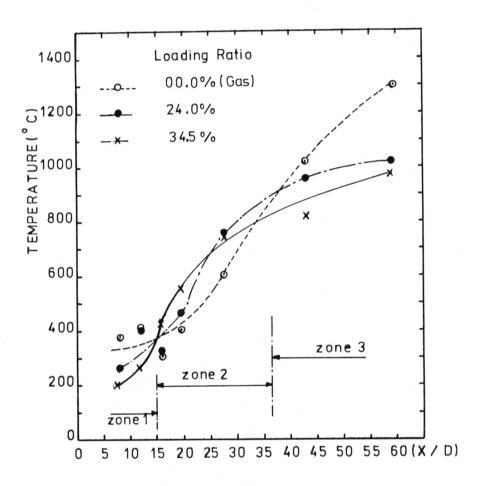

Figure (4) Axial Distribution of Mean Temperature for the Pure Gas and Gas/Coal Flames

that for the pure gas flame. This may be attributed to the energy absorbed by the coal particles. This is supported by the higher temperature drop for the higher load ratios.

Increasing the load ratio results in an increase in the input axial momentum and hence the lift-off distance. Accordingly and due to the higher rate of air entrainment, better partial premixing occurs at the flame base resulting in steeper temperature gradients for higher load ratios. Two contradicting effects may be descerned in this zone; one effect is the larger amount of heat consumed in preheating the particles, the second is the higher heat release due to the better premixing. The temperature distribution shown in Figure 4 suggests that the latter effect surpasses the former. Zone (1) is limited at its upper end at roughly 400 oC, a temperature at which devolatilization of coal particles begins [31].

Zone (2) is characterized by relatively steeper temperature gradients as compared to the pure gas case and extends between X/D=15 and X/D=40.

In this zone rapid devolatilization and volatile combustion take place leading to the observed higher temperatures. The higher load ratio is also seen to yield higher temperatures as expected due to the higher volatile release. Near the end of this zone which is roughly at 800oC devolatilization rate diminishes before the temperature is high enough (=1000oC, Ref. [31])for char to burn. Thus temperature gradients for both load ratios become relatively lower.

In zone (3) volatile release and combustion continues at still diminishing rate and temperatures are seen to drop below those for the pure gas flame. This is believed to be due to the higher radiative heat loss. This is supported by the fact that as the load ratio increases the temperatures become lower due to the higher luminosity of the flame. It is also beleived that a char burn-out analysis in this region would indicate very low char combustion.

Radial profiles of temperature are shown in panels a,b and c of Figures 5 through 7 and assembled as selected contours for the three cases considered in Figures 8 through 10. Figure 8 shows higher rate of spread as compared to Figures 9 and 10 for the coal/gas flames. This agrees well with the findings of Modarress [32] who reported higher spread rates for single phase jet as compared to two-phase jets. However the reaction zone of the gas/coal flames seem to be thicker at their bases. This may be attributed to the volatile combustion and diffusion across the flame envelope.

From visusl observations the flame length was found to increase, as expected, with the load ratio. For the 24% and 34% load ratios, the flames were nearly 31% and 37% respectively

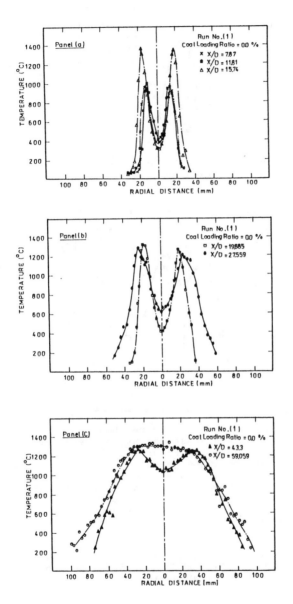

Figure (5) Radial Distribution of Mean Temperature,run(1)
Gaseous Flame

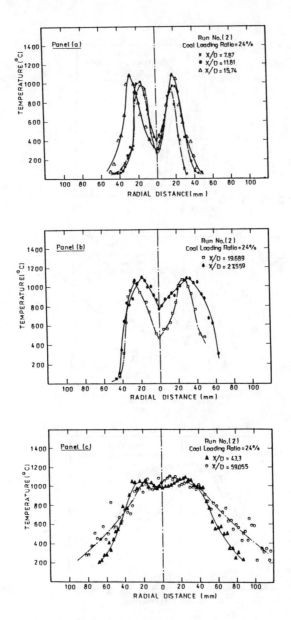

Figure (6) Radial Distribution of Mean Temperature,run(2)
Gas/Coal Flame,Loading ratio = 24%

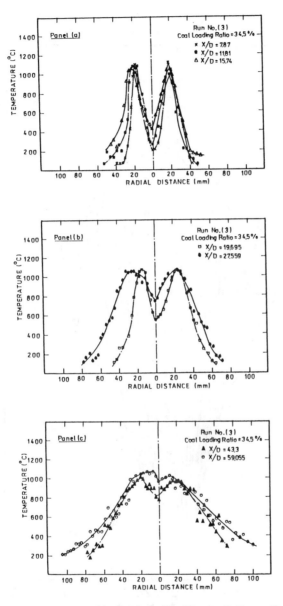

Figure (7) Radial Distribution of Mean Temperature,run (3)
Gas/Coal Flame, Loading ratio = 34.5%

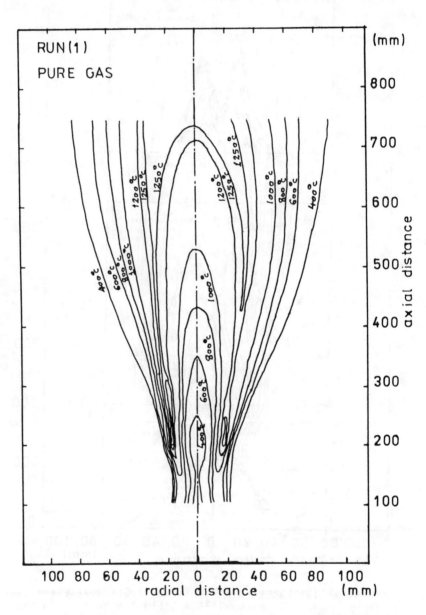

Figure (8) Isothermal Contours,run(1),Gaseous Flame

470

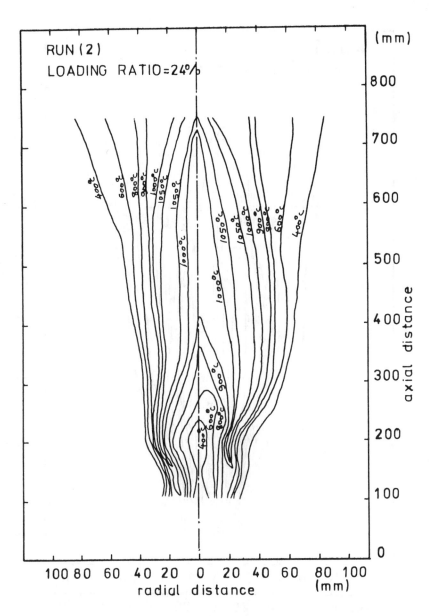

Figure (9) Isothermal Contours,run(2) Gas/Coal Flame
Loading ratio = 24%

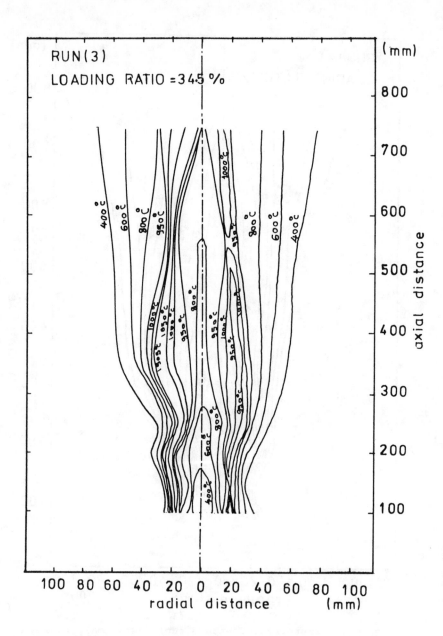

Figure (10) Isothermal Contours,run(3),Gas/Coal Flame
Loading ratio = 34.5%

longer than the pure gas flame. It is, however, seen that the load ratio in this range seem to have little effect on the flame spread rate. The level of the flame envelope contours for the gas flame are seen to be nearly 20% higher than the coal/gas flames. The contours also show that the volume occupied by the reaction zone in the former flame is appreciably larger than that for the latter. This is attributed to the facts that the rate of radial diffusion of the coal particles is lower than that for the gas phase [32] and that the velocity fluctuations (and hence mixing) for the coal/gas flame are smaller than those for the gas flame.

4. CONCLUSIONS

Studies of the mixing pattern and turbulence characteristics of two-phase flows is of great importance in many engineering applications including entrained gasifiers and pulverized coal combustors.

In this paper pulverized coal combustion in a free jet of premixed gas and air has been considered. The experimental set up comprises a gas/coal burner, a coal feeding system, an air supply system, a coal-air mixing section and a gas-(coal/gas) mixing section. It is believed that this simple configuration- apart from being a method of burning coal- is of relevance in:

(i)   providing some basic fundamental understanding of coal combustion.
(ii)   simulating the central primary jet stream in double concentric burners, being widely used in most combustors.
(iii) providing data applicable to the intermediate burning stage that follows gaseous preheating stage of coal fired combustors before turning to pure coal combustion.
(iv)   providing data for validating numerical prediction studies.

At this stage of the experimental program, measurements include flame temperature distribution in both the radial and axial directions, and flame length.

The flame is characterized by three distinct zones explaining coal preheating, coal devoltilization and combustion, and luminous zone. Increasing the load ratio results in a decrease in jet spread rate, increase of flame length and generally lower temperature levels.

Acknowledgements

This project is sponsored by the Foreign Relations Coordination Unit of the Supreme Council of Universities- grant no. FRCU 851010.

473

REFERENCES

1. Hedman, P.O. and Douglas Smoot, L.,"Particle-Gas Dispersion Effects in Confined Coaxial Jets", AIChE Journal, Vol. 21, No. 2, pp. 11-18, March, 1975.

2. Memmott, V.J and Douglas Smoot, L.,"Cold Flow Mixing Rate Data for Pulverized Coal Reactors" AIChE Journal, Vol. 24, No. 3, pp. 466-473, May, 1975.

3. Tice, C.L. and Douglas Smoot, L.,"Cold Flow Mixing Rates with Recirculation for Pulverized Coal Reactors", AIChE Journal, Vol. 24, No. 6, November, 1978.

4. Hedman, P.O, Leavitt, D.R. and Sharp, J.L.,"Cold Flow Mixing Rates in Confined Recirculating Coaxial Jets with Angular Secondary Injection", AIChE Journal, Vol. 31, No. 7, pp. 1105-1112, July, 1985.

5. Shuen, J.S, Chen, L.D. and Faeth, G.M., "Predictions of the Structure of Turbulent, Particle-Laden Round Jets", AIAA Journal, Vol. 21, No. 11, pp. 1483-1484, Nov. 1983.

6. Al-Taweel, A.M. and Landen, J., "Turbulence Modulation in Two-Phase Jets", Int. J. Multiphase Flow, Vol. 3, pp. 341-351, 1977.

7. Shuen, J.S., Solomon, S.P., Zhang, Q-F and Faeth, G.," Structure of Particle-Laden Jets: Measurements and Predictions", AIAA Journal, Vol. 23, No. 3, 1985.

8. Modarress, D., Tan, H. and Elghobashi, S.," Two-Component LDA Measurements in a Two-Phase Turbulent Jet", AIAA Journal, Vol. 22, No. 5, May 1984.

9. Hardalupas, Y., Taylor, A.M.K.P. and Whitelaw, J.H.," Two-Phase Annular Jet Flow", 9th Ausralasian Fluid Mechanics Conference, Auckland, pp. 238-2242, December 1986.

10. Levy, Y. and Lockwood, F.C., "Velocity Measurements in a Particle-Laden Turbulent Free Jet", Combustion and Flame, vol. 40, pp. 333-339, 1981.

11. Field, M.A, Gill, D.W., Morgan, B.B. and Hawksley, P.G.W.,"Combustion of Pulverized Coal", The British Coal Utilization Research Association, Leatherhead, 1967.

12. Thurlow, G.G.,"Research on Pulverized-Fuel Flames by the International Flame Research Foundation: 1. Summary of Work Carried Out", J. of the Institute of Fuel, pp. 366-367, 1960.

13. Loison, R. and Kissel, R.R.,"International Flame Research Foundation: Observations on Experiments Carried Out on the

Combustion of Pulverized Fuel", J. of the Institute of Fuel, pp. 60-73, 1962.

14. Beer, J.M., "International Flame Research Foundation: The Effect of Fineness and Recirculation on the Combustion of Low-Volatile Pulverized Coal", J. of the Institute of Fuel, pp. 286-313, 1964.

15. Chedaille, J. and Hemsath, K.,"Influence of Burner Type, Coal Fineness, the Proportion of Primary Air and Temperature on the Combustion of Pulverized Coal", IFRF Doc. No. F 32/AC/32, Jan. 1965.

16. Kurtzrock, R.C., Bienstock, D. and Field, J.H.,"A Laboratory-Scale Furnace Fired with Pulverized Coal", J. of the Institute of Fuel, pp. 55-61, 1963.

17. Hein, K. and Leuckel, W.,"Further Studies on the Effect of Swirl on Pulverized Coal Flames", IFRF Doc. No. F 32/A/40, 1969.

18. Breen, B.P. and Sotter, J.G." Reducing Inefficiency and Emissions of Large Steam Generators in the United States", Prog. Energy Combustion Sci, Vol. 4, pp. 201-220, 1978.

19. Harding, N.S., Douglass Smoot, L. and Hedman, P.O.,"Nitrogen Pollutant Formation in a Pulverized Coal Combustor: Effect of Secondary Stream Swirl", AIChE J., Vol. 28, pp. 573-580, 1982.

20. Hassan, M.A., Hirji, K.A., Lockwood, F.C. and Moneib, H.A.,"Measurements in a Pulverized Coal-fired Cylindrical Furnace", Experiments in Fluids, Vol. 3, pp. 153-159, 1985.

21. Syred, N. and Beer, J.M.,"Combustion in Swirling Flows- A Review", Combustion and Flame, Vol. 23, pp. 143- 201, 1974.

22. Rao, G.V.S.N. and Sriramulu, V.,"Investigations of a Coaxial Swirl Burner", Combustion and Flame, Vol. 34, pp. 203-207, 1979.

23. Truelove, J.S., Wall, T.F., Dixon, T.F. and Stewart, I.M.,"Flow, Mixing and Combustion within the Quarl of a Swirled, Pulverized-Coal Burner", 19th Symp. (Intl.) on Combustion, The Combustion Institute, pp. 1181-1187, 1982.

24. Dixon, T.F., Truelove, J.S. and Wall, T.F.,"Aerodynamic Studies on Swirled Coaxial Jets from Nozzles with Divergent Quarls", ASME J. Fluid Engineering, Vol. 105, p. 197, 1983.

25. Juniper, L.A. and Wall, T.F., "Combustion of Particles in a Large Pulverized Brown Coal Flame, Combustion and Flame, vol. 39, pp. 69-81, 1980.

475

26. Beer, J.M. and Chigier, N.A., "Stability and Combustion Intensity of Pulverized-Coal Flames- Effect of Swirl and Impingement, Journal of the Institute of Fuel, pp. 443-450, 1969.

27. Krazinski, J.L., Buckius, R.O. and Krier, H., "Coal Dust Flames: A Review and Development of a Model for Flame Propagation", Prog. Energy Combustion Sci., Vol. 5, pp. 31-71, 1979.

28. Csaba, J. and Leggett, A.D., "Prediction of the Temperature Distribution Along a Pulverized-Coal Flame", Institute of Fuels, pp. 440-448, 1964.

29. Moneib, H.A., "Experimental Study of the Fluctuatuion Temperature in Inert and Recting Turbulent Jets", Ph.D. Thesis, University of London, 1980.

30. Hassan, M.M.A., Lockwood, F.C. and Moneib, H.A., "Fluctuating Temperature and Mean Concentration Measurements in a Vertical Turbulent Free Jet Diffusion Flame", Italian Flame Days, June 1980.

31. Essenhigh, R.H., "Combustion and Flame Propagation in Coal Systems: A Review", 16th Symposium (Int'l) on Combustion, The Combustion Institute, p. 353., 1976.

32. Modarress, D., Wuerer, J. and Elghobashi, S.," An Experimental Study of a Turbulent Round Two-Phase Jet", AIAA paper 82-0964, St. Louis, Mo., 1982.

Combustion Processes in Impinging Streams

A. TAMIR, K. LUZZATTO, and A. ZIV
Department of Chemical Engineering
Ben-Gurion University of the Negev
Beer Sheva, Israel

Abstract

Combustion of gas and pulverized coal were carried out in a two-impinging-streams reactor.

For gas, a characteristic behavior was revealed depending on the ratio of gas/air. This behavior consists of three distinct and stable modes of operation as follows: In mode I, no combustion takes place along the path of the jets fed into the reactor and it begins after the impingement plane. In mode II, combustion begins roughly before the impingement plane of the two jets. In mode III, combustion takes place immediately at the entrance of the jets into the reactor. The above findings were modelled quantitatively on the basis of the behavior of the flame front and qualitatively on the basis of exothermic reactions in a mixed reactor.

For combustion of coal, a correlation based on a diffusion-convection model was obtained for prediction of the combustion efficiency as a function of available operational parameters.

1. INTRODUCTION

The technique of impinging streams is an unusually effective tool for performing various types of technological processes either for a single phase or for particle-gas suspensions. The essence of the method, shown in Fig. 1, consists of two coaxial counter-current streams of gas or gas-particle suspension flowing out from closely placed pipelines and brought into impingement. At the impingement zone, heat and mass transfer rates are intensified due to the following effects:
1) Significant increase in the relative velocity, v, between the particles of one stream, v_p, to the relative velocity of the opposite stream, v_a. The relative velocity is given by $v = v_p - (-v_a)$ where under extreme conditions of $v_p = v_a$ the relative velocity may reach twice the gas velocity, and hence it may be significantly increased. Thus, the gas side resistance to heat and mass transfer in systems where the controlling resistance is the external one, is decreased. 2) The mean residence-time of the particles may be increased in contrast to the case where penetration of particles from one stream to the other does not exist. 3) The break-up of liquid fuel droplets in a spray to smaller

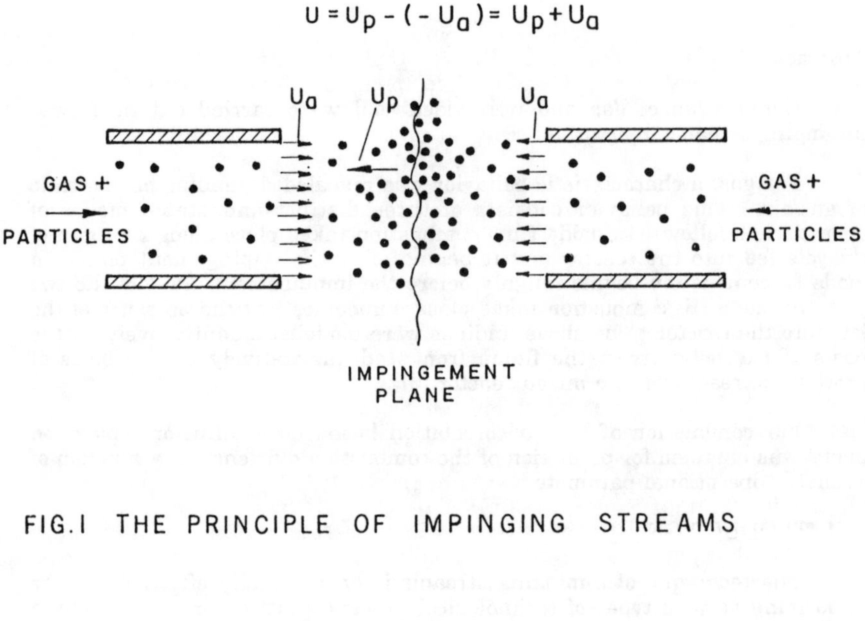

$$U = U_p - (-U_a) = U_p + U_a$$

FIG.I THE PRINCIPLE OF IMPINGING STREAMS

ones, due to shear forces between the continuous phase and the droplets or by interdroplet collisions. Thus an increase in the mass transfer rate is obtained. 4) The efficient mixing between the continuous phases in the impinging zone which homogenizes the temperature and the concentration in the system.

Research on reactors with impinging streams has aimed at demonstrating their superiority over other types of equipment, and at convincing potential users of their great efficiency. Within the framework of the research, two major configurations of the reactor were explored, viz., the tangential mode and the coaxial one. The tangential configuration was explored with respect to its hydrodynamics and residence time [1,2], drying of solids [3], mixing of solids [4,5], scale-up [6] and also for dissolution of solids [7]. In addition, four-impinging-streams and multistage-two-impinging-streams reactor [8] and recently, two impinging streams with primary and secondary air feeds were also thoroughly investigated.

The coaxial configuration of impinging jets was applied for absorption and desorption processes [9-12] and for the production of emulsions [13]. Recently, this configuration was applied to the drying of solids [14,15] and such reactors were also explored with respect to their hydrodynamics, residence-time distribution and scale-up rules for the above characteristics.

Developments until 1972 in the field of impinging streams appeared in Elperin's monograph [16], whereas the more recent research was summarized by Tamir and Kitron [17] and Tamir [18].

Applications of impinging streams in combustion were made with gaseous fuel or with suspensions of solid particles or liquid droplets in a gas stream or with their mixtures. The studies on gaseous fuel, which were summarized by Greenberg et al. [19], provide an understanding of the effect of various parameters, such as fuel/oxidizer concentration or transport properties, on flame location, flame temperature and extinction as well as ignition problems associated with such flames. Early investigations on combustion of liquid fuels in impinging streams in industrial combustors are reported by Elperin [16] and by Enyakim and Dvoretskii [20]. The behavior of the droplets in combustion were treated analytically by Enyakim [21,22], and recently by Greenberg et al. [19] who studied theoretically the effects of initial droplet and vapor mass fractions, droplet rate of vaporization on flame location, temperature and species concentration distribution for an n-decane fuel spray.

Combustion of pulverized coal is a very old process and intensification of its burning has been one of the main research and development goals; to achieve these, the method of impinging streams was applied. In the Koppers-Totzek coal gasification process, two burners are used to produce opposite jets of pulverized coal that impinge in the center of the combustion chamber. High relative velocities between the particulate matter and the gaseous phase in the central area provide good conditions for active diffusion and convection at the particle surface, and together with the high temperature result in fast burning and gasification reactions. Another application of the same principle was tested by Goldberg and Essenhigh [23], who obtained high combustion intensity of coal in a jet-mix stirred reactor, where two or four opposite jets were used (Fig. 2-b,d) to inject pulverized coal into the central area of a spherical cavity. A modification of the four jets configuration is obtained by firing the coal-air suspension where the four burners are arranged at each corner of a square furnace and directed towards an imaginary circle in the center of the furnace (Fig. 2e). This results in the formation of a large vortex which enhances combustion intensity. More developments were made by Goldman et al. [24] and recently by Hazanov et al. [25] who used a unique opposed-flow configuration for the combustion of pulverized coal, in which two streams of the suspension impinge at an incident angle of 45°, and where secondary air is supplied at the final stage of combustion, to improve mixing of fuel and air and the ability to control the motion of pulverized coal in the furnace.

The objective of the present investigation is to explore characteristic behaviors in combustion of gas and of coal in impinging streams. Two configurations of impinging streams are studied: on the first one, two identical streams of gas are tangentially fed into a tubular reactor (Fig. 2a) and flow along its walls where combustion also takes place under some circumstances. In the second configuration, the two opposed coaxial jets of gas, or of a suspension of pulverized coal in air, flow freely into the hollow of the reactor (Fig. 2b) and the combustion takes place in the center of the reactor with minimal contact with its walls.

2. EXPERIMENTAL

The two types of impinging-streams reactors were investigated as shown schematically in Fig. 2-a,b. The first combustor (2a) was constructed of Hastelloy C and characterized by tangential feed of the streams which behave as wall jets inside the reactor. Note also that the inlet pipes to the reactor were of rectangular shape. The experiments were carried out using a 30% - 70% (by volume) mixture of propane and butane respectively, premixed with air. The air feed passed through a gas feeding stage, comprising an ejector which provides the mild suction required for feeding gas to the main air stream, and avoids back-pressure effects at the feed point. When operating the reactor with tangential feed and non-adiabatically, the actual behavior of the gas streams and of the combustion zone were visible on the reactor walls. For more details the attention of the reader is directed to the work of Luzzatto [26].

The second combustor, shown schematically in Fig. 2b, is a coaxial feed type combustor, in which the streams which form two free jets inside the reactor impinge at the center of the combustor space. This configuration was tested because experiments in the tangential feed type reactor indicated burn-up phenomena of the reactor walls in the vicinity of the impingement zone of the streams. The coaxial reactor was applied for combustion of gas and also of pulverized coal where in the latter case it was necessary to heat the air prior to

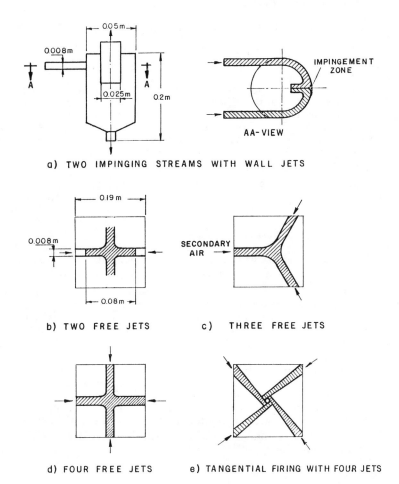

a) TWO IMPINGING STREAMS WITH WALL JETS

b) TWO FREE JETS

c) THREE FREE JETS

d) FOUR FREE JETS

e) TANGENTIAL FIRING WITH FOUR JETS

FIG. 2 COMBUSTION CONFIGURATIONS WITH IMPINGING STREAMS

481

its entrance to the reactor. In addition, the reactor was initially heated to approximately 600°C in order to ignite the coal. For more details, the attention of the reader is directed to the work of Ziv [28].

3. COMBUSTION OF GAS: OBSERVATIONS AND CHARACTERISTIC PHENOMENA

The surprising findings of the experiments were that both configurations, the tangential feed type and the coaxial one, present the same three distinct and different stable modes of operation designated as Mode I, II and III. The observation of the different modes were made as follows. In the coaxial configuration it was possible to view from an opening made in the top of the reactor, where in the tangential feed, the combustion was clearly observed on the outside walls of the reactor. The phenomena are depicted schematically in Fig. 3 but were also photographed [26]. In Mode I, obtained with mixtures which were either very lean, $\phi \sim 0.4$, or very rich, $\phi \sim 1.5$, (where for stoichiometric mixtures $\phi = 1$), no combustion took place along the path of the jets fed into the reactor as well as at the impingement zone. The mixture ignited after leaving the impingement plane and this can be seen schematically at the top of Fig. 3. On the left hand side is depicted the coaxial configuration feed of the streams where combustion may be designated as ring-shaped burning; on the right hand side, the tangential feed of the streams is shown. Mode II was reached by increasing the concentration of the fuel in the air in the zone of $\phi < 1$, or by decreasing its concentration in the zone of $\phi > 1$ when starting with a very rich mixture of gas in air. In this case, combustion did not take place along the jet and it began roughly before the impingement (or stagnation) plane of the two jets. This behavior is shown at the center of Fig. 3 for both types of configurations. When the concentration of the gas in the air attained a stoichiometric concentration, namely $\phi = 1$, Mode III is obtained in which combustion began at the entrance of the jets into the reactor and continued uniformly throughout as shown at the bottom of Fig. 3. In contrast to Modes I and II which existed for two mixtures (lean and rich), Mode III existed only for stoichiometric mixtures. Thus, in practice, the following five clearly distinguishable modes were observed, when operation of the reactor was started from a lean mixture (slightly above the ignition limit of $\phi = 0.4$) and continuously changed: Mode I (lean), Mode II (lean), Mode III, Mode II (rich), Mode I (rich).

A quantitative description of the combustion process of the above modes was obtained by temperature readings taken at different positions along the path of the jet. This was done for the tangential configuration (Fig. 2a) in which 7 thermocouples were positioned along the jet path on the outer wall of the reactor. Fig. 4 shows the temperatures versus the dimensionless position X/D_0 of thermocouples along the combustion path. It contains also initial values of ratios of fuel/air at which the different modes appear. The inlet of the premixed gas-air stream into the reactor is the origin where $X/D_0 = 0$. The temperature profiles, beside differing in the absolute temperatures, also differ

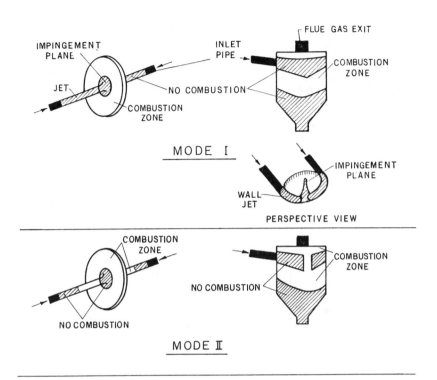

MODE I

PERSPECTIVE VIEW

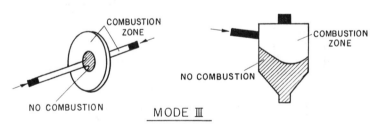

MODE II

MODE III

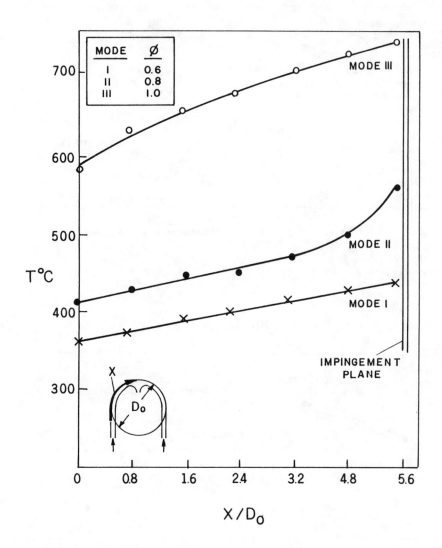

FIG.4 TEMPERATURE PROFILES ALONG THE JET PATH FOR THE DIFFERENT MODES

in their behavior. Mode I, in which no combustion took place along the path of the jet, showed a steady and moderate increase of the wall temperature, due to heating of the gas mixture by the surrounding hot gases. Mode II behaved similarly to Mode I up to the vicinity of the impingement plane, where it ignites and a steep rise of the temperature is seen. Mode III began at a higher temperature which slowly increased towards the impingement plane, probably due to side reactions and diffusive combustion along its path. Another characteristic behavior of Mode III is shown in Fig. 5, where fluctuations of the temperature readings with time, at steady-state conditions, are plotted versus the dimensionless distance X/D_0. As observed, the fluctuations are significant at the inlet to the reactor and decay towards the impingement plane. This behavior is typical of turbulent combustion which takes place at entrance to the reactor. However, the turbulence decreases along the burning jet because of jet homogeinization after the initial zone.

Attention is drawn now to the impingement (or stagnation) plane in which combustion does not take place. In this zone, the pressure in the gas reaches its maximum value and the velocity in the direction of flow becomes zero. However, the flame reappears outside the stagnation plane where pressure in the gas diminishes and the gas velocity in the new direction increases. These situations are clearly shown in Fig. 3.

4. COMBUSTION OF GAS: MODELS

In the following, two approaches will be presented for analysing the combustion of gas in impinging streams. The first approach is based on the flame speed and the second one is based on exothermic reactions in a perfectly mixed reactor.

The Flame Speed Model

A qualitative explanation of the above-described phenomena is based on the flame speed of the air-gas mixture, and on its characteristic dependence on concentration, as shown schematically in Fig. 6. It is clear that by drawing a horizontal line parallel to the ϕ axis, a single flame speed can be obtained for two values of ϕ. Moreover, the maximum flame speed is obtained for a single value of ϕ which is equal to unity. Thus, it is self-evident why for $0<\phi<1$ (lean mixtures) or $\phi>1$ (rich mixtures), four stable modes can exist, two of which, Mode I (lean) and Mode I (rich) as well as Mode II (lean) and Mode II (rich), are identical, where Mode III, for $\phi=1$, is unique. In addition, the condition for a stable combustion to exist at a certain plane is that the velocity of the gaseous mixture be equal to the flame speed. Note also that the gas velocity is maximum at the entrance into the reactor, decreases along the expansion of the jet, and becomes zero at the impingement plane. After the impingement plane, there is a change in the direction of the velocity and it increases. Therefore, the following behavior of the reactor, tangential or coaxial types, is envisaged. In Mode I, which exists after the impingement plane, the gas velocity is very small. Thus, it is obtained either for very lean or very rich gas-

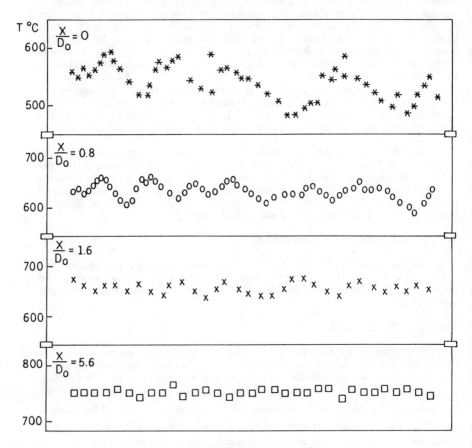

FIG.5   TEMPERATURE   FLUCTUATIONS   AS   A   FUNCTION OF TIME
        FOR MODE Ⅲ

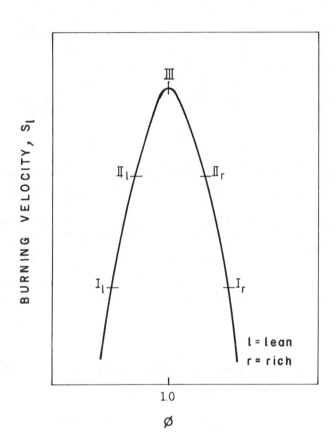

FIG.6 DEPENDENCE OF FLAME VELOCITY ON THE FRACTION
OF THE STOICHIOMETRIC FACTOR

air mixtures (Fig. 6) and near the ignition point. Further increase of gas concentration (on the lean side), or decreasing it (on the rich side), causes an increase of the flame speed as shown in Fig. 6. Thus, the flame front moves back towards the zone of higher flow velocities in the jet, which is located outside the stagnation plane and towards the inlet pipe. This is Mode II where twin flames are obtained on the two sides of the impinging plane as shown in Fig. 3. Further increase of the gas concentration in the range of $0<\phi<1$ or its decrease in the range of $\phi>1$ brings the flame speed to its maximum value at $\phi=1$. Under these conditions, the flame speed is equal to the inlet velocity to the reactor of the gas-air mixture and the mixture is ignited at the feed point and the combustion takes place throughout the jets volume. This is Mode III, shown at the bottom of Fig. 3. Finally, it is noteworthy that the appearance of the various modes took place both when starting from lean mixtures and increasing ϕ, or from rich mixtures and decreasing ϕ. Also, the absolute flow-rates of the gas and air did not change the appearance of the three modes, provided that ϕ was left constant and as long as reasonable total flow-rates which permit combustion were employed.

The possible appearance of different modes is supported by the work of Niioka et al. [29] who studied the effect of flame retardant materials on the change in the laminar burning velocity in two opposed jets. Similarly, Otsuka and Niioka [30] showed in a similar configuration that for a certain gas-air mixture the distance between the twin flames on the two sides of the impinging plane varies with jet velocity (decreases by increasing the velocity).

A quantitative model for the tangential feed type reactor (Fig. 2a, Fig. 3, right), which permits the evaluation of the situation that will exist in the reactor under specific conditions, will be presented in the following where for complete details, the attention of the reader is directed to the work of Luzzatto [26]. As emphasized before, the key quantity in the analysis is the flame speed. On the basis of Fig. 6, which characterizes the behavior of the laminar burning velocity S_l as a function of ϕ, S_l can be approximated by a parabola of the form

$$S_l = a \phi^2 + b\phi + c \qquad (1)$$

Applying the following conditions for evaluation of the constants a, b and c:

$$S_l = S_{l\ max} = 40 \text{ cm/s} \quad \text{for } \phi=1 \qquad (2)$$

where the numerical value is an appropriate value for hydrocarbons,

$$dS_l/d\phi = 0 \quad \text{for } \phi=1 \qquad (3)$$

and that

$$S_l = 0 \qquad \text{for } \phi = 0.4 \qquad\qquad (4)$$

which is the ignition limit of the mixture as measured in the work of Luzzatto [26], yields that

$$S_l \text{ (cm/s)} = -111 \ \phi^2 + 222 \ \phi - 71 \qquad\qquad (5)$$

The conditions inside the reactor are not simply those of laminar combustion because the following effects must be taken into account:

1) The expansion of the jet and its flow along the walls of the reactor; 2) The significant heating of the jets inside the reactor, from approximately 20^0 to 800^0C by the combustion gases; 3) The dilution of the jet due to its expansion and hence reduction of ϕ . Thus, an analogous expression to Eq. 1 for a characteristic burning velocity inside the reactor, S_r, in which the above effects are taken into account will read,

$$S_r = aH_x{}^2 + bH_x + C + fRe_x \qquad\qquad (6)$$

where a, b, c are the constants of the laminar burning velocity in Eq. 1 and 5, and f is an adjustable constant.

In Eq. 6, the Reynolds number accounts for turbulent effects and is defined by

$$Re_x = b_o v_x / \nu_a \qquad\qquad (7)$$

The parameter H_x accounts for the previously mentioned effects and was assumed, as later explained, to be:

$$H_x = \phi_0[1 - k/(b_x/b_o)] \qquad\qquad (8)$$

where k is an adjustable constant which can be determined from prescribed operating conditions of the reactor. Expressions for v_x and b_x are also available elsewhere [26] and their simplified forms are

$$\frac{v_x}{v_0} \cong \frac{3.17}{\sqrt{\theta} \ X/b_0} \sim X^{-1/2} \qquad\qquad (9)$$

where θ ($\cong$ 36-40) is the ratio between the jet temperature and the temperature of the surrounding gas,

$$\frac{b_x}{b_0} \cong 0.74 \sqrt{\theta X/b_0} \sim X^{1/2} \tag{10}$$

As seen, heating of the jet (increase in θ) and proceeding along its path (increase in x), increase the value of b_x which indicates an expansion of the jet, and as a result H_x (Eq. 8) is increased. This trend is in the correct direction because the overall effect of heating, which is very significant in the present experiments, should increase the burning velocity in Eq. 6 as expected.

Equation 6 with the H_x given by Eq. 8 contains two adjustable parameters k and f which are determined from the following conditions obtained from observations in mode II:

at $x = 0$: $S_r = v_0$, $H_x = \phi_0 (1-k)$

In mode II combustion begins at the impingement plane which is a well defined geometrical point X_{II}, namely, at $X = X_{II}$: $S_r = v_x$, H_x is given by Eq. 8, where v_x and b_x are given by Eq. 9 and 10. Applying the following experimental quantities, viz., $X_{II} = 4$ cm, $b_0 = 0.35$ cm, $v_a = 23\times10^{-2}$ cm^2/sec , $v_0 = 340$ cm/s, $\phi_0 = 0.65$, $\theta = 36$ (for a 20°C jet entering the reactor with surrounding gases at 720°C) gives for the present reactor: $f = 0.6$ and $k = 3.9$. Thus Eq. 6 will read

$$S_r = -111\phi_0^2 H_x^2 + 222\phi_0 H_x - 71 + \frac{0.2 v_0 b_0}{v_a \sqrt{X}} \tag{11}$$

where

$$H_x = \phi_0 [1 - 3.9 / (b_x/b_0)] \tag{12}$$

An example of the application of the above results may be as follows. If ϕ_0 is decreased from 0.65 to 0.52, where will the flame front stabilize? Substituting $\phi_0 = 0.52$ in Eq. 11 and equating S_r to v_x from Eq. 9, gives by trial and error, $X = 8$ cm. Thus, decreasing ϕ by 20% causes the flame to move twice as far along the jet path. However, since the above location is behind the stagnation plane, no combustion will take place and Mode I will be obtained. It is clear that Eq. 11 may be similarly used to answer other questions about possible operating conditions of the reactor. In conclusion, to be able to make use of the model, the following data must be available: 1) inlet gas concentration (ϕ_0); 2) mean temperature in the active zone of the reactor for calculating b_x and v_x; 3) physical dimensions of the reactor; and 4) experimental data for the conditions of the existance of Mode II, in order to obtain the constants f and k in Eq. 6 and 8. If such data are not available, the values obtained in the present research may be used as first approximation.

The Autothermic Model

The previous model was based on the effect of the combustion gas concentration in the air on the flame speed. In the following analysis, the effect of the same parameter will be elaborated with respect to the temperature of the gaseous feed to the reactor and its influence on the characteristic behavior of the reactor.

We first consider the behavior of a perfectly mixed reactor in the presence of an exothermic reaction (for example, combustion) and then apply this to the case under consideration. Fig. 7a demonstrates a well known behavior of such a reactor where the S shape curve is the material balance curve and the straight line is the energy balance curve; the intersection of these lines is the operating point of the reactor.

If the feed temperature T_0 varies gradually, the following changes will take place:

a) When the feed temperature changes from 1 to 4, where T_0 is the intersection point between the lines 1 to 4 and the horizontal axis, the reactor will pass through states A, B, C and D. In state D each small addition to the feed temperature will cause a jump to equilibrium state H. An additional increase will cause a movement to point I.

b) When the feed temperature changes from 5 to 2, the equilibrium states will move through the path I-H-G-F as seen in Fig. 7b, and each small additional decrease in the feed temperature will cause the system to reach point B and to quench combustion.

The steady state temperature in the reactor for the above behavior versus the feed temperature is shown in Fig. 7b. A hysteresis effect is revealed when the reactor goes through a thermal cycle in which the feed temperature rises to a value corresponding to point I and then decreases.

The above model for exothermic reactions may be applied to combustion in impinging streams reactor as follows. Previous investigation [1, 4, 17] indicated that the reactor may be described as a combination of mixed vessels

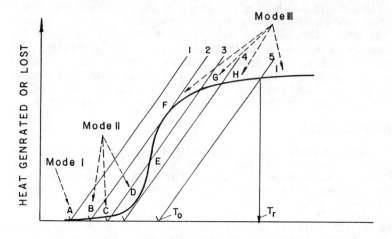

FEED TEMPERATURE T_o AND REACTOR TEMPERATURE T_r

(a)

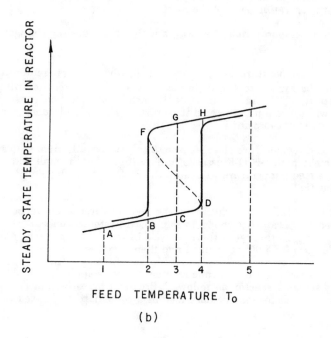

FEED TEMPERATURE T_o

(b)

FIG.7 THE AUTOTHERMAL MODEL AND HYSTERESIS EFFECT

and of recycle streams. Thus, the impingement zone may be visualized as a perfectly mixed reactor where the burnt gases are recycled to mix with the two wall jets, and also to heat them as shown schematically in Fig. 8. It is clear that the situation shown there is essentially autothermal in behavior since the feed is heated by the exhaust gas and the temperature of the hot recycled gas (F) that mix with the cold feed (C) will affect the combustion of the fuel. On the basis of this behavior, the appearance of the various models may be described as follows. Mode I is, in practice, an "extinguished" mode, viz., along the whole jet no combustion takes place and combustion begins only outside the stagnation plane. This situation is described by feed line 1 in Fig. 7a and point A is the extinction point. When the gas concentration in the feed increases, the feed temperature also increases because of the increased temperature of the flue gases which mix with the feed, and such an increase in concentration causes the feed temperature to rise and the equilibrium state will move from line 1 to 4. This movement passes through points B, C and D in which the reaction is possible, but with a low conversion. This is Mode II. When point D is reached, each small addition of temperature (addition of gas to the feed) will cause the reactor to jump to equilibrium state H, which corresponds to Mode III. This is also consistent with the situation existing in the reactor in which it is observed that the change from Mode II to Mode III is very sudden.

When the feed consists of a stoichiometric mixture, each addition of gas will cool the burnt gases which, in turn, will cause a cooling of the feed and a movement from line 5 to line 1. As the mixture becomes richer, the equilibrium state will move from point I to point F, through points H and G, within Mode III. When the system is at point F, each small cooling (addition of gas to the feed) will cause a jump to point B, in Mode II. Additional cooling of the feed will cause a movement to point A, viz., towards extinction or Mode I. The above behavior is entirely consistent with the observed behavior of the reactor.

The results obtained in the experimental system show that the temperature differences between Modes I and II are small (Fig. 4). This result is consistent with a situation in which a slow movement from point A occurs while passing from Mode I to Mode II. Temperature differences between Mode II and Mode III, on the other hand, are large (Fig. 4), as it would be expectable, for instance, from the difference between the heat generated at points D and H (Fig. 7). Furthermore, insulating the reactor causes an immediate homogeinization of temperatures in all ranges, as it would be expected from the passage to adiabatic operation.

It could be argued that the whole TIS reactor should be considered as a well mixed vessel. However, a backmix reactor in which the feed temperature is constant (as it would be in such a case) cannot possess multiple steady states as observed, since this effect takes place only when the feed temperature changes.

The above suggested model would seem to offer a realistic picture of the effects which are responsible for the behavior of the two-impinging streams

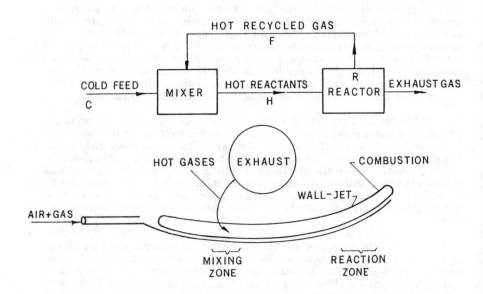

FIG. 8 THE AUTOTHERMAL MODEL

reactor during gas combustion, according to data so far available on the behavior of the gas phase of this reactor.

5. COMBUSTION OF COAL [28]

The coal employed was bituminous with 28% of volatile matter on a weight basis, C = 69.7%, H = 4.8%, S = 0.73%, N = 1.3%, H_2O = 2.4%, ash = 13.2%, O = 7.9% and calorific value of 6730 kcal/kg.

The reactor was operated with its upper and lower openings (Fig. 2b) open. Thus, part of the ash accumulated on the reactor walls and part was discharged from the bottom. The flue gases left from the upper hole carrying, under some circumstances, fly-ash. The most important quantity which was measured was CO_2 concentration in the flue gases. In some experiments, concentrations of CO, NO_2 and SO_2 were also measured as reported below.

The parameters investigated in the combustion of pulverized coal were the mean size of the particles, air and coal flow-rates and the effect of the impinging-streams. The major aim here was to study the combustion efficiency of pulverized coal and its dependence on the above-mentioned operational parameters, where the combustion efficiency η was defined as follows:

$$\eta = \frac{\text{actual concentration of } CO_2 \text{ in flue gases}}{\text{maximal concentration of } CO_2 \text{ in flue gases assuming complete combustion}} \qquad (13)$$

Thus, from measurements of CO_2 in flue gases, air and coal flow rates as well as the composition of the coal, and by assuming that the following reactions take place in the combustion process, namely,

$$C + O_2 \rightarrow CO_2, \ H_2 + 1/2 \ O_2 \rightarrow H_2O \text{ and } S + O_2 \rightarrow SO_2,$$

it was possible to obtain the value of η [28].

In order to express the data in a concise form and to gain more insight about the combustion process of coal and the operational parameters, an attempt was made to develop an appropriate correlation. Since the flow behavior in the impinging-streams reactor is very complicated and there are no governing differential equations which clearly apply, the most feasible approach was by non-dimensional analysis. It was assumed that the measured concentration of CO_2 in the flue gases is a function of the following variables:

$$CO_2 = f(CO_{2,max}, W_p, W_a, k_1, D, \mu_a, \rho_a, d_p, d) \qquad (14)$$

- 1 1 -

495

$CO_{2,max}$ is the maximal value of CO_2 in the flue gases, assuming complete combustion; it characterizes also partially the kind of coal employed. Application of non-dimensional analysis yields the following dimensionless groups expressed in the form of

$$\eta = A\,\mu^\alpha\,(d_p/d)^\beta Re^\gamma Sc^\delta (CO_{2,max}/\rho_a)^\varepsilon (d\sqrt{k_1/D})^\omega \qquad (15)$$

where A, α to ω are adjustable parameters. Equation 15 is a general correlation applicable for correlating data of different coal types. It incorporates the following dimensionless groups, from left to right: the combustion efficiency $\eta = CO_2/CO_{2,max}$, the composition of the feed μ, the size of the particles d_p/d, the hydrodynamics at the inlet pipe expressed by Re, the physical properties of the gas phase given by Sc, the characteristics of the coal $CO_{2,max}/\rho_a$ and the ratio between surface reaction effects to diffusional processes in the gas phase proportional to $d\sqrt{k_1/D}$. Equation 15 was applied to the present experiments by making the following simplifications. The Sc number, the ratio $CO_{2,max}/\rho_a$ as well as $d/\sqrt{D}$ are approximately constant values at some average temperature. The reaction constant k is proportional to $e^{-E/RT}$ and hence Eq. 15 may be reduced to the form of:

$$\eta = \tilde{A}\,\mu^\alpha (d_p/d)^\beta Re^\gamma\, e^{-\tilde{E}/T} \qquad (16)$$

where

$\tilde{A}$, α, β, γ and $\tilde{E}$ are the adjustable parameters determined from the measured quantities.

By fitting the experimental data points, the following correlation was obtained:

$$\eta = 0.131\,\mu^{0.091}\,(d_p/d)^{-0.017} Re^{0.0297}\,\exp(-847.5/T) \qquad (17)$$

where T is in K.

By ignoring the effect of the chemical reaction the following correlation was arrived at:

$$\eta = 0.026\mu^{0.226}(d_p/d)^{-0.055} Re^{0.423} \qquad (18)$$

while by ignoring hydrodynamical effects, the correlation for η reads:

$$\eta = 2.64 \, \mu^{-0.17} \, (d_p/d)^{0.043} \, \exp(-1850/T) \qquad (19)$$

The above correlations were obtained in the following range of the operating parameters:

$d_p = (28,34,69,211) \, \mu m$, $W_p = (0.24 - 1.2) \, kg/h$, $W_a = (3 - 10.8) \, kg/h$,
$\mu = 0.06 - 0.21$, $\%CO_2 = 6 - 17$, $CO = (100 - 2000)ppm$, $SO_2 \approx 10 \, ppm$,
$NO_2 \approx 10 \, ppm$, $U_a = (10\text{-}50)m/s$, $Re = 2800 - 10200$, $\eta = 0.51 - 1$,
$\alpha = 0.5 - 2$, heat output $= (1.3 - 8.6)kw$, maximal temperature at the impingement zone $= (1100 - 1400)K$, flue gases temperature $= (850 - 1200)K$, inlet temperature of the air to the reactor $= (420 - 520)K$.

In Table I, various parameters indicating upon the goodness of the fit by the correlations (17) to (19) are given.

TABLE I: DEVIATIONS OF η FROM EXPERIMENTAL VALUES
(Actual Table Below)

On the basis of the values of the deviations and the sum of squares (S.O.S.) in Table I, the following conclusions may be drawn:

1) Convection plays a dominant role in the combustion process. 2) Comparison between the values of $\bar{D}$ for Eq. 17 and Eq. 19 indicates that the contribution of the chemical kinetic term in Eq. 17 is not significant. This is advantageous because the temperature of combustion, which is not known a priori, is not needed in the calculations. Hence, Eq. 19 is recommended for practical applications for coal having the chemical composition previously detailed and under the operating conditions given above.

Correlations 17 to 19 were obtained in a combustion chamber shown in Fig. 2b and hence possible guidelines for scale-up might be useful. From our recent investigations in drying of solids in a two-impinging-streams reactor similar to that in Fig. 2b, the following rule was deduced: multiplying the geometrical dimensions of a small-scale reactor and keeping the drying air velocity at the inlet pipes identical in the small and the large-scale reactors, yielded similar values of the heat transfer coefficient for identical mass flow-rates of the solids. It has been previously emphasized that in the present combustion experiments, convection has a significant effect on the formation of CO_2 and hence on η. Thus, the above rule might also be applicable for η while keeping, in addition, the temperatures of the hot air in the small and large-scale reactors identical at the inlet to the reactor.

Finally an additional correlation, with no theoretical basis, was obtained by assuming that $\eta/\alpha = A + B\mu$. Fitting of the data to the above equation gave:

$$\eta/\alpha = -0.029 + 7.66\mu \qquad (20)$$

The mean deviation in the prediction of η/α is 12.2% and the maximal deviation is 49.3%.

TABLE I: DEVIATIONS OF η FROM EXPERIMENTAL VALUES

Eq.	Model	$\bar{D}$%	D_{max}%	S.O.S.
17	convection-kinetics	8.7	30.7	1.34
18	convection	8.8	38.8	1.44
19	kinetics	9.5	36.1	1.59

D% = 100 $|\eta_{calc} - \eta_{exp})/\eta_{exp}|$; $\bar{D}$% = $(\Sigma D\%)/m$ where the number of observations m = 221; S.O.S. = sum of squares

NOMENCLATURE

b_0, b_x — the initial half-thickness of the jet; its value at distance x from the inlet, respectively

d, d_p — inlet pipe diameter to reactor; particle diameter, respectively

D_0, D — reactor's diameter; diffusion coefficient, respectively

H_x — parameter defined by Eq. 8

k — adjustable parameter in Eq. 8

k_1 — first order reaction rate constant

Re — Reynolds number at inlet conditions to reactor; dv_0/v_a

498

Sc	- Schmidt number, v_a/D
S_l	- laminar flame speed defined in Eq. 1 and 5
S_r	- flame speed defined according to reactor parameters; Eq. 6
T_0, T	- inlet gas temperature; temperature in reactor, respectively
TIS	- two-impinging-streams
v_0, v_x	- jet velocity at the inlet to the reactor; at distance x, respectively
W_a, W_p	- air and coal mass flow rates, respectively
X	- distance along the jet
α	- coefficient of air excess
ϕ	- fraction of stoichiometric mixture (actual gas concentration of the mixture divided by its concentration in a stoichiometric mixture)
ϕ_0	- value of ϕ at inlet to the reactor
μ	- W_p/W_a
η	- defined in Eq. 13
μ_a, ρ_a, v_a	- mean values of viscosity, density and kinematic viscosity, respectively

REFERENCES

1. K. Luzzatto, A. Tamir and I. Elperin, AIChE J., 30, 600 (1984)
2. A. Tamir, K. Luzzatto, D. Sartana and S. Salomon, AIChE J., 31, 1744 (1985)
3. A. Tamir, K. Luzzatto and I. Elperin, Chem. Eng. Sci., 39, 139 (1984)
4. A. Tamir and K. Luzzatto, AIChE J., 31, 781 (1985)
5. A. Tamir and K. Luzzatto, Journal of Powder and Bulk Solids Technology, 9, 15 (1985)
6. A. Tamir and B. Shalmon, Ind. Eng. Chem. Res., 27, 238 (1988)
7. A. Tamir and M. Grinholtz, Ind. Eng. Chem. Res., 26, 726 (1987)
8. A. Kitron, R. Buchmann, K. Luzzatto and A. Tamir, Ind. Eng. Chem. Res., 26, 2454 (1987)

9. A. Tamir and D. Herskowitz, <u>Chem. Eng. Sci.</u>, 40, 2149 (1985)
10. A. Tamir, <u>Chem. Eng. Sci.</u>, 41, 3023 (1986)
11. D. Herskowitz, V. Herskowitz and A. Tamir, <u>Chem. Eng. Sci.</u>, 42, 2331 (1987)
12. D. Herskowitz, V. Herskowitz, K. Stephan and A. Tamir, <u>Chem. Eng. Sci.</u> (to appear 1988)
13. A. Tamir and S. Sobhi, <u>AIChE J.</u>, 31, 2089 (1985)
14. A. Tamir and Y. Kitron, <u>Drying Tech.</u> (to appear 1989)
15. Y. Kitron and A. Tamir, <u>Ind. Eng. Chem. Res.</u> (to appear 1988)
16. I.I. Elperin, <u>Transport Processes in Impinging Jets</u> (in Russian), Nauka i Tekhnica, Minsk, 213p (1972)
17. A. Tamir and A. Kitron, <u>Chem. Eng. Commun.</u>, 50, 241 (1987)
18. A. Tamir, <u>Chem. Eng. Prog.</u> (to appear 1988)
19. J.B. Greenberg, D. Albagli and Y. Tambour, <u>Combust. Sci. and Tech.</u>, 50, 255 (1986)
20. Y.P. Enyakim and A.I. Dvoretskii, <u>Thermal Engineering</u>, 15, 78 (1968)
21. Y.P. Enyakim, <u>Teploenergetika</u>, 17, 21 (1970)
22. Y.P. Enyakim, <u>Thermal Engineering</u>, 17, 28 (1970)
23. P.M. Goldberg and R.H. Essenhigh, 17th Symposium (Int.) on Combustion, The Combustion Inst., 1978, p. 145
24. Y. Goldman, A. Lipshits and Y.M. Timnat, <u>Israel J. Tech.</u>, 19, 188 (1981)
25. Z. Hazanov, Y. Goldman and Y.M. Timnat, <u>Combustion and Flame</u>, 61, 119 (1985)
26. K. Luzzatto, <u>Investigation of a Two-Impinging-Streams Reactor</u>, Ph.D. thesis, Ben Gurion University (1987)
27. A. Ziv, K. Luzzatto and A. Tamir, <u>Combust. Sci. and Tech.</u>, (to appear 1988)
28. A. Ziv, <u>Combustion of Coal in a Two-impinging-Streams Reactor</u>, Research Report (1986)
29. T. Niioka, T. Mitani and M. Takahshi, <u>Combustion and Flame</u>, 50, 89 (1983)
30. Y. Otsuka and T. Niioka, <u>Combustion and Flame</u>, 19, 171 (1972)

Flash Combustion of Sulfide Mineral Particles in a Turbulent Gas Jet

H. Y. SOHN
Departments of Metallurgy and Metallurgical Engineering
and of Chemical Engineering
University of Utah
Salt Lake City, Utah 84112, USA

K. W. SEO
Department of Chemical Engineering
University of Utah
Salt Lake City, Utah 84112, USA

Experimental measurements and mathematical model predictions for the flash combustion of copper sulfide particles in a turbulent gas jet are described.

A mathematical model has been developed to describe the process taking place in an axisymmetric flash-furnace shaft. The model incorporates turbulent fluid dynamics, chemical reaction kinetics, and heat and mass transfer. The key features include the use of the k - ε turbulence model, incorporating the effect of particles on turbulence, and the four flux-model for radiative heat transfer.

The experiments were carried out in a laboratory flash furnace. Gas temperature, sulfur content in the particles, SO_2 concentration in the gas phase, and particle dispersion during flash-smelting at different locations were measured for various matte grades. Reasonable agreement was obtained between the measured and predicted values.

1. INTRODUCTION

Since the development of flash-smelting process by Outokumpu Oy and INCO in the late 1940s and early 1950s,[1-3] the flash-smelting process has been a dominant process for smelting various sulfide minerals, including those of copper and nickel. The attractiveness of the flash-smelting process has greatly increased with the increasing availability of inexpensive tonnage oxygen and because of major advantages of substantial reduction in fuel requirement, efficient recovery of sulfur dioxide, and rapid smelting rate.

The essential feature of the flash smelting process is that fine particles of dry sulfide concentrate are injected with preheated air mixed with fuel, oxygen- enriched air, or tonnage oxygen into the furnace. The dry charge is dispersed in the furnace flame zone in which sulfide minerals undergo rapid chemical reaction with oxygen, generating a large amount of heat. The high rate of reaction combined with the large exothermicity of the oxidation reaction makes it possible to use the charge as fuel. The use of high concentration oxygen also reduces the amount of energy required to heat the gases. As a result, the smelting process becomes autogeneous.

Chemical reactions taking place in the furnace, using the smelting of chalcopyrite as an example, may be expressed as follows[1,4]:

$$2CuFeS_2 + 2.5O_2 \quad \text{---} \quad Cu_2SFeS + 2SO_2 + FeO \qquad : \Delta H_{298} = -179kcal \quad (1)$$

$$FeS + 1.5O_2 \quad \text{---} \quad FeO + SO_2 \qquad : \Delta H_{298} = -112kcal \quad (2)$$

$$2FeO + SiO_2 \quad \text{--} \quad 2FeOSiO_2 \qquad : \Delta H_{298} = -8kcal \quad (3)$$

Equations (1) and (2) describe the oxidation process, whereas equation(3) describes the slag-forming step. Obviously these equations are a simplification of the overall steps[4]. By controlling the rate of oxygen to sulfur into the furnace, mattes of various grades can be obtained.

When the gas-particle suspension is injected into furnace as a jet, the particles are ignited and undergo the reactions listed above. The rate of these reactions are extremely fast. Reaction times of about 0.1 second have been reported for particles of around 200 mesh size.[5]

The particles go through various stages of reaction after being injected into the furnace. First, they are heated by the hot gas in the furnace as well as by radiation from the furnace walls. The particles then become molten and rather vigorous internal motion is initiated as the molten droplets begin reaction with oxygen.[6] At the same time the particles and surrounding gas exchange heat and mass, while the particles travel within the gas jet. The trajectory and residence time of particles are important factors in determining the degree of oxidation.

In spite of the increasing industrial stature of the process, the design of a flash-smelting furnace remains largely an art. This is mainly due to the difficulty of understanding the complex interactions of the individual subprocesses taking place in a flash furnace. In order to enhance the systematic understanding of the overall process, a reliable mathematical model would be very helpful. Furthermore, the behavior of the complex reacting particle-laden turbulent gas jets can be predicted with a minimum amount of experimental work.

It is only in very recent times that much attention has been focused on the mathematical modeling of the flash-smelting process. The mathematical modeling of the reacting particle-laden turbulent gas flows in confined systems has been a difficult problem not only due to the difficulties of solving the nonlinear partial differential model equations, but also due to the difficulties associated with mathematical description of complex turbulent characteristics of the interaction between the dispersed reacting particle phase and the surrounding gas phase.

Although, since the middle of 1970s, a substantial amount of work has been done on the fundamental mathematical modeling of pulverized-coal combustion process that has some features similar to those of the flash-smelting system, very little work has been done for the flash-smelting process. The previous studies on this process have assumed the flame to be an one dimensional stream[7-8], a two dimensional free jet [9],or a two dimensional confined jet[10].

In the flash-smelting process, the major component is the furnace shaft in which sulfide concentrates go through various stages of reaction after being injected into the furnace. Therefore, it is of great importance to clearly understand the subprocesses taking place in the confined flash-smelting furnace.

In order to predict various major phenomena occurring in the flash-smelting furnace, the comprehensive model must combine the turbulent fluid dynamics of a particle-laden gas jet, chemical kinetics, and the transfer of heat and mass. In this laboratory a comprehensive computer model has been developed by combining all the above mentioned phenomena[11].

In addition, laboratory-scale experiments using the flash system described in the experimental part have been carried out. Measurements of gas temperature, SO_2 concentration, sulfur content in the particles, and the radial distribution of particle flux were made.

In order to verify the mathematical model, the model predictions were compared with experimental data obtained from the laboratory flash furnace.

2. MATHEMATICAL MODEL

The following aspects were included in the mathematical model that was developed in this laboratory:

(1) The two - equation(k-ε) turbulence model that accounts for the dependency of the mixing length on the location in a flow system was used.

(2) The effect of the presence of particles on turbulence was incorporated.

(3) The effect of turbulence on the particle phase, i. e., the dispersion of particles due to the turbulent fluctuations, was considered.

(4) The four-flux model for radiative heat transfer combining the absorption, emission and scattering phenomena was used.

2.1. Model Basis

The flash-furnace shaft can be schematically represented as shown in Figure 1. The primary particle-laden gas stream with or without an oxygen enriched secondary (or process) air stream enters the system through the burner nozzle and expands radially. The modeling equations to describe the overall phenomena of a reacting particle-laden turbulent gas jet in such a confined system can be expressed by the general conservation equations of continuity, momentum and energy for each phase[11].

In the dispersed particle phase, particles may exist locally with different properties due to the varying history of each particle along the trajectories that they are traveling. This different particle histories make the Eulerian treatment of the particle phase quite difficult because of significantly increased computer storage requirement. The

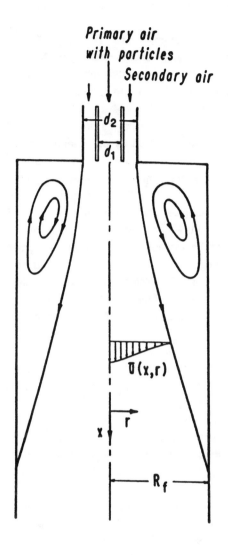

Figure 1. Particle-laden gas jet in a flash-furnace shaft.

Lagrangian treatment of the particle phase, representing the particle field as a series of trajectories, is hence more efficient from the point of view of economical computation.

However, the Lagrangian treatment also has some difficulties in treating the effect of gas turbulence on particle motion and in obtaining the local particle number density. To handle the previous difficulty the particle-phase velocity is broken down into a convective and a turbulent diffusive component that accounts for the effect of turbulence on the particle motion.

The particle dispersion in a turbulent gas flow is described by the local particle number density that can be obtained by an Eulerian equation of continuity if an Eulerian particle velocity field is known. However, with the Lagrangian treatment, the particle velocities are calculated only along trajectories, and the Eulerian particle velocity field is thus not obtained. This difficulty results in an approximation method of obtaining the particle number density by using the gas-phase Eulerian velocity field. The interaction between particles are neglected because the particle concentration in the flash-smelting furnace is very low, and the pressure gradient, virtual mass effect, and Basset force are neglected compared with the aerodynamic drag force.[12,13] In this system, the reaction rate of particles with gaseous reactant is slow compared to the turbulence scale, but fast compared with the mean gas velocity. This assumption allows the particle properties to be calculated from the mean gas properties instead of fluctuating gas properties.[13,14]

Due to the highly emitting, absorbing and scattering nature of a flash-smelting flame, thermal radiation may be a significant mode of heat transfer in a flash-smelting furnace. The radiation field was modeled with the following assumptions:

(1) The particles retain spherical shape.
(2) The temperature field is axially symmetric.
(3) The gas-particle medium is gray.
(4) The gaseous components of the flame are at local thermal equilibrium.

2.2. Gas-Phase Equations

The gas phase was viewed from the Eulerian framework. The continuity and momentum equations combined with the effect of the presence of the reacting particles can, respectively, be expressed as:[13-16]

Continuity:

$$\frac{\partial \rho}{\partial t} + \vec{\nabla} \cdot (\rho \vec{V}) = S_p^m \qquad (4)$$

Momentum:

$$\frac{\partial}{\partial t}(\rho \vec{V}) + \vec{\nabla} \cdot (\rho \vec{V} \vec{V}) = -\vec{\nabla} p - \vec{\nabla} \cdot \overline{\overline{\tau}} + \rho \vec{g} + \vec{S}_p^v + \vec{V} S_p^m \qquad (5)$$

where the arrow and the double over-bar represent a vector and a second-order tensor, respectively.

In Eq. (4), the term S_p^m represents the net rate of mass addition to the gas phase per unit volume due to the reaction of solid particles. In Eq. (5), the fourth and last terms represent momentum sources to the gas phase due to, respectively, the presence of particles and the addition of mass to the gas phase per unit volume by the reaction of particles.

2.3. Turbulence Model

In flash-smelting processes, the turbulent flow confined in a flash furnace has the added complexity due to the presence of particles. This makes it extremely difficult to use fundamental theories to describe the motion of the particle-laden turbulent gas jet. Hence, a great deal of empiricism needs to be included in describing turbulence.

2.3.1. Two-Equation (k - ε) Model

The turbulent fluid dynamics were described by the well-known two-equation(k - ε) model, the details of which can be found elsewhere[11,13-20]

2.3.2. Effect of Particle Phase on Turbulence

The presence of solid or molten particles affects the gas-phase turbulent field. The kinetic energy spectrum of the turbulent gas field is damped by the presence of particles. In addition, the gas phase affects the motion of particles through the phenomena of turbulent fluctuations of flow properties.

Elghobashi and Abou-Arab[21] proposed a two-equation turbulence model for a two-phase flow by incorporating the influence of particles directly into the turbulence model, and obtained good agreement between the predicted results and their experimental data. However, they mentioned that more research was needed to establish the universality of the coefficients used in their turbulence model.[22]

Melville and Bray[23] proposed a semiempirical correlation to account for the effect of particle phase on the gas turbulence. They used the ratio of the mean particle bulk density (i.e., mass of particles per unit volume) to the mean gas density as follows:

$$(v_g^t)_{particles} = (v_g^t)_{no\ particles} \ [1 + (\bar{\rho}_{bp}/\bar{\rho}_g)]^{-0.5} \tag{6}$$

The above equation suggests that the amount of gas-phase turbulence decreases as the particle number density increases. This equation was used throughout for the incorporation of the particulate infulence on turbulence in this study, because no fundamental models are available at the present time.

2.4. Conservation Equation of Gaseous Species

The steady-state continuity equation for gaseous species j in a turbulent flow field can be expressed as:

$$\frac{\partial}{\partial x}(\bar{\rho}\,\bar{u}\,\bar{m}_j) + \frac{1}{r}\frac{\partial}{\partial r}(r\bar{\rho}\,\bar{v}\,\bar{m}_j) - \frac{\partial}{\partial x}(\Gamma_m\frac{\partial \bar{m}_j}{\partial x}) - \frac{1}{r}\frac{\partial}{\partial r}(r\Gamma_m\frac{\partial \bar{m}_j}{\partial r})$$

$$= \pm(\overline{S_p^m})_j \tag{7}$$

where m_j is the time-mean mass fraction of species j. The term in the Eq.(7).represents the time-averaged source (+) or sink (-) per unit volume due to the reactions of particles with gaseous reactants. This time mean source or sink term is completely governed by the mean reaction rate of the individual sulfide particles.

The transport exchange coefficient Γ_m can be expressed as:

$$\Gamma_m = \mu_e / \sigma_m \tag{8}$$

where σ_m is the Schmidt number having a value of about unity.

2.5. Gas-Phase Energy Equation

The steady-state energy equation for the gas phase can be expressed with the Eulerian framework by neglecting viscous dissipation, as follows:

$$\frac{\partial}{\partial x}(\bar{\rho}\,\bar{u}\,\bar{h}_g) + \frac{1}{r}\frac{\partial}{\partial r}(r\bar{\rho}\,\bar{v}\,\bar{h}_g) - \frac{\partial}{\partial x}(\Gamma_h\frac{\partial \bar{h}_g}{\partial x}) - \frac{1}{r}\frac{\partial}{\partial r}(r\Gamma_h\frac{\partial \bar{h}_g}{\partial r})$$
$$= Q_{rg} + \bar{u}\frac{\partial \bar{p}}{\partial x} + \bar{v}\frac{\partial \bar{p}}{\partial r} + S_p^h \qquad (9)$$

On the left-hand side of Eq.(9), the first two terms represent the volumetric heat transfer rate due to fluid convection, and the last two terms are the heat transfer rate due to the gas-phase conduction per unit volume. On the right-hand side, the first term shows the net volumetric heat transfer rate by the gas-phase radiation, the second and third terms represent enthalpy change due to the expansion of fluid which may be neglected for a fluid of constant density, and the last term represents the heat addition to the gas phase due to the reaction of sulfide particles per unit volume.

Γ_h in Eq. (9) represents the transport coefficient for energy that is defined as:

$$\Gamma_h = \mu_e / \sigma_h \qquad (10)$$

where σ_h is the Prandtl number for the gas-phase.

2.6. Particle-Phase Equations

The particle velocities, trajectories and temperatures are obtained by integrating the Lagrangian equations of momentum and energy. The mean particle number density is calculated by an Eulerian equation of continuity by using the Eulerian gas-phase velocity field.

2.6.1. Momentum Equation

The Lagrangian momentum equation for a single particle can be written as:

$$m_p \frac{d\vec{V}_p}{dt} = \frac{1}{2} C_D \rho_g A_p |\vec{V}_g - \vec{V}_p| (\vec{V}_g - \vec{V}_p) + m_p \vec{g} \qquad (11)$$

Equation (11) indicates that the rate of change of the particle momentum is equal to the sum of the aerodynamic drag force and the gravitational force acting on the particle.

The drag coefficient C_D can be calculated by the following correlation:[13,24-26]

$$C_D = \frac{24}{Re} (1 + 0.15 \, Re^{0.687}) \qquad (12)$$

where

$$Re = \rho_g |\vec{V}_g - \vec{V}_p| d_p / \mu_g \qquad (13)$$

One of the important factors involved in the flash-smelting processes is the turbulent dispersion of sulfide particles. Particles tracked through the flow field are considered to have different rate of dispersion from the gas phase.

In order to account for the dispersion of particles due to the turbulent fluctuations, the particle velocity is broken down into a convective and a turbulent diffusive component:[25,27-30]

$$\vec{V}_p = \vec{V}_{pc} + \vec{V}_{pd} \qquad (14)$$

where subscripts pc and pd denote the convective and the diffusive velocities, respectively. The convective velocity of the particle can be defined as the velocity that would result in the absence of turbulence, or that based on the mean gas velocity. This velocity can be obtained by Eq. (14) along a trajectory by numerical integration. The turbulent diffusive velocity accounting for the turbulent fluctuations can be modeled by assuming the turbulent diffusion of particles to be proportional to the gradient of mean particle number density:

$$\vec{V}_{pd}\,\bar{n}_p = D_p^t\,\vec{\nabla}\,\bar{n}_p \tag{15}$$

The transport coefficient D_p^t is defined as the turbulent particle mass diffusivity and can be expressed by:[23,25,27,28,30]

$$D_p^t = v_p^t / \sigma_p^t \tag{16}$$

where v_p^t and σ_p^t are the turbulent particle eddy viscosity and the Schmidt number for particles, respectively. It is noted that, for very small particles completely following the turbulent fluctuations, the turbulent particle Schmidt number can be approximated to unity; for large partilces failing to follow the turbulent fluctuations, it is less than unity. For the present work, the value of 0.35 was used for σ_p^t as recommended [14].

The turbulent kinematic viscosity of particle must account for the degree of turbulence and particle size. Much work is currently being performed by many investigators on how to obtain v_p^t.[16,21-23,25,31] Although no reliable model based on sound theory has been developed yet, a model by Melville and Bray[23] gave saticfactory results for the pulverized-coal combustion process.[14,16,27] For the present work, the following relationship by Melville and Bray[23] was selected for the expression of v_p^t, because it is simpler than other models and is applicable to the particle size range of interest in the flash-smelting processes:

$$v_p^t = v_g^t / [\,1 + (\tau_p / t_t)\,] \tag{17}$$

where τ_p and t_t are the particle relaxation time and the time scale of turbulence, respectively. The particle relaxation time is related to the Stokes particle drag by:[23,25-27,31]

$$\tau_p = m_p / (3\pi\mu_g d_p) \tag{18}$$

The time scale of turbulence can be related to the fluctuating velocity components by:[14,23,25]

$$t_t = v_g^t / (\overline{V' \cdot V'}) \tag{19}$$

By assuming an isotropic turbulence, namely,

$$\overline{u'}^2 = \overline{v'}^2 = \overline{w'}^2, \tag{20}$$

Equation (19) can be expressed as:

$$t_t = 1.5 \, C_\mu \, k / \varepsilon \tag{21}$$

2.6.2. Particle Number Density

Since the local particle number density cannot be calculated from the Lagrangian particle-phase information, it can only be approximated using the Eulerian gas-phase information. The continuity equation for the particle number density of the jth size particle in a turbulent flow at steady state can be expressed as:

$$\frac{\partial}{\partial x}(\overline{u}\,\overline{n}_j) + \frac{1}{r}\frac{\partial}{\partial r}(r\,\overline{v}\,\overline{n}_j) - \frac{\partial}{\partial x}(D_j^t \frac{\partial \overline{n}_j}{\partial x}) - \frac{1}{r}\frac{\partial}{\partial r}(r\,D_j^t \frac{\partial \overline{n}_j}{\partial r}) = 0 \tag{22}$$

The diffusion coefficient D_j^t in Eq.(22) is the same as in Eq.(15).

2.6.3. Particle-Phase Energy Equation

The change of particle temperature is expressed by the following equation written in the Lagrangian framework:

$$\frac{d}{dt}(m_p h_p) = \dot{H}_r + q_{rp} - Q_p - \dot{H}_v - \dot{H}_m \tag{23}$$

On the right-hand side of Eq. (23), the first term represents the heat of reaction of sulfide particles, the second term is the radiation heat transfer between the particles and the surroundings, the third term shows the heat loss to the gas phase by convection, the fourth term is the heat loss due to the volatilization of metal species, and the last term represents the heat loss due to the thermal decomposition or melting of particle. The treatment for the radiation heat transfer is detailed elsewhere.[11,45]

For each particle size, the heat loss due to the gas-phase convection is obtained by the following equations:

$$Q_p = Nu_j \, \pi \, d_j \, k_g \, (T_{pj} - T_g) \tag{24}$$

$$Nu_j = 2 + 0.65 \, Re_j^{1/2} \, Pr_g^{1/3} \tag{25}$$

$$Re_j = d_j \, |\vec{V}_g - \vec{V}_p| \, \rho_g / \mu_g \tag{26}$$

and

$$Pr_g = C_{pg} \, \mu_g / k_g \tag{27}$$

The heat loss due to volatilization of copper species is calculated by:

$$\dot{H}_v = \sum_i r_{vi} \, h_{vi} \tag{28}$$

where subscript i denotes the volatile copper species, r_{vi} is the volatilization rate of species i, and h_{vi} represents the enthalpy required for volatilization. To account for the heat loss due to melting of the particle, the Richards' law was used to obtain the enthalpy of fusion:

$$H_f = \quad 9.24 \text{ (number of atoms)} \, (T_{mp}/M_p) \quad [J/kg] \quad (29)$$

where T_{mp} and M_p are the melting point of particle and the molecular weight of particle, respectively. Once the melting temperature is reached, the heat available to melt the particle is obtained by the following relationship:

$$\dot{H}_m = \dot{H}_r + q_{r_p} - \dot{Q}_p - \dot{H}_v \qquad (30)$$

Before and after the melting is reached, H_m is set to be equal to zero.

2.7. Reactions of Chalcopyrie Particle

In order to develop a comprehensive mathematical model of the flash-smelting process, chemical kinetics must be incorporated with the fluid dynamics of the two-phase turbulent flow and the transfer of heat and mass.

2.7.1. Reaction Kinetics

The data for the intrinsic kinetics of the oxidation of chalcopyrite were obtained by Chaubal and Sohn.[4,32,33] The sulfur removal rate is related to the oxygen consumption by stoichiometric coefficients in oxidation reactions. Below the melting point of concentrate particles, the following equation was suggested for the rate of oxygen consumption:[4,33]

$$N_{O_{2,i}} \left(\frac{dX_i}{dt}\right) = N_{O_{2,i}} \, k_o \exp\left(-E/RT_p\right) f_1\left(p_{O_2}\right) f_2(X) f_3(d_p) \qquad (31)$$

where i is either chalcopyrite or pyrite, and $N_{O2,i}$ is the moles of oxygen required to react completely with either phase in the partcle. X refers to the overall fractional degree of sulfur removed at a certain time.

The values of parameters in Eq.(31) were determined by Chaubal and Sohn,[4,33] and are given in Table 1.

Once the particles become molten, the rate of oxidation of sulfur is considered to be equal to the rate of external mass transfer of oxygen from the bulk. Assuming this mass-transfer control, the reaction rate of the molten particle can be obtained by:

$$N_{O_2} \frac{dX}{dt} = k_m C_{O_2} A_p f_s \tag{32}$$

where k_m and f_s are, respectively, the mass transfer coefficient of oxygen and the fraction of the external surface area occupied by sulfides, which accounts for the fact that the produced oxide phases reduce the available surface area for oxygen transfer from the bulk. f_s is assumed to be the same as the volume fraction occupied by the sulfides. The above equation implies that the reaction between the oxide and the molten sulfides in the particle is negligible compared to the oxidation reaction of sulfides.

2.8. Physical Properties

The transport parameters and heat capacities necessary for solving the gas phase and particle phase modeling equations were obtained using the relationships which can be found in the literature[35,36] and were summarized by Hahn and Sohn[11].

Since changes in particle size are experimentally observed,[8,34,37,38] the particle size is allowed to change linearly with the extent of overall conversion of sulfide, as follows:

$$d_j = (1-X) d_{jo} + d_f X \tag{33}$$

where d_{jo} and d_f are, respectively, the initial size and the final particle size experimentally determined as 25 μm.[34]

3. NUMERICAL METHOD

Many researchers have studied numerical techniques to solve a recirculating turbulent flow system. Roache[40] presented a comprehensive review of available techniques for solving problems related to fluid dynamics. Gosman and Pun[41] developed a well-known numerical technique, which is called TEACH, to solve the gas-phase equations. According to this method, the nonlinear, elliptic, partial differential governing equations are cast into finite difference equations and solved by a line-by-line or tri-diagonal matrix algorithm. Gallagher et al.[42] developed a finite element method. Although this method has certain adventages, especially for arbitrary boundary shapes, it is still in a developing stage and has not been applied extensively to compressible recirculating flows.

The velocity components are governed by momentum equations. The difficluty in solving the velocity field lies in the unknown pressure field. The pressure gradients in the momentum equations become a part of the source term. However, there is no obvious equation for obtaining the pressure field which is only indirectly specified via the continuity equation. To solve this difficulty related to the pressure field, Patankar and Spalding[43,44] developed a semi-implicit method for pressure-linked equations (SIMPLE). Patankar[44] revised the SIMPLE and developed the so-called SIMPLER to improve the rate of numerical convergence.

In this model, based on the PCGC-2 code for coal gasification and combustion systems developed by Smoot and Smith,[28] the TEACH code and SIMPLER algorithm, and the PSI-CELL technique[12] have been used to solve the Eulerian gas field and the Lagrangian particle field, respectively.

4. EXPERIMENTAL

4.1. Experimental Flash Furnace

A laboratory flash furnace was constructed to carry out experiments simulating the flash smelting process. A schematic diagram of the apparatus is shown in Figure 2. It consists of three main units: the solids and gas feeding unit, the reactor shaft, and the off-gas handling unit.

4.1.1. Solid and Gas Feeding Unit

A screw feeder and a vibration feeder connected in series are used to feed solids, primarily chalcopyrite concentrate. The maximum capacity of these feeders are 5.0 kg/h. The vibration feeder minimizes the pulsing of solid feed generated by the screw feeder.

TABLE 1: LIST OF CHEMICAL REACTION KINETICS[*]

	k_o	E	$f_1(P_{O_2})$	$f_2(X)$	$f_3(d_p)$
1. Chalcopyrite					
$T_p < 754K$	2.4×10^8	215	P_{O_2}	$0.07/[\exp(X/0.07)]$	$1/d_p^2$
$754 < T_p < 873K$	0.026	71.4	"	"	"
Sulfur vaporization:					
	2.72×10^9	208	–	$(1-X)^2$	"
2. Pyrite					
Sulfur vaporization:					
	4.5×10^{10}	279	–	$(1-X)^{2/3}$	"

[*] Adapted from reference 4.

Note: k_o in $cm^2/(s \cdot kPa)$, P_{O_2} in kPa, E in kJ/mol, and d_p in cm.

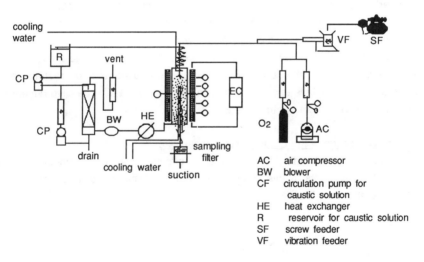

Figure 2. Schematic Diagram of A Flash Reaction System.

AC   air compressor
BW   blower
CF   circulation pump for caustic solution
HE   heat exchanger
R   reservoir for caustic solution
SF   screw feeder
VF   vibration feeder

The solids coming out of the vibration feeder are introduced together with the oxidizing gas as a jet via a T-junction to a water-cooled burner located at the top of the furnace.

The simple burner arrangement used in this study includes a water-cooled chamber. The top half of the unit is the mixing chamber. The bore of the injection pipe(2 cm ID) was so chosen that a linear velocity of 4 m/s could be attained for a gas flow rate of about 70 l/min.

Air and oxygen are used as the oxidizing gases. Air is supplied by an oil-less compressor with a maximum capacity of 252 l/min, and pure oxygen(99.99%) is supplied from an oxygen cylinder.

4.1.2. Reactor Shaft

The reactor shaft is a vertical cylindrical unit externally heated by six 3.175 cmφ silicon carbide electric heaters. The reactor tube was prepared by casting carborundum's carbofrax L-11 silicon carbide mix. The shaft is sealed off at the top with a 2-cm thick silicon carbide disc which has an opening at the center through which a burner is inserted.

This unit rests on a steel frame-work which has a variable opening to facilitate the introduction of various probes into the furnace. At the lower end of this supporter, reacted particles are quenched and deposited.

Two Pt-Pt 10% Rh thermocouples are placed at the outer wall of the reactor and connected to a temperature controller. Six thermocouples are placed at various heights along the outer surface of the reactor and connected to a recorder which monitors the temperature profile of the reactor wall.

The reactor shaft and heating bars are surrounded by a rectangular brickwork of fireclay. Kaowool blanket followed by Kaowool fiber is used as additional insulation, and the entire assembly is encased in a stainless-steel shell. The detailed dimensions of the reactor shaft are shown in Table 2.

4.2. Characterization of the Chalcopyrite Concentrate

Chalcopyrite concentrate from the flotation circuit of Kennecott concentrater was used in all experiments in this study. The results of mineralogical analysis of the sample are shown in Table 3.

TABLE 2: DETAILS OF THE LABORATORY FLASH FURNACE.

Inlet geometry : single entry		
burner tip inside diameter	:	0.02 m
furnace inside diameter	:	0.23 m
furnace length	:	1.32 m
furnace wall thickness	:	0.02 m
Input power to the furnace	:	20 kw
Maximum inlet gas flow rate	:	15 m^3/h
Maximum concentrate feed rate	:	5 kg/h

TABLE 3: ANALYSIS OF CHALCOPYRITE CONCENTRATE.

chemical	% S	: 30
	% Cu	: 26
	% Fe	: 27
	% H_2O	: 0.3
mineralogical		
	chalcopyrite	: 59 %
	bornite	: 13 %
	silicate	: 15 %
	pyrite	: 7 %
	molybdenite	: 3 %
size		cumulative percent
	mesh	passing
	150	95.1
	200	78.4
	270	65.3
	325	57.1
	400	49.2

4.3. Sampling and Measurement

During the operation of the furnace, process measurements are made including gas temperature, SO_2 concentration, sulfur content of the particle, and the distribution of particles at two radial positions as a ratio to those collected at the central position.

4.3.1. Solid Products
4.3.1.1. Collection

In order to collect solid samples at various locations inside the furnace with a minimum disturbance of the flow, a sampling probe which consists of two concentric stainless-steel 316 pipes is introduced through the opening at the bottom of the furnace. Cooling water flows through the annular space between the pipes whereas the gas-particle sample is collected through the inner tube.

The bottom end of the probe is connected by a flexible tubing to a filter unit to which suction is applied by a vacuum pump. The pumping rate is adjusted so that the sampling conditions are close to isokinetic with a minimum disturbance of the flow. The two main requirements of the probe are that it should be able to quench the particles close to the entry point and that the sample should be representative of the particles at the location in the reactor. Once the particle temperature falls below the melting point, very little reaction occurs and it can be assumed that the chemical and mineralogical make-up of the collected particles is close to that at the probe tip. In this study, a flow rate of 3.2 m/s(at 285K) is used for the maximum cooling of particles as close to the entry point as possible.

4.3.1.2. Analysis

The sulfur contents of the collected particles and the feed concentrate were measured by a high frequency combustion titration method. For this purpose, a LECO sulfur analyzer was used.

For the non-reacting system, the size analysis of collected particles and the feed is carried out using a Sedigraph 5000 particle size analyzer. The changes in particle size during flash smelting was investigated by means of microscopy. Particles were collected for 5 minutes at 5 cm radial positions along seven axial positions. Collection of particles were performed for reacting and non-reacting systems. The non-reacting system was at room temperature.

4.3.2. Gas Phase

The sulfur dioxide concentration and the temperature of the gas are the two variables to be measured in the gas phase.

The sulfur dioxide content of the gas at various positions is measured by an infrared SO_2 analyzer. The gas sample is sucked at a linear velocity of 8 cm/s through a 0.635cm ID pipe attached to the vacuum line after a filter. This pipe runs into the inlet of the infrared SO_2 analyzer, the exit end of which is connected to an aspirator.

In order to measure the gas temperature, a simple suction probe as shown in Figure3 was constructed. The tube is made of stainless steel 314. The tip of the probe is wrapped in a kaowool refractory blanket to reduce the effect of radiation from the walls. A Pt-Pt 13%Rh thermocouple is positioned inside the tube so that the tip of the thermocouple is about 0.5 cm below the tip of the tube. Gas is sucked at a linear velocity of 5 cm/s through the probe during measurement.

5. RESULTS AND DISCUSSION

The model predictions were compared with experimental data with a laboratory flash furnace. Both the experimental data and the predictions are for gas-phase temperature, SO_2 concentration, sulfur remaining, and the particle mass flux normalized by that at the centerline. The conditions for experimental runs are shown in Table 4.

By using a laboratory flash furnace some experimental data along the centerline of the furnace were obtained and compared with the predictions.

Figure 4. shows the effect of inlet gas velocity on the gas temperature for non-reacting heating system(exp. nos. 1 to 3). As shown in this figure, measured experimental data are a little higher than the predictions at the upper position of the furnace due to radiative heat transfer at the tip of temperature probe. The lower value of the measured is due to leakage of surrounding air from the bottom of the furnace.

Figures 5 and 6 show the furnace wall temperature, gas temperature, and model predictions obtained at various axial and radial positions with the experimental conditions of nos. 7 and 9. Although some discrepancy exits between the predictions and the measurements, there is a reasonably good overall agreement.

521

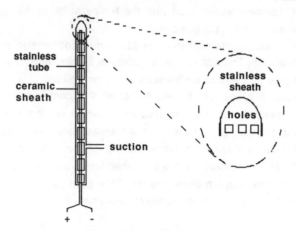

Figure 3. Temperature Probe.

TABLE 4: CONDITIONS FOR EXPERIMENTAL RUNS.

exp. no.	Lv[a]	MT[b]	%O_2	G[c]	S[d]	remark
1	2.8	"	21	0.76	0.	g
2	4.4	"	21	1.19	0.	g
3	9.8	"	21	2.66	0.	g
4	4.0	50	21	4.57	4.8	e f
5	4.0	60	21	4.57	3.8	e f
6	4.0	70	21	4.57	3.3	e f
7	4.0	70	30	4.57	4.7	e f
8	3.4	65	21	3.79	3.0	f
9	3.4	55	21	3.79	3.5	f
10	5.4	50	21	0.76	0.79	e
11	2.8	60	21	0.76	0.64	e
12	2.8	70	21	0.76	0.57	e

a	linear velocity of gas at the burner tip(m/s)
b	target matte grade(% Cu) assuming Fe_3O_4 production
c	volumetric flow rate of gas(m^3/h) at STP
d	mass flow rate of solid feed(kg/h)
e	central position only
f	central,5 and 10 cm radial positions
g	central,gas temperature measurement

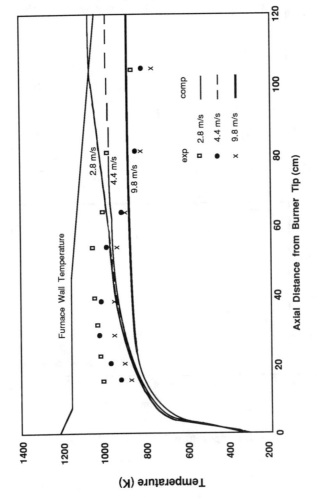

Figure 4. Effect of Inlet Gas Velocity on Gas Temperature for Non - Reacting Heating System.

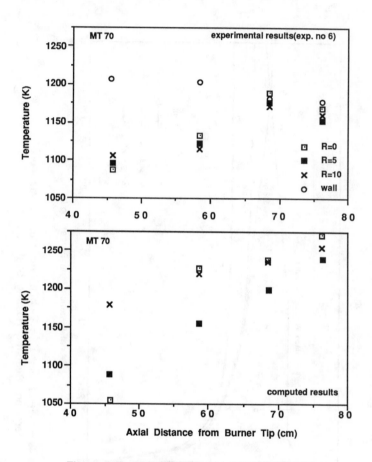

Figure 5. Temperature Profile: MT 70 , 21% Oxygen.

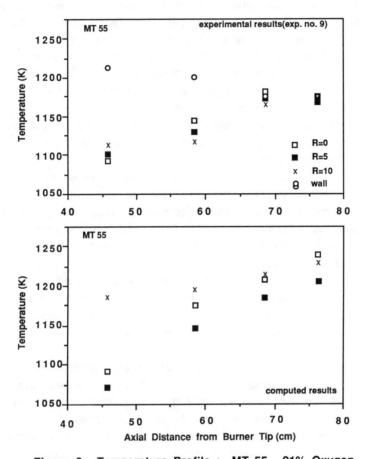

Figure 6.   Temperature Profile :   MT 55 , 21% Oxygen.

In Figures 7 and 8, the particle mass flux normalized by that at the centerline with the experimental conditions of Nos. 4,7,8 and 9 are shown. The centerline mass fluxes for the reacting and non-reacting(303K) systems are shown in tables 5 and 6. From the figures and tables, the distribution of particles can be predicted.

The experimental measurements show that the particles are evenly dis-persed after 85 cm from the burner tip, and the dispersion of reacted particles is more even than the unreacted ones, especially at the upper section of the furnace. The effects of gas inlet velocity and matte grade on the distribution of particles are significant as shown in these figures. The particles are dispersed much better at low inlet gas velocity, especially for the reacted particles. For the same gas inlet velocity, the dispersion is better at higher target matte grade.

Model predictions for the same experimental conditions are also shown in Figures 7 and 8. There is some discrepancy between the predictions and measurements, which needs further investigation.

Figures 9 to 12 show the comparison between model predictions and measurements obtained by varying the target matte grade between 50% and 70% with 21% oxygen. Figure 9 is for the injection velocity of 4 m/s (exp. nos. 4 , 6) and Figures 10 to 12 are for that of 2.8 m/s (exp. nos. 10,11,12). In Figure 9, the center-line profile of the SO_2 and O_2 concentrations and the amount of sulfur remaining are plotted against the axial distance from the top.

As shown in Figure 9, reasonable agreement between the predicted results and measurements was obtained except near the furnace bottom. However, as shown in Figures 10 to 12, for lower inlet gas velocity, there is some discrepancy between the predictions and measurements. This is due to the air leaking into the furnace from an opening at the bottom for the sample probe.

The effect of oxygen concentration at the inlet is also shown in Figure 13. The oxygen content in the inlet stream was varied from 21 to 30% with the injection velocity of 4 m/s and 70 % matte grade(exp. nos. 6,7). The predicted and measured results show that the reaction rate increases somewhat as the O_2 concentration increases.

Figure 14 shows the amount of sulfur remaining from both the predictions and measurements for different oxygen concentrations of gas at inlet, but with the same target matte grade of 70 at three different radial positions(exp. nos. 6,7). The reaction extent of particles near the wall is somewhat greater than those near the centerline. From this figure, we can conclude that under our experimental conditions, most of the

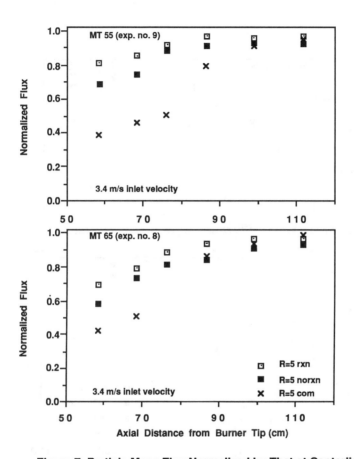

Figure 7. Particle Mass Flux Normalized by That at Centerline.

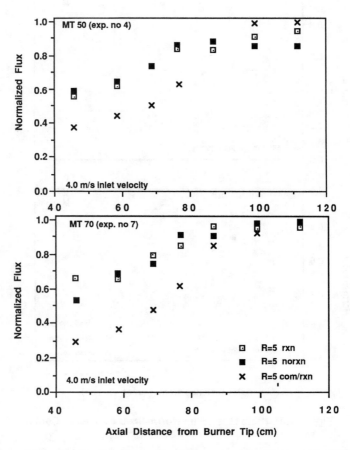

Figure 8. Particle Mass Flux Normalized by That at Centerline.

TABLE 5: PARTICLE MASS FLUX[a] AT THE CENTERLINE ;
FOR NON-REACTING SYSTEM

location[b]	experiment no.			
	4	7	8	9
46	5.75×10^{-6}	5.69×10^{-6}	3.60×10^{-6}	3.75×10^{-6}
58	4.92 "	4.49 "	2.86 "	2.89 "
69	4.12 "	3.97 "	2.22 "	2.49 "
76	3.29 "	3.05 "	1.94 "	2.00 "
86	3.05 "	3.02 "	1.82 "	1.94 "
99	3.02 "	2.83 "	1.63 "	1.85 "
112	3.01 "	2.80 "	1.57 "	1.78 "

a : $kg / cm^2/s$
b : axial distance from the burner tip (cm)

TABLE 6: PARTICLE MASS FLUX[a] AT THE CENTERLINE ;
FOR REACTING SYSTEM

location	experiment no.			
	4	7	8	9
46	4.83×10^{-6}	4.12×10^{-6}	2.83×10^{-6}	3.05×10^{-6}
58	4.06 "	3.73 "	2.21 "	2.22 "
69	3.32 "	3.11 "	1.72 "	2.00 "
76	2.83 "	2.49 "	1.42 "	1.63 "
86	2.58 "	2.46 "	1.35 "	1.54 "
99	2.49 "	2.40 "	1.32 "	1.51 "
112	2.34 "	2.34 "	1.26 "	1.48 "

a : $kg / cm^2/s$
b : axial distance from the burner tip (cm)

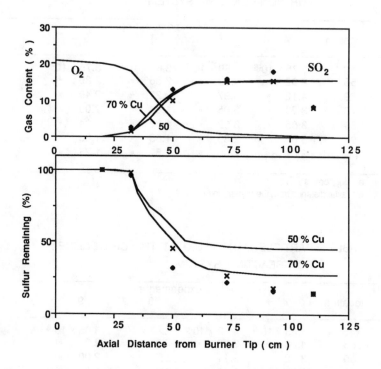

Figure 9. Effect of Matte Grade on Reaction of Chalcopyrite Concentrate. Measured Data (• 70 & × 50 % Cu ; 21 % O_2), -- Predictions.

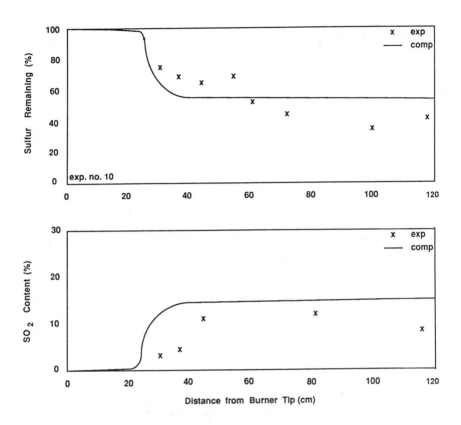

Figure 10. Reaction of Chalcopyrite Concentrate for Target
Matte Grade of 50 : Exp. No. of 10 .

531

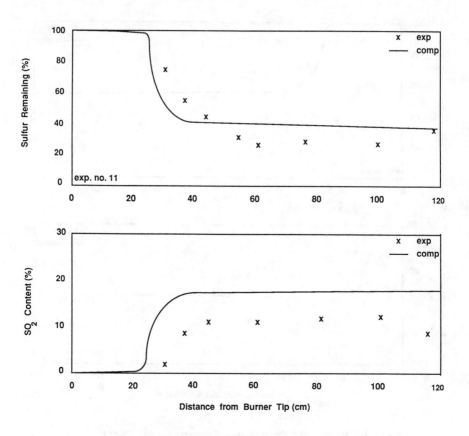

Figure 11. Reaction of Chalcopyrite Concentrate for Target
Matte Grade of 60 : Exp. No. of 11 .

532

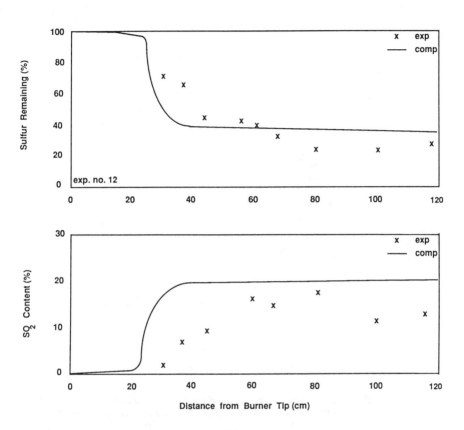

Figure 12. Reaction of Chalcopyrite Concentrate for Target
Matte Grade of 70 : Exp. No. of 12 .

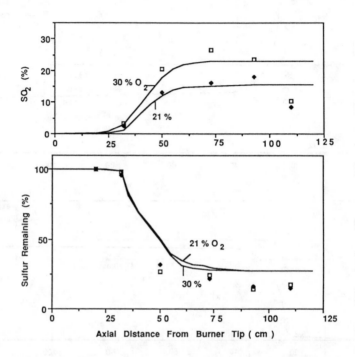

Figure 13. Effect of Oxygen Content on Reaction of Chalcopyrite Concentrate.
Measured Data (◆ 21 & ◻ 30 % O$_2$; 70 % Cu), -- Predictions.

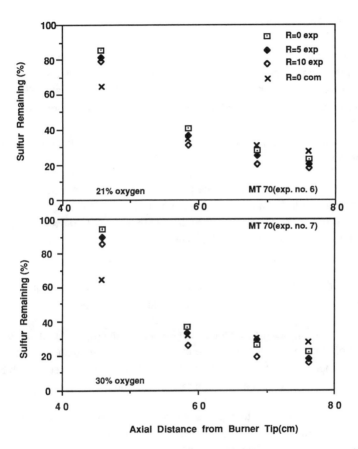

Figure 14. Reaction of Chalcopyrite Concentrate for 70 PCT
            Matte Grade.

reaction of chalcopyrite concentrate is finished within 60 cm from the burner tip, and the effect of oxygen content at the inlet on the sulfur remaining is not large for the same target matte grade.

The mathematical model developed in this work was also used to predict the performance of the Outokumpu pilot flash furnace, the results of which can be found elsewhere[11,45].

6. CONCLUDING REMARKS

The numerical predictions of the flash-smelting furnace show overall satisfactory agreement with experimental data in spite of complex nature of the system and mathematical models, This leads to the conclusion that the two-equation (k-ε) turbulence model incorporating the effect of particles on turbulence can adequately be used for the modeling of the flash-smelting furnace.

Model predictions of the performance of the flash smelting furnace point to the following conclusions:

1 For the effect of inlet gas velocity on the gas temperature for non-reacting heating system(exp. nos. 1 to 3), measured experimental data are a little higher than the predictions at the upper position of the furnace probably due to radiative heat transfer at the tip of temperature probe. The lower values of the measured data is due to leakage of surrounding air from the bottom of the furnace.

2. For the centerline profiles of SO_2 and O_2 concentrations, reasonable agreement between the predicted results and measurements was obtained.

3. Particles were evenly dispersed after 85 cm away from the burner tip, and the dispersion of reacted particles was more even than in a non-reacting system, especially in theupper section of the furnace. Particles were dispersed much better at low inlet gas velocities, especially in a reacting system.

4. For the profile of gas temperature and sulfur remaining in particles at different radial positions(5 and 10 cm) of the furnace, the predictions were in good agreement with the measurements up to 80 cm from the burner tip.

NOMENCLATURE

A_p projected area of a particle

C_D drag coefficient, defined

C_p specific heat capacity

d_p particle diameter

D diffusivity

$D_p{}^t$ turbulent particle diffusivity

E activation energy

g gravitational acceleration

h enthalpy

H_m rate of heat loss due to melting

H_r rate of heat production by reaction

H_v rate of heat loss due to volatilization

k turbulent kinetic energy

k_g gas-phase heat conductivity

k_m mass transfer coefficient

k_o pre-exponential factor

m_j mass fraction gas species j

m_p mass of a particle

n particle number density

Nu Nusselt number, defined

p pressure

Pr Prandtl number, defined

q_{rp} radiative heat-transfer rate for the particle phase

Q_p rate of heat loss due to gas-phase convection

Q_{rg} volumetric heat-transfer rate by gas-phase radiation

r radial distance from the axis of symmetry

R universal gas constant

Re Reynolds number

S source or sink term in conservation equations

Sc Schmidt number

Sh Sherwood number

t	time
t_t	turbulent time scale
T	temperature
u	axial velocity
v	radial velocity
w	tangential velocity
w_k	mass of gas species k
X	overall fraction of sulfur removed

GREEK SYMBOLS

β	scattering angle
Γ_e	effective transport exchange coefficient
ε	dissipation rate of turbulent kinetic energy or emissivity
μ	viscosity
υ	kinematic viscosity
ρ	density or reflectivity
ρ_{bp}	bulk particle density (mass of particles per volume)
$\sigma_\varnothing$	Prandtl-Schmidt number for ø
τ	shear stress
τ_p	particle relaxation time
ø	general dependant variable

SUBSCRIPT

e	effective value
g	gas
p	particle
t	turbulent

SUPERSCRIPT

t	turbulent
'	fluctuation component

OVERLINE

--   time-averaged

=   second-order tensor

Acknowledgment

The authors wish to express their appreciation to Dr. K.J. Richards, Mr. D. B. George and Dr. L.K. Bailey of Kennecott Process Technology for helpful discussions, and to Dr. P. J.Smith of Brigham Young University for providing their computer program for a gas-particle jet.

This work was supported in part by the National Science Foundation under Grant No. CPE-8204280, and by the Department of the Interior's Mineral Institute program administered by the Bureau of Mines through the Generic Mineral Thechnology Center for Pyrometallurgy under allotment Grant No. G1125129.

REFERENCE

1.   The Staff, Mining and Smelting Div., INCO Ltd., "Oxygen Flash Smelting Process of the International Nickel Company", J. Metals, 1955, vol.7, pp. 742-751.

2.   P. Bryk, J. Ryselin, J. Honkasalo, and R. Malmstrom, "Flash Smelting Copper Concentrates", J. Metals, 1958, vol. 10, pp. 395-400.

3.   S. V. Harkki and J. T. Juusela, "New Development in Outokumpu Flash Smelting. Method", TMS paper selection A74-16, TMS-AIME, Warrendale, PA, 1976.

4.   P. C. Chaubal, "The Reaction of Chalcopyrite Concentrate in a Flash Furnace Shaft", Ph.D Dissertation, University of Utah, 1986.

5   INCO Research, "Oxygen Flash Smelting in a Convertor", paper presented at the 106th Annual Meeting of AIME, Atlanta, Georgia, March 7-11, 1977.

6.   J. C. Yannopoulous, C. E. Swanson, and J. W. Ahlrichs, "Oxidation Reactions in a Dispersed Copper Smelting System", Extractive Metallurgy of Copper, J. C. Agarwal and J. C. Yannopoulous eds., vol.1, TMS-AIME, New York, 1976, pp. 49-65.

7. N. J. Themelis, J. K. Makinen, and N. D. H. Munroe, "Rate Phenomena in the Outokumpu Flash Smelting Reaction Shaft", the Symposium on "Physical Chemistry of Extractive Metallurgy", V. Kudryk and Y. K. Rao eds., TMS-AIME, Warrendale, PA, 1985, pp. 289-309.

8. Y. H. Kim and N. J. Themelis, "Effect of Phase Transformation and Particle Fragmentation on the Flash Reaction of Complex Metal Sulfides", The Reinhardt Schuhmann International Symposium on Innovative Technology and Reactor Design in Extraction Metallurgy, D. R. Gaskell, J. P. Hager, J. E. Hoffmann, and P. J. Mackey eds., TMS-AIME, Warrendale, PA, 1986, pp. 349-369.

9. Y. Fukunaka, S. Nakashita, Z. Asaki, and Y. Kondo, "A Modeling Study on the Pyrite Smelting Process", World Mining and Metals Technology, A. Weiss ed., AIME, New York, 1976, vol.1, pp. 481-504.

10. S. Ruottu, "The Description of a Mathematical Model for the Flash Smelting of Cu Concentrate", Combustion and Flame, 1979, vol. 34, pp. 1-11.

11. Y.B. Hahn and H.Y. Sohn, "Mathematical Modeling of the Combined Turbulent Transport Phenomena, Chemical Reactions, and Thermal Radiation in a Flash-Furnace Shaft," Mathematical Modeling of Materials Processing Operations, Proceedings of a Symposium on "Extractive and Process Metallurgical Meeting," J. Szekely et al. eds., TMS, Palm Spring, CA,1987, pp.799-834.

12. C.T.Crowe,M. P. Sharma, and D. E. Stock, "The Particle-Source-In-cell (PSI-CELL) Model for Gas-Droplet Flows", Journal of Fluid Engineering, Trans. ASME, 1977, pp. 325-332.

13. L. D. Smoot and D. T. Pratt, Pulverized Coal Combustion and Gasification, Plenum Press, New York, 1979, pp. 57-64, 83-104, 217-231.

14. L. D. Smoot and P. J. Smith, User's Manual for a Computer Program for 2-Dimensional Coal Gasification or Combustion(PCGC-2), Combustion Laboratory, Brigham Young University, 1983.

15. D. B. Spalding, Numerical Computation of Multiphase Flows, Lecture Notes, Thermal Science and Propulsion, Purdue University, 1979, pp. 161-190.

16. L. D. Smoot and P. J. Smith, Coal Cumbustion and Gasification , Plenum press, New York, 1985, pp. 245-264, 349-371.

17. B. E. Launder and D. B. Spalding, Mathematical Models of Turbulence, Academic Press, London, 1972.

18. E. E. Khalil, D.B. Spalding and J. H. Whitelaw, "The Calculation of Local Flow Properties in Two-Dimensional Furnaces", Int. J. Heat Mass Transfer, 1975, vol. 18, pp. 775-790.

19. H. Schlichting, Boundary-Layer Theory, Mc Graw-Hill Inc., New York, 1979, pp. 473-480, 578-595.

20. B. E. Launder and D. B. Spalding, "The Numerical Computation of Turbulent Flows", Computer Methods in Applied Mechnics and Engineering, 1974, vol. 3, pp. 269-289.

21. S. E. Elghobashi and T.W. Abou-Arab, "A Two-Equation Turbulence Model for Two-Phase Flows",Phys. Fluids,1983, vol. 26, pp. 931-938.

22. S. E. Elghobashi, T. W. Abou-Arab, M. Rizk, and A. Mostafa, "Prediction of the Particle-Laden Jet with a Two-Equation Turbulence Model", Int. J. Multiphase Flow, 1984, vol. 10, pp. 697-710.

23. E. K. Melville and N. C. Bray, "A Model of the Two-Phase Turbulent Jet", Int. J. Heat Mass Transfer, 1979, vol. 22, pp. 647-656.

24. P. J. Smith and L. D. Smoot, "One-Dimensional Model for Pulverized Coal Combustion and Gasification", Combustion Science and Technology, 1980, vol. 23, pp. 17-31.

25. A. S. Abbas, S. S. Koussa, and F. C. Lockwood, "The Prediction of the Particle Laden Gas Flows", Eighteenth Symposium (International) on Combustion, the Combustion Institute, 1980, pp. 1427-1437.

26. F. C. Lockwood, A. P. Salooja, and S. A. Syed, "A Prediction Method for Coal-Fired Furnace", Combustion and Flame, 1980, vol. 38, pp. 1-15.

27. P. J. Smith, T. H. Fletcher, and L. D. Smoot, "Model for Pulverized Coal-Fired Reactors", Eighteenth Symposium (International) on Combustion, the Combustion Institute, 1980, pp. 1285-1293.

28. L. D. Smoot and P. J. Smith, User's Manual for a Computer Program for 2-Dimensional Coal Gasification or Combustion(PCGC-2), Combustion Laboratory, Brigham Young University, 1983.

29. J. O. Hinze, "Turbulent Fluid and Particle Interaction", Progr. Heat Mass Transfer, 1971, vol. VI, pp. 433-452.

30. T. H. Fletcher, "A Two-Dimensional Model for Coal Gasification and Combustion", Ph. D. Dissertation, Brigham Young University, 1983.

31. G. P. Lilly, "Effect of Particle Size on Particle Eddy Diffusivity", Ind. Eng. Chem. Fundam., 1973, vol. 12, pp. 268-275.

32. P. C. Chaubal and H. Y. Sohn, "Intrinsic Kinetics of the Oxidation of Chalcopyrite Particles under Isothermal and Nonisothermal Conditions," Metall. Trans. B, 1986, vol. 17B, pp. 51-60.

33. P. C. Chaubal and H. Y. Sohn, "Combustion and Ignition of Chalcopyrite Particles under Suspension Smelting Conditions", submitted to Metall. Trans. B, 1987.

34. F. R. A. Jorgensen, "Single Particle Combustion of Chalcopyrite", Proc. Australas. Inst. Min. Metall., 1983., vol. 288, pp. 37-46.

35. R. B. Bird, W. E. Stewart, and E. N. Lightfoot, Transport Phenomena, John Wiley & Sons, Inc., N. Y. , 1960.

36. J. Szekely, J. W. Evans, and H. Y. Sohn, Gas-Solid Reactions, Academic press, New York, 1976.

37. T. Kimura, Y. Ojima, Y. Mori, and Y. Ishii, "Reaction Mechnism in a Flash Smleting Reaction Shaft," the Reinhardt Schumann International Symposium on Innovative Technology and Reactor Design in Extraction Metallurgy, D. R. Gaskell, J. P. Hager, J. E. Hoffmann, and P. J. Mackey eds., TMS-AIME, Warrendale, PA, 1986, pp. 403-418.

38. E. Partelpoeg, personal communication, Phelps Dodge Corp. Hidalgo Smelter, Playas, New Mexico, June, 1987.

39. P. J. Smith, "Theoretical Modeling of Coal and Gas Fired Turbulent Combustion or Gasification", Ph. D. Dissertation, Brigham Young University, 1979.

40. P. J. Roache, Computational Fluid Dynamics, Hermosa publishers, Albuquerque, New Mexico, 1976.

41. A. D. Gosman and W. M. Pun, Lecture Notes for Course Entitled "Calculation of Recirculating Flows", Imperial College, London, 1973.

42. R. H. Gallagher, J. T. Oden, C. Taylor, and O. C. Zienkiewicz, Finite Elements in Fluids, vols. 1 and 2, John Wiely & Sons, London, 1975.

43. S. V. Patankar and D. B. Spalding, "A Calculation Procedure for Heat, Mass and Momentum Transfer in Three-Dimensional Parabolic Flows", Int. J. Heat Mass Transfer, 1972, vol.15, pp.1787-1806.

44. S. V. Patankar, Numerical Heat Transfer and Fluid Flow, Mc Graw-Hill Book Co., New York, 1980.

45. Y. B. Hahn, "Mathematical Modeling of Chalcopyrite Concentrate Combustion in an Axisymmetric Flash-Furnace Shaft", Ph.D Dissertation, University of Utah, 1988.

The Influence of Spray Characteristics and Air Preheat on the Flame Properties in a Gas Turbine Combustor

A. M. ATTYA and M. A. HABIB
Mechanical Power Engineering Department
Cairo University
Giza, Egypt

ABSTRACT:

A computational procedure for the calculation of the flow, temperature and species concentration in a model of a gas-turbine combustor is provided and analysed. The equations which represent the transport of mass, momentum, enthalpy, unburnt fuel and mixture fraction were solved together with the equations representing the turbulence, combustion, radiation and spray models. A two equation model which comprises the solution of the turbulence kinetic energy and its dissipation rate was used to present the Reynolds stresses in the momentum equations. A single step reaction model in which the rate of reaction is connected to the rate of mixing between fuel and oxygen was used. The model used to represent the fuel spray is of the tracking type and is based on the assumption that the spray is represented by a finite number of groups of droplets rather than by a continuous distribution function. The results of the solution procedure were compared to the corresponding experimental data for different cases of preheated combustion air and different spray mean droplet diameters. The comparison indicated reasonable agreement between the calculated temperature, oxygen, carbon dioxide and unburnt fuel concentrations, but near the centerline high discrepancies were observed. The reasons for these discrepancies were provided and analysed.

The results indicated that the spread of the injection points of the spray resulted in an enhanced improvement of the flame properties. The results indicated also that increasing the combustion air preheat temperature from 125 to 225 °C resulted in a considerable improvement of the turbulent mixing rates between the fuel and air. The decrease in the droplet diameter from 115 to 54 um was shown to shift the flame upstream of the combustor.

1. INTRODUCTION:

In the majority of gas turbine combustion chambers, the fuel is introduced in the form of a spray of liquid fuel droplets. The combustion process of the spray involves complex, chemically reacting, multi-component two phase flows with phase change which encompass fluid mechanic, chemical and heterogeneous effects. The fluid mechanic and chemical effects are well known to be controlled by the turbulent mixing between

543

the fuel and the oxidant and the chemical kinetics. However, the heterogeneous effects are linked to the spray characteristics including evaporation rate, droplet size and velocity and fuel properties. The increasing need for the fuel economy and to establish design criteria for efficient and stable combustion has generated fresh interest in both experimental and theoretical studies of spray flames. Recent experimental studies including Attya [1], Attya and Whitelaw [2, 3], Elbanhawy and Whitelaw [4] and Yule et al.[5] concentrated on the study of the detailed structure of spray flames while Miyasaka and Law [6], Sangiovanni and Labowsky [7], Okajima [8], Karasawa et al. [9] and Cho and Law [10] examined evaporating sprays.

On the theoretical side, variants of two models are currently employed, namely the Williams [11] statistical spray model as used by Westbrook [12], Haselman and Westbrook [13], Cliffe et al. [14] and Jones and McGuirk [15]. The second model is called the "discrete droplet" model used by Crowe et al. [16], Elbanhawy and Whitelaw [17], Gosman and Ioannides [18], Jones and McGuirk [15] Attya and Whitelaw [2], Okajima [8], Sirignano [19] and Sellens and Brzustowski [20].

The statistical spray model considers a generalized spray distribution function originally defined in eight-dimensional space of droplet diameter, location, velocity and time. Conservation principles yield a partial integro-differential equation for this function and the solution of this equation , together with the gas conservation equations, provides the required model for the spray. This approach, as applied, is costly in terms of computer storage and time unless simplifications are introduced. Furthermore, due to the limited resolution especially in the vicinity of the atomizer, this method may introduce substantial spurious numerical diffusion into the calculation.

In the "discrete droplet" model, the spray is represented by a finite number of groups of droplets rather than by a continuous distribution function. Each of these groups is characterised by the same initial size, velocity and temperature. The motion, heating and evaporation of each group as it traverses the gas are computed by solving numerically the Lagrangian ordinary differential equations which govern the mass, momentum and energy conservation. The effects of the droplets on the gas phase are introduced into the Eulerian equations representing the gas by feeding in the local rates of heat, mass and momentum exchange deduced from the analysis of the droplet group trajectories. The solution is obtained by iterating between the calculation of the two phases.

In the present paper the "discrete droplet" model was used to predict the properties of preheated kerosene spray flames within a two-dimensional combustor model. The present mathematical formulation comprises the application of the Eulerian conservation equations to the gas phase and Lagrangian equations of droplet motion and heat balance to the droplets representing the spray.

The two sets of equations are coupled through the droplet tracking technique of Crow et al. [16] to allow for the determination of the droplet location and properties within the flow field.

The eddy dissipation model of Magnussen et al. [21] was used to calculate the mean rate of burnt fuel. The radiative heat transfer between the combustion gases and combustor walls was calculated using the "four flux" equation radiation model of DeMacro and Lockwood [22]. The influence of the spray on the gas field was considered through source terms representing different exchange processes included in the conservation equation of the gas field.

The paper describes the mathematical method together with its application for the calculation of spray flame properties. The combustion air was preheated to 125°C and 225°C with a swirl number of 1.4. The combustor was equipped with a rotating cup atomizer which was capable of producing mono-disperse droplet size. The sauter mean droplet diameters of the sprays used were 54 μm and 115 μm. The predicted results are compared with the measurements given by Attya [1] and the results of the comparison are analysed and appraised.

2. THE MATHEMATICAL PROCEDURE:

The Governing Equations:

The flow field inside the combustor may be described by a set of equations governing the transport of mass, momentum, turbulence kinetic energy and its dissipation rate, enthalpy and species concentration. The general form of these conservation equations can be written in a general form as follows:

$$\frac{\partial}{\partial x}(\rho U \phi) + \frac{1}{r}\frac{\partial}{\partial r}(r\rho V \phi) = \frac{\partial}{\partial x}\left(\Gamma_{\phi_{eff}}\frac{\partial \phi}{\partial x}\right) + \frac{1}{r}\frac{\partial}{\partial r}\left(r\Gamma_{\phi_{eff}}\frac{\partial \phi}{\partial r}\right) + S_\phi + S_d \quad (1)$$

This equation presumes that the mass flow is two-dimensional, axi-symmetric and that the turbulent diffusion can be represented through the gradient of the mean quantities and exchange coefficients $\Gamma_{\phi eff}$. The source term S_ϕ represents the generation or destruction of the variable ϕ by processes concerned with the gas phase. The additional source term S_d represents the generation or destruction due to the fuel droplets and will be described later. These gas phase equations with their models for turbulence, combustion, and radiation were solved by a finite difference technique described by Gosman and Pun [23]. Table (1) presents the exact expressions for the exchange coefficients and the source term S corresponding to each variable.

The Turbulence Model:

A two-equation turbulence model was used to represent

the turbulence characteristics as introduced by Jones and Launder [24] for recirculating flows. The model involves an assumed linear relationship between the Reynolds stress and the rate of strain.

$$\rho.\overline{u_i.u_j} = \frac{2}{3}\,\sigma_{ij}\,(\rho k + \mu_{eff}\,\frac{\partial u_1}{\partial x_1}) - \mu_{eff}\,(\frac{\partial u_i}{\partial x_j} + \frac{\partial u_j}{\partial x_i}) \tag{2}$$

For turbulence fluxes of scaler quantities a gradient diffusion model is used:

$$\rho.\overline{u.\phi'} = - (\frac{\mu_t}{\sigma_t}).\frac{\partial \phi}{\partial x_i} \tag{3}$$

The effective viscosity, μ_{eff}, is defined as:

$$\mu_{eff} = \mu_t + \mu_1$$

where μ_1 is the laminar viscosity and μ_t is the turbulent viscosity defined by:

$$\mu_t = C_\mu.\rho.k^2/\varepsilon. \tag{4}$$

The k and ε equations have been defined by many modellers, see for example Jones and Launder [24];

k-equation:

$$\frac{\partial(\rho Uk)}{\partial x} + \frac{1}{r}.\frac{\partial(r\rho Vk)}{\partial r} = \frac{\partial}{\partial x}\,(\frac{\mu_{eff}}{\sigma_k}.\frac{\partial k}{\partial x}) + \frac{1}{r}\frac{\partial}{\partial r}(r\,\frac{\mu_{eff}}{\sigma_k}.\frac{\partial k}{\partial r}) + \mu_{eff}.G - \rho\varepsilon \tag{5}$$

ε- equation:

$$\frac{\partial(\rho U\varepsilon)}{\partial x} + \frac{1}{r}.\frac{\partial(r\rho V\varepsilon)}{\partial r} = \frac{\partial}{\partial x}\,(\frac{\mu_{eff}}{\sigma_\varepsilon}.\frac{\partial \varepsilon}{\partial x}) + \frac{1}{r}\frac{\partial}{\partial r}(r\,\frac{\mu_{eff}}{\sigma_\varepsilon}.\frac{\partial \varepsilon}{\partial r}) + \mu_{eff}C_1\,(\frac{\varepsilon}{k}).G$$

$$- C_2\rho\,(\frac{\varepsilon^2}{k}) \tag{6}$$

where G is the production of the turbulence kinetic energy and is given by;

$$G = 2\,\{(\frac{\partial U}{\partial x})^2 + (\frac{\partial V}{\partial r})^2 + (\frac{V}{r})^2\} + (\frac{\partial W}{\partial r})^2 + \{r\,\frac{\partial}{\partial r}\,(\frac{W}{r})\}^2 + \{(\frac{\partial U}{\partial r}) + (\frac{\partial V}{\partial x})\}^2 \tag{7}$$

The used values of the turbulence model constants and turbulent Prandtl numbers are given in table (2).

The Combustion Model:

The eddy dissipation model of Magnussen et al. [21, 25] was used to evaluate the mean formation rate of the species. The model assumes that the reaction rate takes place as a global one-step, infinitely fast chemical reaction between the fuel and

TABLE I: EXCHANGE COEFFICIENTS AND SOURCE TERMS EXPRESSIONS

OF VARIABLES

(ϕ)	Γ_ϕ	S_ϕ
U	μ_{eff}	$\frac{\partial}{\partial x}(\mu_{eff}\frac{\partial U}{\partial x}) + \frac{1}{r}\frac{\partial}{\partial r}(\mu_{eff}r\frac{\partial V}{\partial x}) - \frac{\partial P}{\partial x}$
V	μ_{eff}	$\frac{\partial}{\partial x}(\mu_{eff}\frac{\partial U}{\partial r}) + \frac{1}{r}\frac{\partial}{\partial r}(\mu_{eff}r\frac{\partial V}{\partial r}) - \frac{2\mu_{eff}V}{r^2} + \frac{\rho W^2}{r} - \frac{\partial P}{\partial r}$
W	μ_{eff}	$-(\frac{\mu_{eff}}{r^2} + \frac{\rho V}{r} + \frac{1}{r}\frac{\partial\mu_{eff}}{\partial r})W$
k	$\frac{\mu_{eff}}{\sigma_k}$	$\mu_{eff}\cdot G - \rho\epsilon$
ϵ	$\frac{\mu_{eff}}{\sigma_\epsilon}$	$\frac{\epsilon}{k}(\mu_{eff}\cdot C_1\cdot G - C_2\cdot\rho\cdot\epsilon)$
f	$\frac{\mu_{eff}}{\sigma_f}$	0
h	$\frac{\mu_{eff}}{\sigma_h}$	$\frac{16}{9}\cdot K_g(R_x + R_y - 2\sigma T^4)$
m_{fu}	$\frac{\mu_{eff}}{\sigma_{fu}}$	$23.6 \; (\frac{\mu_1\epsilon}{\rho k^2})^{0.25}\cdot(\epsilon/k)\cdot m_{min}\cdot F\cdot\rho$

TABLE II: VALUES OF THE TURBULENCE MODEL CONSTANTS AND TURBULENT
PRANDTL NUMBERS

Constant	C_μ	C_1	C_2	σ_k	σ_ϵ	σ_f	$\sigma_{m_{fu}}$	σ_h	$\not{z}$	E
Value	0.09	1.44	1.92	1.0	1.22	0.9	0.9	0.9	0.419	9.8

547

oxidant where both combine in stoichiometric proportions. The model also assumes that the rate of reaction is determined by the rate of intermixing of fuel and oxygen on the molecular scale or effectively by the rate of dissipation of the eddies. However, in turbulent flows, this molecular mixing is similar to the dissipation of turbulent kinetic energy and the rate of fuel consumption can be expressed as given by Magnussen et al. [21, 25]:

$$R_{fu} = 23.6 (\mu \, \epsilon / \rho k^2)^{0.25} . (\epsilon / k) . m_{min} . F . \rho . \tag{8}$$

where m_{min} is the smallest of m_{fu} and (m_{ox}/i) and F is defined as the ratio between the local concentration of reacted fuel and the total fuel;

$$F = \frac{(1 - m_{fu} - m_{ox}) / (1 + i)}{\{(1 - m_{fu} - m_{ox}) / (1+i)\} + m_{fu}} \tag{9}$$

The mass of the fuel, m_{fu}, is obtained from the solution of the conservation equation using the combustion rate expression (equation 8). The unburnt oxygen concentraction, m_{ox} is obtained from the following expression;

$$J = m_{fu} - m_{ox}/i \tag{10}$$

where J is a passive scalar of the flow and is defined by;

$$f = (J - J_A)/(J_{fu} - J_A) \tag{11}$$

with J_A and J_{fu} are the scalar values of the air and fuel streams at the inlet to the combustor, respectively. The mixture fraction of fuel, f, is defined as the mass fraction of fuel both burnt and unburnt and is obtained from the solution of the transport equation for mixture fraction with no source term.

The local values of the mean enthalpy were calculated from the solution of the transport equation and were used to calculate the local gas temperatures from the following expression;

$$T = \frac{h - m_{fu} H_{fu} + \sum_{all \; i} m_i C_{pi}(T_{ref}) . T_{ref}}{\sum_{all \; i} m_i C_{pi}(T)} \tag{12}$$

The Spray Model:

The model used to represent the fuel spray is based on the assumption that liquid droplets act as distributed sources of fuel vapour. The model is of the tracking type described by Crowe et al. [16] which assumes that the fuel is injected into the combustion domain as a fully atomized spray of spherical droplets and represented by a finite number of size ranges; each of which characterizes droplets of the same initial size, velocity and temperature.

548

For each droplet size range the momentum conservation equations for axial, radial and circumferential directions may be written as:

$$\frac{\partial U_d}{\partial t} = \frac{-1}{\rho_d}\left(\frac{\partial P}{\partial x}\right) + \frac{3}{4}\frac{\mu C_D}{\rho_d D_d^2}\,(U_f - U_d)\,Re_d \qquad (13)$$

$$\frac{\partial V_d}{\partial t} = \frac{-1}{\rho_d}\left(\frac{\partial P}{\partial r}\right) + \frac{3}{4}\frac{\mu C_D}{\rho_d D_d^2}\,(V_f - V_d)\,Re_d + \frac{W_d^2}{r_d} \qquad (14)$$

$$\frac{\partial W_d}{\partial t} = \frac{3}{4}\frac{\mu C_D}{\rho_d D_d^2}\,(W_f - W_d)\,Re_d - \frac{V_d W_d}{r_d} \qquad (15)$$

where Re_d is based on the droplet diameter and its relative velocity. The drag coefficient C_D is calculated from the correlations given by Williams [26] and applied by many investigators, for example, Attya and Whitelaw [2], Boysan [27] and El-Banhawy and Whitelaw [17];

$$C_D = 27\ Re_d^{-0.84} \qquad\qquad\qquad 0 < Re_d < 80$$

$$C_D = 0.271\ Re_d^{0.27} \qquad\qquad\quad 80 < Re_d < 10^4$$

$$C_D = 2.0 \qquad\qquad\qquad\qquad\quad Re_d > 10^4$$

The rate of change of droplet diameter with respect to time is calculated from this correlation given by Agoston et al. [28] under forced convection conditions:

$$\frac{dD_d}{dt} = (4\lambda_f / \rho_d Cp_f D_d).(1 + 0.24\ Re_d^{0.25})\ \ln[1 + Cp_f (T_f - T_s)/L] \qquad (16)$$

To allow for the droplet heating up time after entering into the gas stream, it is assumed that the evaporation begins when the droplet temperature reaches the boiling point and the temperature is uniform across its radius. The equation representing the temperature change of the droplet can be expressed as;

$$\frac{dT_d}{dt} = (6\lambda_f / \rho_d Cp_d D_d).(T_f - T_d).Nu \qquad (17)$$

The Nusselt number, Nu, is calculated from the Ranz and Marshall [29] expression;

$$Nu = (2 + 0.6\ Re^{0.5}\ Pr^{0.33}) \qquad (18)$$

The solution of the momentum equations together with equations (16) and (17) provide the variation of the droplet velocity, diameter and temperature in the time domain. The droplet tracking technique transfers this to the space domain and allows the calculation of the droplet source terms.

549

The Radiation Model:

The radiant heat transfer between the combustion gases and the combustor walls was calculated using the "four flux" radiation model of DeMarco and Lockwood [22]. For the axi-symmetric cylindrical geometry, the flux equations are:

for radial direction

$$\frac{1}{r} \frac{\partial}{\partial r} \left(\frac{1}{K_g} \cdot \frac{\partial (r R_r)}{\partial r} \right) = 1.33 K_g (2R_r - R_x) - 1.33 K_g \sigma T^4 \qquad (19)$$

for axial direction

$$\frac{\partial}{\partial x} \left(\frac{1}{K_g} \cdot \frac{\partial R_x}{\partial x} \right) = 1.33 K_g (2R_x - R_r) - 1.33 K_g \sigma T^4 \qquad (20)$$

The radiation fluxes R_r and R_x depend on four terms of Taylor's series representing the radiation intensity. K_g is represented by one clear and two gray gases as proposed by Truelove [30].

The radiant energy contribution to the source term in the enthalpy equation is:

$$S_h = (16/19) \, K_g \, (R_x + R_r - 2\sigma T^4) \qquad (21)$$

3. THE SOLUTION PROCEDURE:

The gas phase equation represented by equation (1) together with the models for turbulence, combustion and radiation were solved by a finite difference procedure using a computer program developed on a VAX 11/780 computer. The solution of the equations representing the fuel spray model were obtained with the Runge-Kutta-4 method and the droplet source terms were obtained through the tracking technique.

The GAS Field Solution:

A grid of 22x25 nodes was overlayed over the calculation domain of figure (1) and the dependent variables W, h, m_{fu}, m_{ox}, f, k, ϵ, R_x and R_r were evaluated at each node. The axial and radial velocities U and V were obtained at the nodes staggered with respect to the grid for the scalar variables. The solution of the partial differential equations of the dependent variables at every grid node was found by the finite difference approximation. The finite difference equations were obtained by integrating the respective differential equations over the control volume surrounding each node. The finite difference equations for all variables can be written in the following form:

$$(A_p - S_p + S_{dm}) \, \Phi_p = \sum A_i \phi_i + S_u + S_d \qquad (22)$$

where A_i is the coefficient in the i-direction and A_p is their sum. $(S_u + S_p \Phi_p)$ is the linearized form of the gas source term and $(S_d - S_{dm} \Phi_p)$ is the linearized form of the droplet source term.

The resulting finite difference equations were solved using the iterative method SIMPLE algorithm described by Caretto et al. [31]. The droplet gas coupling is incorporated in the numerical procedure as follows:

1. An isothermal droplet-free solution of the gas field is obtained.
2. The droplet mass, momentum and heat transfer equations are integrated using the Lagrangian approach to yield the droplet trajectories and diameter and temperature history along each trajectory.
3. The droplet source terms are calculated for each cell in the calculation domain and supplied to the finite difference equation to obtain adjusted values for the dependant variables.
4. These new dependant variable values are used again in the solution of the droplet equations.
5. This process is repeated until a preset convergence criterion is met.

Droplet Field Solution:

The differential equations of the droplets were solved using the Runge-Kutta-4 method. The obtained droplet velocity components were used through the tracking technique described by Crowe et al.[16] and El-Banhawy and Whitelaw [4] to determine the instantaneous locations of the droplets relative to the gas field as follows;

$$X_{do} = X_{di} + (U_{do} + U_{di}).(\Delta t/2) \tag{23}$$

$$Y_{do} = Y_{di} + (V_{do} + V_{di}).(\Delta t/2) \tag{24}$$

The tracking procedure allows the size, velocity and position of the droplets to be determined and the source terms are obtained by calculating the loss or gain of the droplet mass and momentum within each cell. Summing up the respective source terms for the droplets representing different size ranges gives the total droplet source term.

The net rate of the droplet mass for the cell is:

$$S_{dm} = \sum_{I=1}^{N} [(m_{di} - m_{do})\dot{n}]_1 \tag{25}$$

The net enthalpy flux for the cell is:

$$S_{dh} = \sum_{I=1}^{N} [(m_{di} - m_{do})\dot{n}]_1 . (H_{fu} - L) \tag{26}$$

The net rate of momentum for the cell is:
   axial direction:

$$S_{dU} = \sum_{I=1}^{N} [((mU)_{do} - (mU)_{di})\dot{n}]_1 \tag{27}$$

   radial direction:

$$S_{dV} = \sum_{I=1}^{N} [((mV)_{do} - (mV)_{di})\dot{n}]_1 \tag{28}$$

tangential direction:

$$S_{dW} = \sum_{1=1}^{N} [((mW)_{do} - (mW)_{di})\dot{n}]_1 \qquad (29)$$

4. BOUNDARY AND INLET CONDITIONS:

Boundary Conditions:

At the axis of symmetry, the fluxes of convection and diffusion were set to zero. This is incorportated by setting the velocity component normal to the axis of symmetry as well as the diffusion flux coefficient to zero. At the exit plane where the flow becomes parabolic or boundary layer in behaviour, the exit values of the flow variables, with the exception of the axial velocity, have no influence on the upstream calculation. The conservation of mass over the whole solution domain was checked by adding an incremental value to the velocity of the plane before the outflow plane. This incremental velocity is given by:

$$U_{inc} = (m_{in} - {}_A\int \rho U dA) / {}_A\int \rho dA \qquad (30)$$

Where $\dot{m}_{in}$ and ${}_A\int \rho U dA$ represent the net mass flow across the entry plane and outlet plane of the solution domain, respectively. As the solution proceeds to convergence U_{inc} tends to zero. Similar procedures were applied to ensure overall conservation of heat and mass of the fuel.

Close to the wall where the turbulence model is invalid, the logarithmic wall function proposed by Patankar and Spalding [32] was used to overcome the problem of viscous effects.

For the flux momentum:

$$(\rho U_p / \mathcal{T}_w) \, C_\mu^{0.25} \, K_p^{0.25} = (1/\mathcal{Z}) \, \ln(E \, Y^+) \qquad (31)$$

$$\text{where } Y^+ = Y_1 \, \rho \, C_\mu^{0.25} \, k_p^{0.5} / \mu_1 \qquad (32)$$

Inlet Conditions:

The inlet conditions were taken from the experimental data of Attya [1] or were estimated from suitable correlations.
Table (3) summarizes the inlet conditions for the predicted flames.

5. RESULTS AND DISCUSSION:

The above calculation procedure was applied to predict the local flame properties in different confined preheated spray flames whose characteristics are given in table (3). The geometry of the combustion chamber is shown in fig.(1). The calculation procedure was evaluated by comparing its results with the experimental data given by Attya [1]. In these experiments the liquid fuel was injected inside the combustion

TABLE III: EXPERIMENTS INLET CONDITIONS

	Flame1	Flame2	Flame3	Flame4
Combustion air mass flow rate (Kg/hr)	266	266	266	266
Combustion air swirl number.	1.4	1.4	1.4	1.4
Combustion air preheat temp.(°C)	225	125	225	125
Atomizer cooling air (Kg/hr)	12.8	12.8	12.8	12.8
Fuel mass flow rate (Kg/hr)	8.66	8.66	8.66	8.66
Sauter mean droplet diameter (μm)	54	54	115	115
Satellite droplet diameter (μm)	30	30	60	60
%age mass of mean droplet.	80	80	75	75
%age mass of satellite droplet.	20	20	25	25
Spray mean axial velocity (m/s)	5-10	5-10	3-9	3-9
Spray mean radial velocity (m/s)	6.5-10	6.5-10	5-8.2	5-8.2
Spray mean tangential vel. (m/s)	6.5-9	6.5-9	5-6.2	5-6.2

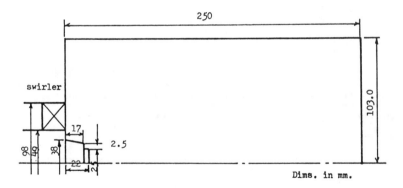

Fig. 1 Geometry of the Combustor.

chamber through a rotating cup atomizer. This atomiser produces a near mono-disperse droplet size which makes possible the representation of the spray with a few droplet size ranges. The detailed description of the experiment and measurements are given by Attya [1]. Due to the difficulties encountered with the measurement of the spray velocities at its injection point, the measurements were specified near the atomizer. Therefore, in order to represent the spray characteristics in the model it was necessary to make some trials to correlate the best injection locations with the measured velocity distribution and proposed droplet size ranges. The spray was represented by six size ranges to allow for the velocity distribution. Three of them represent the main droplet diameter and the others correspond to the satellite droplets. In test case (1), it was assumed that all size ranges are injected from one point of X= 26 mm and Y= 25 mm while in case (2), there were three injection points of X= 26 mm and Y= 10, 15 and 20 mm as shown in fig. (2) of the droplet trajectories.

The comparison of the predicted results of the temperature and species concentrations of flame 1 and 2 with the corresponding measurements are shown in figures 3 through 6; the measured CO profiles are presented to assist in the analysis. It is appreciable that the spread of the injection points of the spray has resulted in better distribution of the flame properties. As observed from the trajectories of the droplets in figure (2), the spread of the injection points of the spray has improved the distribution of the trajectories and thus the flame properties.

Figures 3a to 6a indicate that the temperature distributions are predicted quite reasonably by the present calculation procedure. The temperature is more accurately presented by the multi-point fuel spray injection which, also, gives, better predictions of CO_2 concentrations as shown in figures 3b to 6b. The two maxima near the centerline and the wall of figures 3b and 5b are clearly predicted by the model. Figures 3c to 6c present a comparison of the calculated O_2 and unburnt hydrocarbon (UHC) concentrations with their corresponding experimental data. The trend of the distributions is again presented quite reasonably. In the vicinity of the centerline and the atomizer, the figures indicate that the temperature and carbon dioxide are under-predicted by the present calculation procedure while the oxygen concentrations are overpredicted. This indicates that the combustion rate of fuel is underpredicted by the present model near the centerline. This is mainly due to the slow diffusion rate of the fuel towards the centerline.

The present turbulence model presumes that the Reynolds stresses are connected with the rate of strain through the Boussinesq equation. As shown by Bradshaw [33], the Reynolds stresses are observed to change by values which are much larger than the direct effect of strains because of the importance of the extra strain rates. With the presence of the swirling flow, and at high swirl numbers in particular, the Reynolds stresses

554

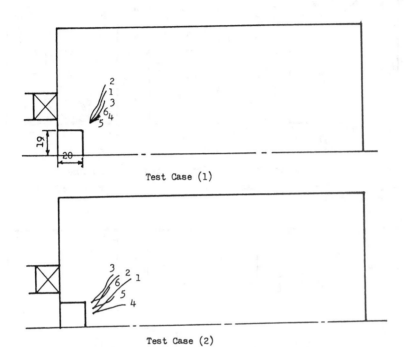

Test Case (1)

Test Case (2)

Fig. 2 Droplet Trajectories of Flame 2 Through the Combustor.

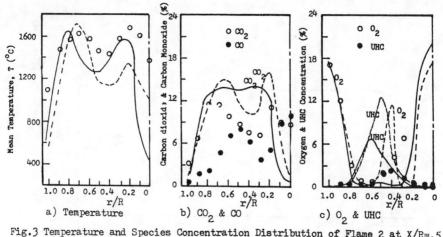

Fig.3 Temperature and Species Concentration Distribution of Flame 2 at X/R=.5
————— ,Calculations of Test Case 1 ;— — —,—·—·— Calculations of Test
Case 2 ; ○, ● Measurements.

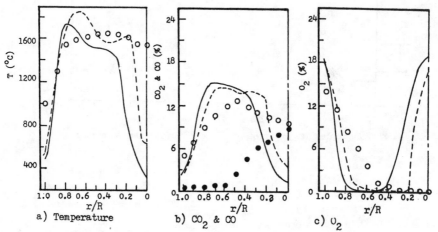

Fig.4 Temperature and Species Concentration Distributions of Flame 2 at X/R=1.0
Notation as in Fig. 3

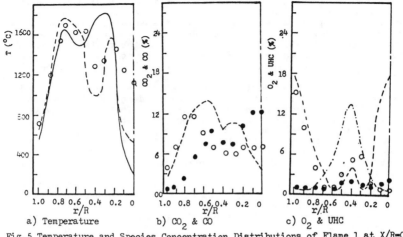

a) Temperature   b) CO_2 & CO   c) O_2 & UHC

Fig.5 Temperature and Species Concentration Distributions of Flame 1 at X/R=0.5
Notation as in Fig. 3.

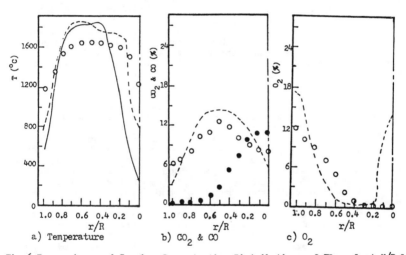

a) Temperature   b) CO_2 & CO   c) O_2

Fig.6 Temperature and Species Concentration Distributions of Flame 1 at X/R=1.0
Notation as in Fig. 3.

are shown to be much larger than what is predicted by the Boussinesq equation and therefore the calculation of the Reynolds stresses using the Boussinesq equation results in large discrepancies (Habib and Whitelaw, [34]). Near the centerline, the effect of the extra strain rates is specially important as shown by Spalding [35]. These effects are expected to cause underpredicted diffusion in the vicinity of the centerline as a result of the underprediction of the turbulence kinetic energy. A direct effect of this is the decrease of the rate of combustion as given by equation 8.

Away from the centerline, the predicted temperature and carbon dioxide are higher than the corresponding measured values by about 200 ^{o}C and around 4% respectively. This is mainly due to the deficiency in the combustion model used which assumes single step reaction and neglect the formation of carbon monoxide. Near to the wall, the prediction results is in quite agreement with the experimental data due to the validity of this assumption since measured CO concentrations are zero near the walls.

Fig. (7) presents the comparison between the contours of the experimental data and the calculated temperature and CO_2 and oxygen concentrations of flame 1. Generally, the contour maps of the predicted results are in qualitative agreement with the corresponding measurements. However, close to the centerline and fuel injection locations, the contour lines show that the temperature is underpredicted by up to 200°C due mainly to uncertainties associated with the single step chemical reaction combuston model and the neglection of soot formation and intermediate species. At small radii of the combustion chamber and around r/R= 0.5, the predicted temperatures and carbon dioxide are higher than the measurements by about 230°C and 3%, respectively. On the other hand, calculated oxygen concentrations are lower than the measurements by about 3 to 4%. These discrepancies may be explained by the high CO concentrations measured at this region and due to the deficiencies in the coupled turbulence/chemistry model.

The effect of increasing the combustion air preheat temperature on the combustion characteristics is shown in figures 7 and 8 for the spray mean droplet diameter of 54 μm and figures 9 and 10 for 115 μm droplet diameter. In general, the results indicate that the increase of the air inlet temperature from 125 to 225°C has resulted in the onset of combustion closer to the combustor and in the propagation of the flame towards the wall of the combustor. The temperature contours of figures 7 to 10 show that the rise of the inlet combustion air temperature has resulted in higher local gas temperatures and expansion of the high temperature region. However, the species concentrations of CO_2 and O_2 indicate improvement of the combustion intensity and chemical reaction through the combustor. These observations reveal that the rise of the preheat temperature has improved the turbulent mixing rates between the fuel and oxidant with consequent enhancement of combustion. The main reason is the acceleration of the evaporation rate nearer to the atomizer which has resulted in early feeding of the combustion domain with higher rates of fuel vapour than liquid droplets.

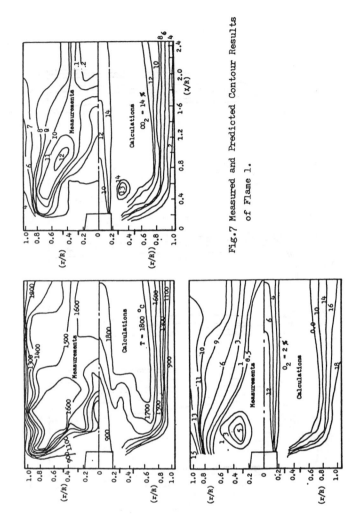

Fig.7 Measured and Predicted Contour Results
of Flame 1.

559

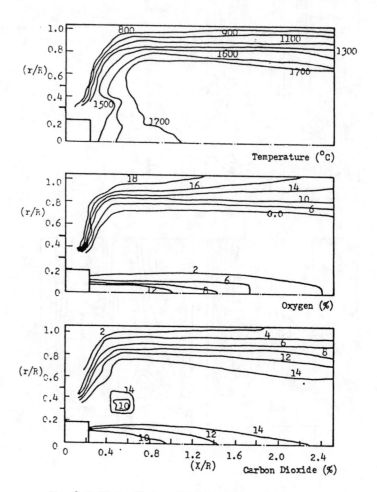

Fig. 8 Predicted Contour Results of Flame 2.

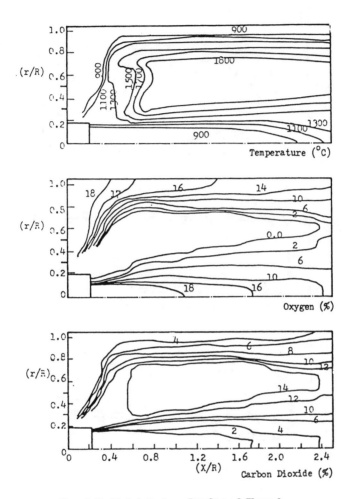

Fig. 9 Predicted Contour Results of Flame 3.

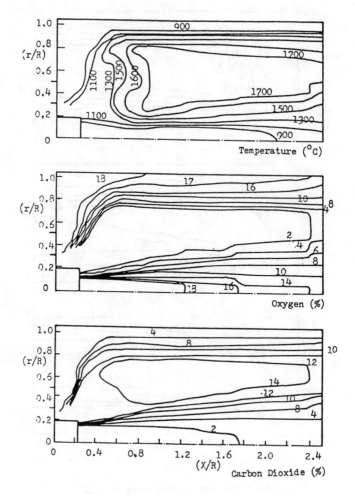

Fig. 10 Predicted Contour Results of Flame 4.

The comparison between fig. 7 and fig. 9 and, also, between fig. 8 and fig. 10 presents the influence of increasing the initial spray mean droplet diameter from 54 μm to 115 μm on the flame properties keeping the inlet combustion air temperature unchanged. It is seen that the reduction of the spray mean droplet diameter shifted the high temperature region, and consequently, the main flame region upstream of the combustor with relatively higher temperatures. These high temperatures are associated with relatively high CO_2 concentrations and lower O_2 values which imply an increase in the combustion intensity. This is influenced by the spreading and increased rate of fuel evaporation associated with small droplets. As shown from equations 16 and 17, the heating period of the droplet and the rate of change of its diameter with time are both inversely proportional to its diameter. This means that a reduction in the spray mean droplet diameter will be associated with earlier spray evaporation and consequently shorter droplet trajectories. The increase in temperature values of flames 1 and 2 (spray diameter = 54 μm) at upstream locations than flames 3 and 4 (spray diameter = 115 μm) is mainly influenced by this rise in the evaporation rate as more fuel vapour is available for combustion in the initial part of the flame. This increase in the evaporation rate in the upstream region is also associated with the reduction in the amount of fuel vapour evolved from the droplets during their subsequent movement and helps to explain the lower temperature and carbon dioxide concentration closer to the wall associated with flames 1 and 2.

6. CONCLUSIONS:

A computational procedure for the calculation of the flow, temperature and species concentration in a model of a gas-turbine combustor was developed and evaluated. The procedure involved a two-equation turbulence model and a "discrete droplet " model in which the spray was represented by a finite number of groups of droplets. Comparison was made between single and multi-point fuel spray injection.

The calculation procedure was evaluated through the comparison of the results with the experimental data for two different preheat temperatures of combustion air and for two different mean droplet diameters of the fuel spray. The comparison between the single and multi-point fuel spray injection indicated that the multi-point spray injection resulted in better predicted values for the temperature and concentration of species.

Apart from the centerline, reasonable agreement between the experimental and calculated results was obtained. In the vicinity of the centerline and the atomizer, the temperature and carbon dioxide were underpredicted while the oxygen concentration was overpredicted. This was attributed to the underprediction of the combustion rate of fuel by the present model and to the slow diffusion rates calculated by the turbulence model.

The calculations showed also that increasing the

combustion air preheat temperature from 125 to 225 C has resulted in an improvement of the turbulent mixing rates between the fuel and oxidant and thus to a higher combustion intensity throughout the combustor. On the other hand, decreasing the mean droplet diameter from 115 μm to 54 μm while keeping the inlet combustion air temperature unchanged resulted in an earlier spray evaporation and consequently shorter droplet trajectories with more fuel vapour available for combustion. This appeared in the form of shifted regions of high temperature and flame location upstream of the combustor with relatively higher temperature, higher carbon dioxide concentration and lower oxygen concentration.

7. NOMENCLATURE:

C_D	Drag coefficient.
C_p	Specific heat.
H^p	Calorific value.
h	Enthalpy.
i	Stoichiometric ratio.
k	Turbulent kinetic energy.
K_g	Local gas mixture absorption coefficient of the gas medium.
L	Latent heat of evaporation of fuel.
m	mass of droplet, mass of fuel, unburnt oxygen or mass of species.
N	Number of size ranges.
$\dot{n}$	Number of droplets per unit time.
Nu	Nusselt number under forced convection motion of droplet.
P	Pressure.
Re	Reynolds number.
R_x	Axial total radiation flux.
R_r	Radial total radiation flux.
S_{dm}	Droplet mass source term.
S_{dh}	Radiation source term.
T	Temperature.
t	Time.
U	Mean axial velocity.
u	Fluctuating velocity component.
V	Radial velocity component.
W	Tangential velocity component.
δ_{ij}	Kronecker-delta (= 1 for i = j; = 0 for i ≠ j)
ϕ	General dependent variable.
ϵ	Dissipation rate of turbulent kinetic energy.
ρ	Density.
τ	Wall shear stress.

Subscripts:

A	Air.
d	Droplet.
di	Inlet of the droplet to the cell.
do	Exit of the droplet from the cell.
fu	Fuel.
f	Flow of gas.
i	Axial direction.
j	Radial direction.
ox	oxygen.
p	Near wall node.
s	Saturation.
w	Wall node.

REFERENCES

1. Attya,A.M. (1983) "Kerosene Spray Flames." Ph.D. Thesis, University of London.
2. Attya, A.M. and Whitelaw, J.H. (1984) "Measurements and Calculations of Preheated and Unpreheated Confined Kerosine Spray Flames." Combustion Science and Technology, V40, pp 193-215.
3. Attya, A.M. and Whitelaw,J.H. (1981) "Velocity, Temperature and Species Concentrations in Unconfined Kerosine Spray flames." ASME Winter Annual Meeting Nov., Paper No. 81-WA/HT-43.
4. El-Banhawy Y. and Whitelaw, J.H.(1981) " Experimental Study of Interaction Between a Fuel and Surrounding Air." Combustion and Flame, V42, pp 253-275.
5. Yule, A., Seng,.C, Felton, P., Ungut, A. and Chigier, N.A. (1982) "A Study of Vaporizing Fuel Sprays by Laser Techniques." Combustion and Flame, V44, pp 71-84.
6. Miyasaka,K. and Law, C.K. (1981) "Combustion of Strongly Interacting Linear Droplet Arrays." 18th Symp. Int. on Combustion, pp 283-292.
7. Sangiovanni,J.J. and Labowsky,M. (1982) "Burning Times of Linear Fuel Droplet Arrays: A Comparison of Experiment and Theory." Combustion and Flame, V47, pp 15-30.
8. Okajima, S. (1985) "Experimental Investigation of Flame Propagation in Liquid Fuel Droplet Arrays Under Zero Gravity." Arch. Combustion, V5 n3-4.
9. Karasawa, T., Shiga, S. and Kurabayashi, T. (1985) "Ignition Phenomena of The Fuel Droplet Impinged Upon a Hot Surface." Japanese Bull., V51 n465, pp 1725-1730
10 Cho,P. and Law,C.K. (1985) "Pressure/Temperature Ignition Limits of Fuel Droplets Vaporizing Over a Hot Plate" Inst. Heat and Mass Transfer,V28 n11, pp 2174-2176.
11. Williams, F.A. (1965) "Combustion Theory" Addison Wesley Pub. Co. Inc., London.
12. Westbrook, C.K. (1977) "Three Dimensional Numerical Modelling of Liquid Fuel Sprays." 16th Symp. Int. on Combustion, pp 1517-1526.
13. Haselman,L.E. and Westbrook,C.K.(1978) . ASME Paper 780138.
14. Cliffe,K.A., Lever,D.A. and Winters,K.(1979) International Conference on Numerical Methods in Thermal Problems, Swansea U.K.
15. Jones, W.P. and McGuirk, J. (1981)." A Comparison of two Droplet Models for Gas-Turbine Combustor Flows" 5th International Symposium on Airbreathing Engines.
16. Crowe, C.T., Sharma,M. and Stock,D.1(1977) "The Particle Source in Cell (PSI_ Cell) Model for a gas Droplet Flows." ASME J. Fluid Eng., V99, pp 325-332.
17. El-Banhawy,Y. and Whitelaw, J.H. (1980)" The Calculation of the Flow Properties of a Confined Kerosene Spray Flame." AIAA Journal, V18, pp 1507-1510.
18. Gosman, A.D. and Ioannides, E. (1981)". Aspects of Computer Simulation of Liquid Fuelled Combustors." AIAA 19th Aerospace Science Meeting.
19. Sirignano, W.A. (1986) "Formulation of Combustion Models Resolution Compared to Droplet Spacing." J. Heat Transfer Trans. ASME, V108 n3, pp 633-639.

20. Sellens, R. and Brzustowski, T.A. (1986) "Simplified Prediction of Droplet Velocity Distributions in a Spray." Combustion and Flame, V65 n3.

21. Magnussen,B.F., Hjertager,B., Olsen,J.G. and Bhaduri,D. (1979) " Effects of Turbulent Structure and Local Concentrations on Soot Formation and Combustion in C2H2 Diffusion Flame" 17th Symp. Int. on Combustion, pp 1383-1393.

22. DeMarco,A.G. and Lockwood,F. (1975) "A New Flux Model For the Calculation of Radiation in Furnaces.", Italian Flame Days, p184.

23. Gosman,A.D. and Pun, W.M.(1974) Lecture Notes for Course Entitled : Calculation of Recirculating Flows." Imperial College , Mech. Eng. Report, HTS/74/2.

24. Jones.,W.P. and Launder, B.E.(1972) "The Prediction of Laminarization with a Two-Equation Turbulence Model." Int. J. of Heat and Mass Transfer, V15, pp 301

25. Magnussen,B.F. and Hjertager, B. (1977) "On Mathematical Modelling on Turbulent Combustion With Special Emphasis on Soot Formation and Combustion." 16th Symp. Int. on Combustion, pp 719-729.

26. Williams, A. (1973) "Combustion of Droplets of Liquid Fuels: A Review." Combustion and Flame, V21, pp 1-13.

27. Boysan, F., Ayers, W.H., Swithenbank, J. and Pan, Z. (1982) "Three Dimensional Model of Spray Combustion in Gas Turbine Combustor" J. Energy, V6, pp 368 - 375.

28. Agoston, G.A., Wise, H. and Rosser, W.A. (1957) "Dynamic Factors Affecting the Combustion of Liquid Fuel Spheres" Sixth Symp. Int. on Combustion, pp 708 - 717.

29. Ranz,W.E. and Marshall,W.R. (1952) Chemical Engineering Progress, V84, pp 141-173.

30. Truelove, J.S. (1976) "A Mixed Gas Gray Model for Flame Radiation." UKAEA Harwell Report AERE-R84494.

31 Caretto,L.S., Gosman,A.D., Patankar,S.V. and Spalding, D.B.(1972)"Two Calculation Procedures for Steady Three Dimensional Flows with Recirculation." Proceedings of the 3rd International Conference on Numerical Methods in Fluid Mechanics pp 60-68.

32. Patankar,S.V. and Spalding, D.B. (1970) "Heat and Mass Transfer in Boundary Layer." 2nd Edition, Intertext Books, London.

33. Bradshaw,P., (1973) "Effect of Streamline Curvature on Turbulent Flow" AGARDograph, 169.

34. Habib,M.A. and Whitelaw,J.H. (1980) "Velocity Characteristics of a Confined Coaxial Jets With and Without Swirl" J. Fluid Eng., V102, p47.

35. Spalding,D.B. (1980) "Turbulence Modelling: Solved and Unsolved Problems; Turbulent mixing in non-reactive and reactive flows." Ed. by S.N.B. Murthy.

Study of Combustion Processes through Imaging Techniques

CHOWEN C. WEY, BAHMAN GHORASHI,
and C. JOHN MAREK*
Cleveland State University
Cleveland, Ohio USA

In order to better understand the physics of combustion processes, one must utilize a technique to visually observe the complete flowfield and simultaneously extract quantitative information from the flow. At present, many well developed techniques are available for flow visualization and flow measurements, and numerous laser diagnostic and particle-tracking concepts are at the early stages of development. Several of these techniques are selectively chosen for discussion, however, it should be noted that this presentation is not exhaustive and many other techniques are available. Reference is also made to works of many researchers and the development of new diagnostic techniques. Furthermore, as the nature of a specific study and the system requirements would limit the choice of diagnostic techniques, different methods and their applications are discussed separately. Each method is preceeded by a very brief introduction to its basic principals.

I. LASER DIAGNOSTICS

Laser diagnostic techniques offer the advantage of in-situ, non-intrusive and remote measurements and allow precise spatial and temporal measurements. A grouping of these techniques may be based on the nature of the scattering process, i.e. elastic versus inelastic scattering. The examples of the former would be Rayleigh and Mie scattering of light quanta from molecules or particulate matter, and those of the latter would be Raman and Near Resonant Raman scattering [1]. Fluorescence is the emission of light from an atom or molecule following promotion to an excited state [1] Laser Induced Fluorescence Spectroscopy (LIF). LIF spectroscopy may be used to detect flame radical species such as OH and NH at the ppm level since fluorescence is many times stronger than Raman scattering. Two conditions, among others, have to be met for the utilization of this technique. First, the molecule must have a known emission spectrum and second, it must have an absorption wavelength which would be accessible to a tunable laser source [1].

Among the recent LIF works, Gross and McKenzie [2] developed a LIF method to measure the instantaneous temperature in low temperature turbulent flows. They reviewed the capabilities of

---

*NASA-Lewis Research Center

the method and reported its application to a simple two-dimensional Mach 2 turbulent boundary-layer flow which was seeded with NO. Gross, et al. [3] extended their LIF method to provide simultaneous measurements of temperature, density and their fluctuations owing to turbulence in unheated compressible flows. Pressure and its fluctuations were also deduced using the equation of states for a non-reacting perfect gas mixture. The method was applied to a supersonic turbulent boundary-layer flow with free-stream Mach number of 2.06. In another study, Cattolica and Vosen [4] observed the interaction of a methane-air flame with a vortex-ring structure by means of quantitative imaging of the OH concentration using LIF. A planar sheat of UV laser light from a Nd: YAG-pumped dye laser was used to excite fluorescence in the OH molecules produced by the spark-ignited flame. A two-dimensional image of the OH fluorescence was obtained with a gated image-intensified vidicon camera. The two-dimensional imaging results show the effect on the OH concentration distribution and the fluid mechanic features of the vortex-ring structure. Also, Cattolica and Vosen [5] studied the temporal and spatial development of the OH concentration during the ignition of a lean methane-air mixture by a combustion torch. The fluid physics of the combustion torch has a significant influence on its chemical structure and the development of the subsequent ignition process in the main combustion chamber. Observations of the chemical structure of the combustion-torch ignition process were made by quantitative imaging of the OH concentration using LIF.

Seitzman, et al. [6] described a single-pulse, LIF diagnostic for the measurement of two-dimensional temperature fields in combustion flows. The method uses a sheet illumination from a tunable laser to excite PLIF in a stable tracer molecule (NO), seeded at constant mole fraction into the flow field. The temporal resolution of this technique is determined by the laser pulse length.

Hiller, et al. [7] described a nonintrusive technique for multiple-point velocity measurements in subsonic flows. The technique was based on the detection of fluorescence from a Doppler-shifted absorption line of seeded iodine molecules excited at a laser frequency fixed in the wing of the line. Counterpropagating laser sheets were used to illuminate the flow, thereby eliminating the need for an unshifted reference signal. The fluorescence was detected simultaneously at 10000 points in a plane of the flow using a 100x100 element photodiode-array camera. The velocity at each point was computed from four successive camera frames, each recorded with a different beam direction.

Kychadoff, et al. [8] described a quantitative flow visualization technique for combustion research. The method used sheet illumination from a tunable laser to excite PLIF which was detected using an image-intensified 2-D detector. They discussed the detectivity, dynamic range, spatial and temporal resolution of the technique, and an application to view the OH concentration in a rod-stabilized flame.

Hanson [9] established a technique for visualizing velocity in gaseous flows. The technique was based on the Doppler effect, and used an intensified photoiode array camera and a planar form of LIF to detect 2-D velocities of I_2 (in I_2 - N_2 mixtures) via Doppler-shifted absorption of narrow-linewidth laser radiation at 514.5 nm. The flowfield was probed with two fixed laser frequencies so that the spatially varying slope of the absorption line could be measured directly at each point. The directly determined slope provides additional information on the pressure distribution in the flow field.

Allen and Hanson [10] demonstrated the ability to make single-shot, two-dimensional measurements of radical species and fuel pyrolysis product distributions in a turbulent burning spray. OH and CH images were acquired using PLIF; the fuel vapor distribution was imaged by planar multiphoton dissociation (PMPD) of C_2H_2.

Also, Allen and Hanson [11] described a technique for the instantaneous measurement of the two-dimensional distribution of the OH radical in spray flames based on PLIF. Sheet illumination was used with the resulting fluorescence imaged at 90° onto an intensified photodiode array camera. They demonstrated the ability to make OH concentration distribution images in two-phase combustion systems.

Lee, et al. [12] studied PLIF images of O_2 in flames using an ArF excimer laser as the excitation source. The resultant fluorescence was detected with an image-intensified 100x100 photodiode array camera. The real-time evolution of discrete flame structures was displayed by a high-speed sequence of images that was acquired using the high repetition rate capability of the excimer laser.

Bowman, et al. [13] reported an ongoing experimental and computational investigation of supersonic combustion flows which included an experimental study of mixing and combustion in a supersonic plane mixing layer; development of LIF techniques for time-resolved two-dimensional imaging of species concentration, temperature, velocity and pressure; and numerical simulations of compressible reacting flows.

Hiller, et al. [14] and Hiller and Hanson [15] presented an optical technique for combined, spatially resolved measurements of 2-D velocity and pressure field in compressible flows. The single-mode frequency of an argon laser is fixed in the wing of an absorption line of iodine molecules and the emitted fluorescence is detected with an intensified 100x100 photodiode array camera. Three components of the velocity vector in a cross-sectional plane were sequentially probed with four laser sheets from three different directions. By shifting the laser frequency in one pair of the sheets, the slope of the absorption line could be measured in situ to provide the required scaling factor for the velocity measurement. The slope was also used to infer pressure.

Also, Hiller and Hanson [16] reported the development of a non-intrusive technique which permits simultaneous spatially resolved measurements of two velocity components and pressure in a plane of compressible gaseous flow field. The technique was based on the detection of fluorescence from an absorption line excited with a narrow-bandwidth laser. Doppler shift and pressure broadening of the line were exploited to extract velocity and pressure information, respectively. The fluorescence was detected at a 90° angle with an image-intensified 100x100 element photodiode-array camera which was interfaced with a laboratory computer.

Hassa, et al. [17] used fluorescence of iodine, excited at 514.5 nm by a single-mode argon-ion laser tuned to the quasi-linear part of an absorption line to detect the Doppler shift and hence the velocity of iodine molecules seeded in a nitrogen jet flow. The slope of the absorption line profile was measured directly using a frequency shift introduced by acoustooptic modulators (AOMs). To reduce experimental noise, the laser beams were switched on and off at a fast rate to allow the use of ac-coupling and signal-tuned amplification.

II.   HOLOCINEMATOGRAPHIC VELOCIMETRY (HCV)

Weinstein, et al. [18] proposed the use of a dual view (orthogonal), high speed, holographic instantaneous movie technique for the study of the evolution of instantaneous, 3-D velocity profiles in turbulent flow fields. An imaging system using far-field holography is used to provide full field of view and tracking of individual seed particles. Velocity information is determined by measuring particle displacements of sequential hologram reconstruction which provides each particle's relative position.

Liburdy [19] examined the resolution limits of farfield holography as applied to the Holocinematographic Velocimeter (HCV). Analytical and experimental results show the relative insensitivity of particle detection to interference by nearby particles either in or out of the image plane. A straightforward enhancement technique provides a means to eliminate noise and reduce out of image plane ambiguity.

Weinstein and Beeler [20] used a small prototype water tunnel to demonstrate proof-of-concept for the HCV. After utilizing a conventional flow visualization apparatus with a laser light sheet to illuminate tracer particles in order to evaluate flow quality, a simplified version of the HCV was employed to demonstrate the capabilities of the technique.

In another study, Beeler and Weinstein [21] used two simultaneous, orthogonal-axis holographic movies that were made of tracer particles in a low-speed water tunnel to determine the time-dependent, three-dimensional velocity field. The holographic movies were reduced to the velocity field with an automatic data reduction system. Automatic fiducial mark location was still required for a fully automatic Image Analysis System (IAS).

III. PARTICLE IMAGE TRACKING

Chang and Tatterson [22] used a standard stereoscopic motion picture technique to track neutrally buoyant tracer particles, which yield three-dimensional particle paths. An automated method was developed for data acquisition. A software package was constructed to perform particle image identification (no enhance algorithm used), stereo pair image matching (using color, brightness, size, y-coordinates), and particle path tracking (constraints on displacement and direction) on digitized film.

Chang, et al. [23] improved the software package utilizing thresholding and pattern matching to perform particle image identification; particle image tracking in consecutive frames; stereo pair matching of particle trajectories; and velocities evaluation in three-dimensional space evolving in time. Combination of image processing and stereoscopic motion pictures was used to provide sufficient data to characterize the flow field from a Lagrangian/Eulerian point of view.

Racca and Dewey [24] recorded simultaneous orthogonal views of the tracer-seeded flow by a single high speed cine camera through a split field mirror system which was subsequently converted to machine readable form by a video digitizer. Digital enhancement was used to separate the tracers from the contrasting background. Also, algorithms were developed to (1) correct rotation and linear distortion in the image, (2) match the projections of individual tracers in the two views (geometrical considerations), (3) obtain the three-dimensional coordiantes, (4) follow the tracers from frame to frame and compute the velocity vectors along the particle trajectories. Eulerian information was derived from the pooled velocity data points by interpolation on a regular spatial grid.

Kobayashi, et al. [25] developed a technique to derive the two-dimensional velocity distribution around a circular cylinder from photographs of the trace particles. The system consisted of digitization of pathline, derivation of velocity vector and prediction of velocity distribution. The problem of the determination of flow direction was solved by using a two-camera system to obtain both short-exposure and long-exposure photographs in order to define the starting points.

Gharib, et al. [26] developed a computerized flow visualization technique capable of quantifying the flow field automatically. This technique used afterglowing effect of optically activated phosphorescent particles to retrieve vectorial information on each trace. By defining starting and ending points, which is done by applying certain constraints on eight different square masks which are centered on the point, the direction of the movement is determined.

Lewis, et al. [27] descussed the restriction on the particle type and size as they relate to the particle's ability to follow the flow and to withstand combustion temperature. They developed the technique to obtain instantaneous two-dimensional velocity measurements from multiple-exposed photographic images (taken at a right angle to a thin laser sheet on the diametric center of the jet) of scattered light from seed particles in the flow. The interrogation software that identify particle patterns is not developed. The task was performed manually.

Landreth, et al. [28] constructed an automated pulsed-laser particle-image velocimeter (PIV) to determine fluid velocity. Special image shifting techniques, which offsets all negative displacements of particle image pairs on a PIV photograph, resolve directional ambiguity in reversing flow fields.

IV. OTHER IMAGING TECHNIQUES

Rockwell, et al. [29] developed techniques for implementing and interpreting time lines (lines of elements marked at specific times). Their emphasis was on: the correlation of the visualized flow structure and the surface pressure; three-dimensional reconstruction of flow structure from dual views of flow field; and interactive interpretation of flow images with basic classes of thoretical simulations.

V. FLOW VISUALIZATION-HYDROGEN BUBBLE TIMELINES

In another study, Rockwell, et al. [30] recently discussed the techniques for curve-tracking of images; visual correlation and ensemble-averaging both within a given image and between images; determination of three-dimensional flow structure from single and dual-view images; estimates of velocity eigenfunctions from images; determination of flow structure-surface pressure relations from images and pressure measurements which were acquired simultaneously.

Utami and Ueno [32] took successive pictures of flow patterns in two horizontal cross-sections at different levels. The pictures were then digitized and analysed by a computer to calculate the distributions of the three components of the velocity vectors. They proposed a conceptual model of turbulence structure in which the elementary unit of coherent structure in the buffer layer is presumed to be a horseshoe vortex.

Ongoren, et al. [33] used a single bubble wire and video camera in conjunction with a phase reference from the oscillating body to acquire two-dimensional projections of the flow at desired spanwise locations to reconstruct (Computer-Aided-Design) the large-scale, time-dependent flow structure. Appropriate generation of bubble timelines leads to three-dimensional "time-surfaces" at selected instants of time, and thereby a three-dimensional space-time estimate of the macroscale features of the flow.

VI. VELOCITY MEASUREMENTS

Talbot, et al. [34] performed a LDV study of velocity profiles in the laminar boundary layer adjacent to a heated flat plate. They noted that the seed particles used for the LDV measurements were driven away from the plate surface by thermophoretic forces. A fitting formula for the thermophoretic force was proposed which reasonably agreed with the majority of the available data.

Lusseyran and Rockwell [36] showed that use of the hydrogen bubble timeline method, for the case of quasi-periodis flow, leads to reasonable estimates of the eigen-function of the streamwise velocity fluctuation. Both amplitude and phase distributions across an unstable wake flow were well-approximated. The vorticity extrema, as well as the degree of concentration of vorticity, were in good agreement with those calculated from the linear stability theory.

Smith and Paxson [37,38] described a system which allows the recreation of the three-dimensional motion and deformation of a single hydrogen bubble time-line in time and space. By digitally interfacing dual-view video sequences of a bubble time-line with a computer-aided display system, the Lagrangian motion of the bubble-line can be displayed in any viewing perspective that is desired. The flow structure in the nearwall region of a turbulent boundary layer was illustrated.

Kychakoff, et al. [39] discussed the obstacles in rapid multiple-frame imaging which was required for observation of the temporal evolution of flowfields (movies) or of 3-D structures. Some promising approaches were suggested. Example movies and 3-D images were displayed, but the technology was still in its infancy by contrast with the mature phase of the single-plane digital flowfield imaging technology.

Lu and Smith [40] described a technique which employs automated image processing of hydrogen-bubble flow visualization pictures in order to establish local, instantaneous velocity profile information. Hydrogen bubble flow visualization sequences were recorded using a high speed video system then digitized, stored, and evaluated by a VAX 11/780 computer. Employing special smoothing and gradient detection algorithms, individual bubble-lines were computer identified, which allowed local velocity profiles to be constructed using time-of-flight techniques.

VII. STEREO (THREE-DIMENSIONAL) IMAGING AND SUMMARY

Ghorashi, et al. [41] have described a three-dimensional (stereo) image analysis technique for the study of turbulent structures in a combustion process. The technique requires the illumination of a section of the flow via a laser beam while two cameras capture a stereo image of the flow. The visual images can then be digitized and each pair may be analyzed through a particle-matching process based on shape, geometry, orientation and position (x,y) in each view. Quantitative information can then be extracted from the digitized images, e.g., vector velocity field and subsequently the vorticity field. Figures (1) and (2) are the proposed schematic drawings of the equipment. It should be noted that due to the significant amount of data that

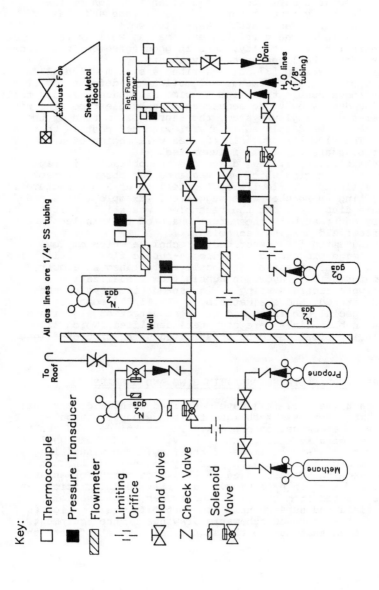

Fig.1.–Flow System Schematic

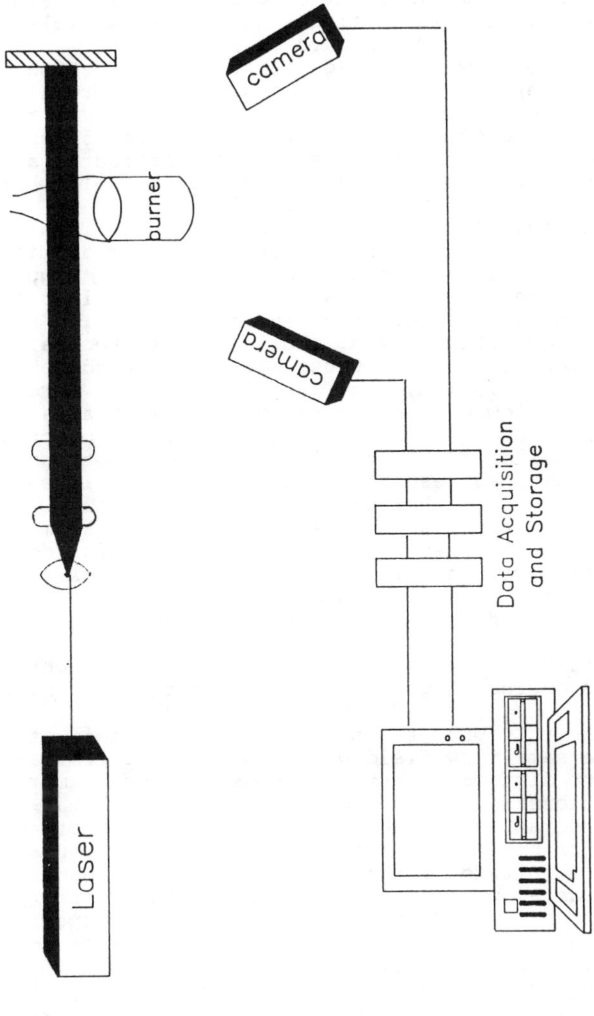

Fig.2.-Flow Imaging System

would be generated in such an undertaking, a totally automated
data reduction and acquisition system would be required. The
stereo image technique is based on the premise that the tracer
particles in the flow would have two positions (x,y) on each
frame of the stereoscopic pair. The particles that exist in both
pairs would be identified and matched to a standard grid scale
according to their size, shape, reflected light intensity, and
three-dimensional orientation. A frame-by-frame comparison of
the stereo images would then reveal information regarding the
direction of motion and displacement of each particle. Figure
(3) is a two-dimensional illustration showing the particles in a
spray-nozzle type of flow. The figure is from a 16mm
computer-generated movie which was digitized and a section of the
flow (AB) was arbitrarily selected for analysis (Figure 4).
Then, the same segment was identified in consecutive views and
the images were subtracted from each other. Figure (5) shows one
such subracted image. Ghorashi, et al. [41] constructed a
computer-generated stereo pair of a vortical flow field where
each tracer particle on one of the stereo images was
appropriately shifted. Furthermore, in order to distinguish
different particles from each other, different color densities,
geometries and sizes were assigned to each particle. Figure (6)
shows photographs of a sequence of stereo images obtained from
the computer screen at different time intervals. Motion picture
photography of the computer screen, showing the stereo pairs at
different time intervals enabled these workers to produce stereo
movies of the vortical motion, i.e.; as if the movies were taken
from a segment of a flow with a camera equipped with a pair of
stereo lenses. By viewing the movies via a stereo-projector, one
would learn the details of the individual particle motions. Once
the movies were digitized, each color density was assigned a
certain shade of grey which helped in its identification. Figure
(7) illustrates a digitized and magnified image of a segment of
the flow. It should be noted that this test was performed under
an inverse-condition with a limited number of particles that were
computer generated and by design showed no overlap. In reality,
particles do overlap and further complications arise when size
and shape similarities as well as equal amounts of light
reflection, due to particle orientations towards the light, pose
challenges in the course of particle identification and matching
process.
     Note that in this proposed method [41], Ghorashi, et al.
would observe the same flow field with two separate cameras,
positioned a short distance apart, as opposed to one camera with
a pair of stereo lenses. It should be further noted that
stereoscopic photography provides valuable spatial information
which at present cannot be extracted from other techniques.
Nonetheless, the magnitude of the problems that are encountered
due to excessive data, particle overlap and tracking are
enormous. Therefore, depending upon the nature of a specific
experiment, one should make a decision as to whether or not a
three dimensional approach to flow imaging is necessary. Many
two-dimensional techniques, such as the ones previously
described, would provide 2-D slices of the flow pattern at
different locations that may provide sufficient information

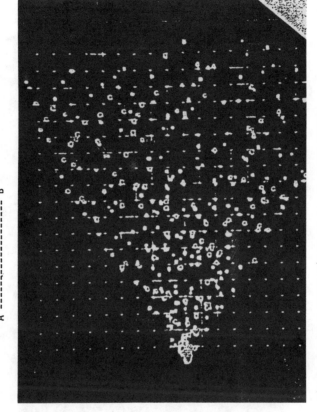

A -------------- B

Fig. 3.—Computer Generated Particles in a Flowfield

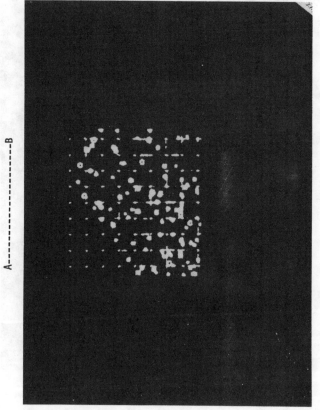

A----------B

Fig.4.-Digitized Image of the Region of Interest (ROI) in the FLOW

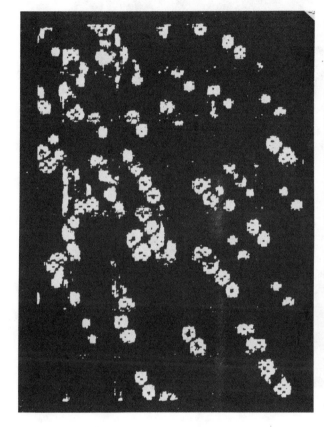

Fig.5.—Determination of Particle Motion by Subtraction of Consecutive ROI Images (Amplified Three Times)

579

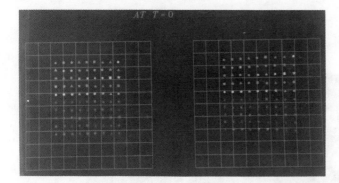

6a. Stereo Pairs
at T=0

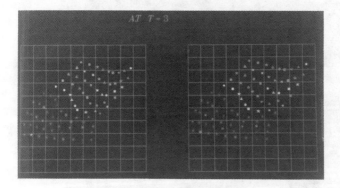

6b. Stereo Pairs
at T=3

Fig.6.-Segments from a Motion Picture Photography
of a Sequence of Stereo Images Obtained from
the Computer Screen at Different Time Intervals

580

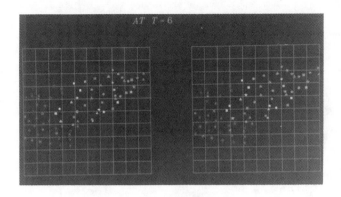

6c. Stereo Pairs
    at T=6

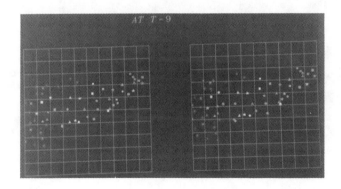

6d. Stereo Pairs
    at T=9

581

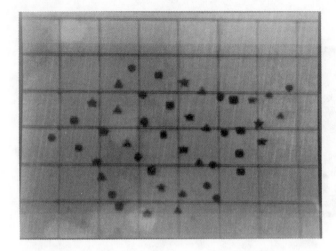

7a. Region of
    Interest(ROI)

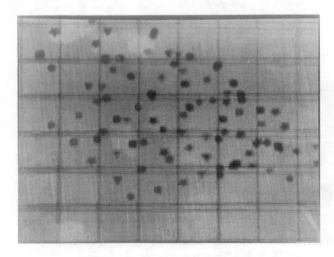

7b. Subtraction
    of Consecutiv
    ROI Images

Fig.7.-A Digitized and Magnified Image of a Segment
        of the Flow

without the need to resort to more complicated three-dimensional techniques.

REFERENCES

[1] Eckbreth, A.C., "Laser Diagnostics for Combustion Temperature and Species", Abacus Press, 1988.

[2] Gross, K.P.; McKenzie, R.L. "Optical Measurements of Fluctuating Temperatures in a Supersonic Turbulent Flow Using One- and Two-Photon, Laser-Induced Fluorescence". NASA Technical Memorandum 85949, 1984.

[3] Gross, K.P.; McKenzie, R.L., Logan, P., "Measurements of Temperature, Density, Pressure, and Their Fluctuations in Supersonic Turbulence Using Laser-Induced Fluorescence", Experiments in Fluids, Vol. 5, pp. 372-380, 1987.

[4] Cattolica, R.J.; Vosen, S.R., "Two-Dimensional Fluorescence Imaging of a Flame-Vortex Interaction", Combustion Science and Technology, Vol. 48, pp. 77-87, 1986.

[5] Cattolica, R.J.; Vosen, S.R., "Combustion-Torch Ignition: Flourescence Imaging of OH Concentration", Combustion and Flame, Vol. 68, pp. 267-281, 1987.

[6] Seitzman, J.M.; Kychakoff, G.; Hanson, R.K., "Instantaneous Temperature Field Measurements Using Planar Lasar-Induced Fluorescence", Optics Letters, Vol. 10, No. 9, pp. 439-441, 1985.

[7] Hiller, B.; McDaniel, J.C.; Rea Jr., E.C.; Hanson, R.K., "Laser-Induced Fluorescence Technique for Velocity Field Measurements in Subsonic Gas Flows", Optic Letters, Vol. 8, No. 9, pp. 474-476, 1983.

[8] Kychakoff, G.; Howe, R.D.; Hanson, R.K., "Quantitative Flow Visualization Technique for Measurements in Combustion Gases", Applied Optics, Vol. 23, No. 5, pp. 704-712, 1984.

[9] Hanson, R.K., "Flow Visualization in Gaseous Flows", NASA Contractor Report 174954, 1985.

[10] Allen, M.G.; Hanson, R.K., "Digital Imaging of Species Concentration Fields in Spray Flames", Twenty-First Symposium (International)/on Combustion/The Combustion Institute, pp. 1755-1762, 1986.

[11] Allen, M.G.; Hanson, R.K., "Planar Lasar-Induced-Fluorescence Monitoring of OH in a Spray Flame", Optical Engineering, Vol. 25, No. 12, pp. 1309-1311, 1986.

[12] Lee, M.P.; Paul, P.H.; Hanson, R.K., "Laser-Fluorescence Imaging of O_2 in Combustion Flows Using an ArF Laser", Optic Letters, Vol. 11, No. 1, pp. 7-9, 1986.

[13] Bowman, C.T.; Hanson, R.K.; Mungal, M.G., Reynolds, W.C. "Turbulent Reacting Flows and Supersonic Combustion" Annual Report, 1987.

[14] Hiller, B.; Cohen, L.M.; Hanson, R.K., "Simultaneous Measurements of Velocity and Pressure Fields in Subsonic and Supersonic Flows Through Image-Intensified Detection of Laser-Induced Fluorescence", AIAA-86-0161, 24th Aerospace Sciences Meeting, Reno, Nevada, Jan. 6-9, 1986

[15] Hiller, B.; Hanson, R.K., "Two-Frequency Laser-Induced Fluorescence Technique for Rapid Velocity-Field Measurements in Gas Flows", Optic Letters, Vol. 10, No. 5, pp. 206-208, 1985.

[16] Hiller, B.; Hanson, R.K., "Simultaneous Planar Measurements of Velocity and Pressure Fields in Gas Flows Using Laser-Induced Fluorescence", Applied Optics, Vol. 27, No. 1, pp. 33-48, 1988.

[17] Hassa, C.; Paul, P.H.; Hanson, R.K.,"Laser-Induced Fluorescence Modulation Techniques for Velocity Measurements in Gas Flows", Experiments In Fluids, Vol. 5, pp. 240-246, 1987.

[18] Weinstein, L.M.; Beeler, G.B.; Lindemann, A.M., "High-Speed Holocinematographic Velocimeter for Studying Turbulent Flow Control Physics", AIAA-85-0526, Shear Flow Control Conference, Boulder, CO, March 12-14, 1985.

[19] Liburdy, J.A.; "Resolution Limits for a Holocinematographic Velocimeter (HCV)", L.I.A., Vol. 58, ICALEO, pp. 97-105, 1986.

[20] Weinstein, L.M.; Beeler, G.B., "Flow Measurement in a Water Tunnel Using a Holocinematographic Velocimeter", Conference Proceedings No. 413, Aerodynamic and Related Hydrodynamic Studies Using Water Facilities, Oct. 20-23, 1986.

[21] Beeler, G.B.; Weinstein, L.M., "Holocinematographic Velocimeter for Measuring Time-Dependent, Three-Dimensional Flows", Presented at the 12th International Congress on Instrumentation in Aerospace Simulation Facilities, Williamsburg, VA, June 22-25, 1987.

[22] Chang, T.P.; Tatterson, G.B., "An Automated Analysis Method for Complex Three Dimensional Mean Flow Fields", Flow Visualization III, ed. Yang, W.J., pp. 236-243, 1983.

[23] Chang, T.P.; Wilcox, N.A.; Tatterson, G.B., "Application of Image Processing to the Analysis of Three-Dimensional Flow Fields", Optical Engineering, Vol. 23, pp. 283-287, 1984.

[24] Racca, R.G.; Dewey, J.M., "A Method for Automatic Particle Tracking in a Three-Dimensional Flow Field", Experiments in Fluids, Vol. 6, pp. 25-32, 1988.

[25] Kobayashi, T.; Ishihara, T.; Sasaki, N., "Automatic Analysis of Photographs of Trace Particles by Micro-computer Systems", Flow Visualization III, ed. Yang, W.J., pp. 231-235, 1983.

[26] Gharib, M.; Hernan, M.A.; Yavrouian, A.H.; Sarohia, V., "Flow Velocity Measurement by Image Processing of Optically Activated Tracers", AIAA-85-0172, 23rd Aerospace Sciences Meeting, Reno, Nevada, Jan. 14-17, 1985.

[27] Lewis, G.S.; Canmtwell, B.J.; Lecuona, A.,"The Use of Particle Tracking to Obtain Planar Velocity Measurements in an Unsteady Laminar Diffusion Flame", Combustion Inst., Western States Section Annual Meeting, Paper 87-35, April 1987.

[28] Landreth, C.C.; Adrian, R.J.; Yao, C.S., "Double Pulsed Particle Image Velocimeter with Directional Resolution for Complex Flows", Experiments in Fluids, Vol. 6, pp. 119-128, 1988.

[29] Rockwell, D.; Atta, R.; Kramer, L.; Lawson, R.; Lusseyran, D.; Magness, C.; Sohn, D.; Staubli, T., "Flow Visualization and Its Interpretation", AGARD Symposium on Aerodynamic and Related Hydrodynamic Studies Using Water Facilities.

[30] Rockwell, D.; Gumas, C.; Kerstens, P.; Backenstose, J.; Ongoren, A.; Chen, J.; Lusseyran, D., "Computer-Aided Flow Visualization", Sixteenth Symposium on Naval Hydrodynamics.

[31] Zucherman, L.; Kawall, J.G.; Keffer, J.F., "Digital Image Analysis of a Turbulent Flame", Experiments in Fluids, Vol. 6, pp. 16-24, 1988.

[32] Utami, T.; Ueno, T.,"Experimental Study on the Coherent Structure of Turbulent Open-Channel Flow Using Visualization and Picture Processing", J. Fluid Mech., Vol. 174, pp. 399-440, 1987.

[33] Ongoren, A.; Chen, J.; Rockwell, D.,"Multiple Time-Surface Characterization of Time-Dependent, Three-Dimensional Flows", Experiments in Fluids, Vol. 5, pp. 418-422, 1987.

[34] Talbot, L.; Cheng, R.K.; Schefer, R.W., Willis, D.R., "Thermophoresis of Particles in a Heated Boundary Layer", J. Fluid Mech., Vol. 101, pp. 737-758, 1980.

585

[35] Hiller, B.; Cohen, L.M.; Hanson, R.K., "Simultaneous Measurements of Velocity and Pressure Fields in Subsonic and Supersonic Flows Through Image-Intensified Detection of Laser-Induced Fluorescence", AIAA-86-0161, 24th Aerospace Sciences Meeting, Reno, Nevada, Jan. 6-9, 1986

[36] Lusseyran, D.; Rockwell, D., "Estimation of Velocity Eigenfunction and Velocity Distributions from the Timeline Visualization Technique", Experiments in Fluids, Vol. 6, pp. 16-24, 1988.

[37] Smith, C.R.; Paxson, R.D., "A Technique for Evaluation of Three-Dimensional Behaviour in Turbulent Boundary Layers Using Computer Augmented Hydrogen Bubble-wire Flow Visualization", Experiments in Fluids, Vol. 1, pp. 43-49, 1983.

[38] Smith, C.R.; Paxson, R.D., "Computer Augmented Hydrogen Bubble-wire Flow Visualization of Turbulent Boundary Layers", Flow Visualization, III Ed., Yang, W.J., pp. 185-189, 1983.

[39] Kychakoff, G.; Paul, P.H.; Cruyningen, I.V.; Hanson, R.K. "Movies and 3-D Images of Flowfields Using Planar Laser-Induced Fluorescence", Applied Optics, Vol. 26, No. 13, pp. 2498-2500, 1987.

[40] Lu, L.J.; Smith, C.R., "Image Processing of Hydrogen Bubble Flow Visualization for Determination of Turbulence Statistics and Bursting Characteristics", Experiments in Fluids, Vol. 3, pp. 349-356, 1985.

[41] Ghorashi, B.; Wey, C.C., and Marek, C.J., "An Image Analysis Technique for Identifying Time-Dependent, Three-Dimensional Flow Structures in Combustion Processes", Twenty-Second Symposium International on Combustion, Abstracts, The Combustion Institute, p. 181, pp. 278, August 1988.

Intensification of the Burning Process
with the Use of Swirling Jets

V. SH. KAKABADZE
Institute of Natural Resources
Academy of Sciences of the Georgian SSR
Tbilisi, 380030, Paliashvili str. 87, USSR

Abstract

The swirling jet, unlike the straight-flow one has a wide angle of expansion and high ejection capability, causing quick ignition and stable burning of fuel. It has been established that the use of powerful swirling jets strongly increases the pulsating component of the velocity, thereby increasing the speed of flame spreading in a gas flow.

Since increasing the tubulence of a flow increases the number of turbulent eddies size, a remarkable decrease in chemical reaction time is expected in swirling jets as compared to straight-flow ones. In addition, the relative length of the mixing path in a swirling jet is almost three times greater than that in a strangth-flow jet, indicating a greater capacity for heat and mass transfer in the swirling jet.

I. INTRODUCTION

It is well known that up to-date the thermal power stations (TPS) produce the main part of the electric energy. Taking into consideration the production volume of the electric energy it becomes clear that TPS should be thouroughly studied in order to find out and to decrease the negative effect of the energetics on the environment. According to the existing data TPS consume more than one third of the mineral fuel produced in the world and therefore TPS along with motorized transport and industry are the basic source of the environment pollution in the total balance of the negative effects of the industry on the nature as a whole.

At present most of the TPS of high and medium capacity are operated with the use of coal. According to data of /1/ it is expected that by 2025 the coal mining and production of the gaseous and liquid fuel out of it will considerably exceed the volumes of production of the natural gas and petroleum. It therefore follows that after 2000 the mining and production of the coal and its artificial derivatives should be essentially increased.

One of the most perspective trends of the high-effective burning of the coal is utilization of the water-coal suspensions in combination with the natural gas as a fuel and the heated air as the sprayer. The full burning of the fuel is provided similar to that, when dust-coal fuel is used. For this process high-boost tangential burners are used in the boiler furnaces. The flow velocities of the water-coal suspension and the air in the burner jets is 20 to 20 m/s. The

control range of the air consumption by the swirl is 6:1.
A recirculation zone, which stabilizes the flame is formed
in front of the burner. Such technology of the coal burning
gives an economy of the fuel by 15-20 %, the temperature control in the furnace with protection against overheating of
the screened tubes, and decrease of the number of the formed
nitric oxides in the burnt products by 25-30%, which drastically affect the ozonic layer of the earth atmosphere /2-4/.

2. ANALYSIS

When studying the burning processes in high temperature
furnaces and jet engine chambers, one of the most important
problem is the finding out of the interrelation between the
ignition and burning processes in the flow with aerodynami-
cal state of the latter. This fundamental problem is essential for consideration of the extent of the practical effect
on the burning of such factors, as the flow velocity, the
size of the combuction chamber, turbulence, etc. and for obtaining basic representations on the physical mechanism of
the process. Formation of mixes are preparative stages for
burning. They can be isolated from the burning processes,
whereas the ignition processes in the flow, propagation of
the flame and subsequent burning is the single process, Up-
to-date there is no precise analytical theory of diffusive
flame. However, by means of simplified theories one can describe certain properties of the diffusive flames. One of the
most important assumptions used in these theories is the hy-
pothesis on infinitely great value of the chemical reaction,
and therefore, on the infinitely small thickness of the burning zone, as well as on the propagation velocity of the
flame, its location and sizes, which depend upon the conditions of mixing of the fuel and oxidizer only. Calculation
of the actual process of mixing on account of the density difference and chemical reactions is very complicated.
Unlike straight-flow jet, the swirled one has a large
angle of broadening and accordingly the less aerodynamic and
thermal long-range, and the inner ejective capacity of the
swirled jet with intensive counterflow increases remarkably
the flame stabilization and intensity of the heat and mass
exchange. As it is known, the swirled jet is characterized
aerodynamically by axial, radial, tangential velocities as
well as by pulsation velocity being the component of the axial velocity /5/. In the front processes of the chemical reaction, which are realized in the gaseous and liquid media the
role of the pulsation velocity is rather high. For instance,
when the gaseous mixtures are burning, the flame velocity increases portional to the pulsation velocity of the turbulent
flow. The turbulent state of the swirled jets is characterized
by higher intensity of the turbulence as compared with
straight-flow jet. Besides, the relative length of the mixing
in the swirled jet is almost three times of the unswirled one,
that indicates at the higher heat and mass transfer in the
swirled jet. With the increase of the intensity criterion of
the turbulence in the spectrum of turbulent state of the flow

the number of turbulent moles of less size increases. Taking
into mind that the time of annihilation of molecular inhomo-
geneties in the flow depend strongly upon the size of the
turbulent moles of the mixed masses (proportional to the
square size) it is expected that in the swirled jets there
should be remarkable decrease of the time of chemical reaction as compared with straight-flow jets. Therefore, less
time needed for annihilation of molecular inhomogeneities in
the swirled jets as compared with straight-flow ones under sufficient temperatures with high diffusion factor allows to
make combustion chambers without preliminary mixing of reacting
mases.

3. CONCLUSION

We have studied experimentally (Fig.1) a noncatalytic
conversion of the natural gas in the swirling jet reactor
without prior mixing of the reacting masses /6/. Tests were
carried out on the laboratory unit, capacity of which was
7-12 nm^3/h of the basic mixture /7,8/. As a result, main characteristics were found: thermal strength of the reactor
reached $160x10^6 kcal/m^3 h$. If we assume that the thermal
strength of the industrial reactor is $60x10^6 kcal/m^3 h$ then in
this case the given volume of the reactive space will provide
high output too. With consumption of the natural gas of
40.000 nm^3/h, the necessary volume of the reactive space will
be only about 1m. In the laboratory reactor with swirling
jets a possibility of decreasing of the oxygen consumption
per unit of the converted gas by 10-14% averagely was found
as compared with conversion in straight flow jet reactor under similar conditions. The process of high temperature conversion in the swirling jet reactor excludes in full the
flame breakthrough into the device for swirling of jets or
its tear off. Therefore, there is no need in the mixer that
simplifies the design of the reactor. An aerodynamic speci-
ficity of the swirling flows provides a wide range of the
control of the basic reagent consumption - the fuel and the
oxidizer and the high level of forcing. Following from the
obtained experimental data, it may be recommended to use the
swirling jets for the high effective burning of both dust
coal and water-coal suspensions in combination with natural
gas and utilization of the heated air as a sprayer.

REFERENCES

1. T.N.Veziriglu, Hydrogen technology for energy needs of
   humen settlements, I.G. of Hydrogen energy, V.12,No2,1987.
2. J.Makensi, Coal water fuels, "Power", v.129, No7, p.17-
   24, 1985.
3. D.Liu, High intensity, combustion of coal with water injection, "Combustion and flame", v.63, No1-2, 1986.
4. M.S.Anand, F.C.Gouldin. Combustion Efficiency of a
   Premixad Continuous Flow Combustor, Jornal of Engineering for Gas Turbines and Power, v.107, No3, 1985.

5. V.Sh.Kakabadze and S.N.Shorin, High-temperature conversion of natural gas in swirling jet reactor. Bull. Georgian Acad. Sci. 38, 1965.
6. V.Sh.Kakabadze, Application of method of swirling jets for conversion of natural gas. Bull. El-Tabbin Inst. Metall. Studies, 21, Cairo, ARE, 1976.
7. V.Sh.Kakabadze, Experimental investigation of the conversion of natural gas in a reactor with swirling jets, J. Hydrogen energy, Vol. 11, N. 8, USA, 1986.
8. V.Sh.Kakabadze, Analysis of energy balance and role of thermal losses in high-temperature reactor of natural gas, Conf. on Altern. Energy Sources, VII, Vol. 5, USA, 1985.

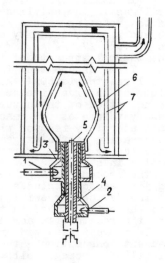

Fig.1. Shematic of the reactor with a nozzle
for swirling jets
1- tube of natural gas
2- tube of oxidizer
3-4 - cyclone;5- electric ignition
device
6- reactor; 7-thermal insulator

Index